New Approaches in Social, Environmental Management and Policy to Address SDGs

New Approaches in Social, Environmental Management and Policy to Address SDGs

Editors

Margarita Martinez-Nuñez
Mª Pilar Latorre-Martínez

MDPI • Basel • Beijing • Wuhan • Barcelona • Belgrade • Manchester • Tokyo • Cluj • Tianjin

Editors

Margarita Martinez-Nuñez
Organization Engineering,
Business Administration and
Statistics
Technical University of Madrid
Madrid
Spain

Mª Pilar Latorre-Martínez
Management and Organisation
University of Zaragoza
Zaragoza
Spain

Editorial Office
MDPI
St. Alban-Anlage 66
4052 Basel, Switzerland

This is a reprint of articles from the Special Issue published online in the open access journal *Sustainability* (ISSN 2071-1050) (available at: www.mdpi.com/journal/sustainability/special_issues/New_Approaches_SDGs).

For citation purposes, cite each article independently as indicated on the article page online and as indicated below:

LastName, A.A.; LastName, B.B.; LastName, C.C. Article Title. *Journal Name* **Year**, *Volume Number*, Page Range.

ISBN 978-3-0365-2951-6 (Hbk)
ISBN 978-3-0365-2950-9 (PDF)

Contents

About the Editors

Margarita Martinez-Nuñez

Margarita Martinez-Nuñez is an Associate Professor who teaches in the areas of Economy and Business Management, Project Management, Information and Communication Technology Management in Organizations and Business Administration in the Technical University of Madrid (Spain). Her research interests are focused on forest and environmental economy, business organization and economic models and information systems within the framework of sustainability, life cycle assessment, efficiency and technological innovation, knowledge management and data analysis. She has published more than 30 articles in international and national journals of recognized prestige. She has presented more than 40 communications in numerous scientific international and national conferences and congresses. She has also published several books and chapters in the field of knowledge management, environmental and economic models and research methodologies. She has been a participant or director of more than 20 competitive research or teaching innovation projects funded by national and international organizations like European Commission (Erasmus + Program), EIT Climate-KIC, Regional Government of Madrid (Spain), Spanish Ministry of Education and Science, or Spanish Ministry of Science and Innovation. She has been a member of the scientific committees of several international conferences and guest editor of special issues and frequent reviewer of high impact scientific journals.

Mª Pilar Latorre-Martínez

Maria Pilar Latorre Martinez is an Associate Professor in the Department of Management and Organisation at the University of Zaragoza (Spain). She specializes in Industrial Engineering and she has a Phd in Economics and Managing Organisations. Her research interests are in the areas of business analytics and network sciences. She teaches in the areas of Human Resource Management , Strategic Management , Digital Transformation in Business Model and Team Management and Innovation. She has published 13 papers in international and national journals of renowned prestige. She has presented more than 20 communications in numerous scientific international and national conferences and congresses. She has also published several books chapters in the field of gender equality and relations of organizations. She has been participant of several competitive research or teaching innovation projects funded by national and international organizations. She has been reviewer of high impact scientific journals.

Preface to "New Approaches in Social, Environmental Management and Policy to Address SDGs"

The SDGs (Sustainable Development Goals) are principles created by the UN, with the ultimate goal of guaranteeing well-being on and off the planet. The importance of these objectives is essential, not only for promoting environmental education and citizen science, but also for the commitment to the solutions at different levels of governments, companies, and civil society.

Among the SDGs, there are goals that both public and private organizations must meet before 2030, all with the aim of causing a positive impact on society. Policy makers, companies, and organizations are challenged to cooperate to define new regulations, change business models, and promote the adaptation of society to the challenges posed by the 17 SDGs.

This Special Issue has welcomed papers, research works, and investigations on new approaches to address SDGs from a more macro perspective—environmental, social, and governance (ESG) criteria.

For this purpose, the book comprises a selection of papers addressing some of the most relevant challenges and opportunities for addressing SDGs from many different perspectives. Papers in this collection cover the most recent lines and approaches of research in addressing SDGs and are all novel propositions that deepen the analysis of environmental, social and governance strategies in the adaptation of the society to meet the 17 SDGs.

In total, a selection of 14 papers forms this book, covering topics such as ecosystem services, urban-industrial systems, governance, institutional quality, waste shipment and disposal, sustainable urban drainage systems, green infrastructures, entrepreneurship, European funds, forestry, wood production, carbon storage, forest biodiversity, agro-food system, environmental performance of small and medium enterprises, business incubators, maturity models, regional and local planning, project management, public infrastructures, local level SDGs implementation, stakeholders synergies, ecological footprint, community empowerment, rural depopulation, etc.

Contributors to the book represent a wide spectrum of nationalities from all over the world. Papers come from Canada, China, Estonia, Finland, Japan, Hungary, Italy, Latvia, Lithuania, Norway, Poland, Spain, Sweden, Thailand and the United States of America (presented in alphabetical order).

Therefore, this book represents an excellent forum and contribution to present some of the latest lines of research about the impacts of SDGs in society.

Finally, the Guest Editors are grateful to all the people that have helped us to complete this Special Issue. We would like to thank all the authors who answered our invitation and submitted their papers to this Special Issue and also to those authors that have made the effort to complete a paper, but it was impossible for them to submit it. We would like to send a special thanks to Mavis Li, our Special Issue's Managing Editor, for her continuous support and assistance, kindness and patience, she is part of the successful completion of this Special Issue. Another special thanks go to the academic editor of *Sustainability* that have supported each submission as well as the reviewers who have generously dedicated part of their valuable time to reviewing papers for this Special Issue.

Margarita Martinez-Nuñez, Mª Pilar Latorre-Martínez

Editors

sustainability MDPI

Perspective

Implementing SDGs to a Sustainable Rural Village Development from Community Empowerment: Linking Energy, Education, Innovation, and Research

Isabel del Arco [1,*], Anabel Ramos-Pla [1], Gabriel Zsembinszki [2], Alvaro de Gracia [2] and Luisa F. Cabeza [2,*]

1 Organisational Development Team (EDO-UdL), University of Lleida, 25001 Lleida, Spain; anabel.ramos@udl.cat
2 GREiA Research Group, University of Lleida, 25001 Lleida, Spain; gabriel.zsembinszki@udl.cat (G.Z.); alvaro.degracia@udl.cat (A.d.G.)
* Correspondence: isabel.delarco@udl.cat (I.d.A.); luisaf.cabeza@udl.cat (L.F.C.)

Abstract: Rural depopulation is a worldwide fact and has a domino effect on medium and small cities, which act as a nucleus of reference for small towns. Moreover, the United Nations (UN) stressed that disparities between rural and urban areas are pronounced and still growing over time. Globally, people in rural areas lack access to modern energy services, which affects productivity, educational and health services, exacerbating poverty, among other things. Given this reality, the following research questions arise: how can we act to reverse this reality? Are there examples of transformation in rural contexts where community empowerment is a key strategy? This paper aims at describing the transformation process of a small rural municipality towards a sustainable development, in parallel to the activation of the local productivity that helps to eliminate the effects of rural depopulation. Therefore, the project ALMIA was established as an example of a sustainable village that is Almatret (Catalonia-Spain). The backbone of such project is the commitment to community empowerment, where the main results are the generation of networks with experts and researchers to help the municipality's energy transition, the involvement of the local administration, the commitment to technological development, as well as the socio-community development. Moreover, the activities developed within the project ALMIA are aligned with the UNs Sustainable Development Goals, alignment that is analyzed in detail. Thus, this paper aims to further highlight existing sustainable development practices related to community empowerment in order to promote similar practices.

Keywords: community empowerment; energy transition; rural depopulation; sustainable development; Sustainable Development Goals (SDGs)

Citation: del Arco, I.; Ramos-Pla, A.; Zsembinszki, G.; de Gracia, A.; Cabeza, L.F. Implementing SDGs to a Sustainable Rural Village Development from Community Empowerment: Linking Energy, Education, Innovation, and Research. *Sustainability* **2021**, *13*, 12946. https://doi.org/10.3390/su132312946

Academic Editors: Margarita Martinez-Nuñez, Mª Pilar Latorre-Martínez and Marc A. Rosen

Received: 19 August 2021
Accepted: 18 November 2021
Published: 23 November 2021

1. Introduction

In 2015, the United Nations (UN) and its member states adopted by resolution the 2030 Agenda for Sustainable Development [1] with the overall goal of providing a path to peace and prosperity. This agenda ensures that all countries develop in relation to economic growth. At the same time, it aims to improve health and education, reducing inequality and tackling climate change.

All of this is implemented in 17 Sustainable Development Goals (SDGs) that include targets and actions to achieve this ambitious overarching goal.

These objectives are based on a modern conception of human development that gives relevance to health and education, overcoming the old conception that considered development as "a process of continuous economic growth, which ensures a lasting surplus of all kinds of goods, which can be used to meet human needs and enhance greater well-being" [2]. Thus understood, human development seeks the welfare and quality of life of people.

We speak of Sustainable Development because the person is at the center of all policies and efforts to improve conditions in each social context. To achieve this, three key principles must be taken into account: (i) the promotion of remote human and socially equitable development for all humanity based on inclusion, (ii) the guidance of economic development at the service of human development, and finally, (iii) the promotion of responsible use of the planet's natural resources [3].

The 17 SDGs are inter-related and mark inclusion, security, production, and consumption as core issues for the sustainable development of human settlements (cities, towns and villages) [4]. The creation of viable human settlements becomes a clear indicator of socially and economically sustainable development.

This article focuses on rural settlements and how the SDGs can be applied to these currently vulnerable populations. It is interesting to know how we can act in these populations to promote their sustainable development by basing their transformation on community empowerment as a key strategy.

Thus, the aim of this paper is to describe the transformation process of a rural municipality towards sustainable development in parallel with the activation of local productivity that helps to reduce or eliminate, among others, the effects of depopulation. In order to do so, the global situation of rural settlements, the problems they face and the consequences that arise from them will be contextualized. Subsequently, it will be argued that it is important to focus on these settlements in order to make the development of the 2030 Agenda globally effective.

The municipality of Almatret is presented as a current example of a positive and sustainable village and the ALMIA project as a strategic plan designed for the transformation of the municipality. It is important to understand how the ALMIA project is aligned with the global implementation of the 17 SDGs, and specifically with some of them.

2. Towards a Sustainable Village

The dramatic rapid growth of cities has been a constant during the last years. It is estimated that by 2030, almost 60% of the world population will live in urban areas. This strong expansion will occur in 95% of developing countries.

This has generated problems such as high levels of exclusion and citizen insecurity, pressure on water supply, waste generation and, in short, uncontrolled production and over-consumption. These problems are even expected to become even more accentuated in the coming years.

These are the questions that SDGs eleven and twelve of the 2030 Agenda seek to address. Both goals propose that cities and human settlements in general should be inclusive, safe, resilient, and sustainable by promoting controlled production and moderate consumption.

On the other hand, rural depopulation is a fact worldwide and has a domino effect also for medium and small cities that act as a nucleus of reference for small towns [2]. A reduction of inhabitants in the region means that the supply of services, leisure, health, or education becomes meaningless, and with them, the corresponding jobs disappear. All this generates a chain effect that leads the population to seek opportunities in large cities [5]. These large concentrations of population can become unviable and unsustainable.

It is essential to have better and more sustainable cities, but it is also important to have an impact on rural populations, which are generally limited in terms of roads, electricity supply, telecommunications and other types of infrastructure that are of key importance for their development and growth. All these factors mean that the rural population receives more limited educational, health and social services, which has a negative impact on their well-being.

Let us remember that agriculture is still the main economic activity in rural territories, and despite being an activity with a high potential to support environmentally sustainable growth, there is evidence that the current form of food production is one of the main drivers of climate change [6,7]. In addition, the economic income it generates is low, and gender inequality is exclusionary [8,9].

The 2030 Agenda recognizes that rural settlements are determinant for the achievement of its goals and that public policies cannot forget that development also involves rural areas. According to UN-Habitat (2006), sustainable human settlements involve providing adequate shelter for all inhabitants, and this involves improving management by:

- Promoting sustainable land-use planning and management.
- Promoting the integrated provision of environmental infrastructure: water, sanitation, drainage, and solid waste management.
- Promoting sustainable energy and transportation systems
- Addressing planning and management in disaster-prone areas
- Encouraging sustainable industrial and construction activities.
- Promoting sustainable human resource development.

SDG two devotes specific attention to rural areas by proposing increased investment in infrastructure, agricultural research and extension services, technological development and plant and livestock gene banks. This is the only way to reduce inequality between rural and urban areas.

We must prevent the population of rural settlements from lacking access to modern energy services because it affects productivity, educational services and even the health of the population, aggravating poverty [10].

Recently, new approaches to sustainable rural development have emerged, emphasizing the nexus between energy, health, education, water, food, gender, and economic growth.

The focus is on inclusive and sustainable development. It is about increasing productivity, creating jobs and generating income. All of this is based on social inclusion, gender equality and decent employment for young people. Development must focus on the creation of added value through the application of research and innovation (R&D&I).

On the other hand, there is a growing concern about economic and environmental development, intensified by the global energy, food and financial crises, and the COVID-19 pandemic. All this translates to a continuous warning about the transgression of planetary or ecological limits, reflected in climate change. The green economy is the proposed alternative to catalyse new national policies in support of sustainable development.

Access to affordable, reliable, sustainable, and modern energy for all would create new job opportunities, empowerment of women and youth, better education and health, and ultimately more sustainable settlements using a powerful climate change strategy.

Education is an integral element of sustainable development, not only ensuring quality, inclusive, and equitable education for children, but also including lifelong learning opportunities. Education can build sustainable societies by actively integrating citizens' education into sustainable development.

Isolation, lack of infrastructure, scarce diversification and access to services can affect the health of the rural population.

Finally, we must focus on sustainable consumption and production as key factors for the proper management of natural resources. The rational use of aquifers, the reduction of waste production and its recyclability and reuse are part of what is being called circular economy, which affects sustainable development.

At this point, is it possible to reverse reality in order to achieve sustainable development? How can we act in rural settlements to promote their development? Are there examples of transformation of rural contexts where community empowerment is a key strategy?

There are some examples of sustainable cities and towns that are making the achievement of the SDGs possible: from the Middle East (Education City to Qatar and Masdar City in Abu Dhabi), the Far East (Dongtan Institute in China, Nova Songdo City in Korea South), Ruhr, Feldheim, and Freiburg (Germany), Vancouver (Canada), Brighton (United Kingdom), Genk (Belgium), Stockholm (Sweden) [11–14]. Considering the number of existing large cities (400) and mega cities (23) [2], it is necessary to rethink their sustainability. There are also energy transition initiatives towards sustainability in small towns such as in Poland (Biskupiec, Nowe Miasto Lubawaskie, Nowy Dwór Gdánski, Górowo Ilaweckie, Goldap or Ryn), Scandinavia (Danishh Svendborg, Norwegian Skotterutd, or Swedish Falkoping), and Germany (Marihn, Meldorf, or Penzlin) [15]. In Spain, the island of El

Hierro has become, through renewable energies, the first self-sufficient island in terms of energy [16]. These initiatives seek to achieve local sustainable development [17,18].

An example of interest is the case of "East Village at Knutsfod" (EVK) [19], located 1.5 km from the central business district of Fremantle (Western Australia). The project aims to create semi-detached houses that incorporate solar energy storage through photovoltaic solar systems (designed to produce more energy than is consumed in the domestic environment), a battery deposit for charging electric vehicles and integration of the urban water management (through stormwater infiltration, alternative urban supplies, and biophilic urban designs) to reduce dependence on centralized water sources. With this, it is intended that the town ends up using 100% renewable energy for daily consumption [20].

In this context, the municipality of Almatret has developed the project ALMIA, with the aim of increasing its sustainable development by becoming a net positive energy village. This document presents the process that the village is undertaking and how it links all its actions (past, present, and future) to the SDGs. The objective of this article is to present the ALMIA project and the processes of transformation into a sustainable development village.

3. Almatret, an Example of a Sustainable Village

3.1. Description of the Municipality

Almatret is a small village located in the southwest of Catalonia, Spain (Figure 1) [21], next to the river Ebro in the limit with Aragon. Today, it has around 300 inhabitants in a surface area of 56.83 km^2. Almatret has a very rich history because Iberians, Romans, and Arabs lived in its area. The village is considered to be founded in 1301.

Figure 1. Almatret map (Google Maps, 2021) [21].

Today, Almatret's main industry is agriculture, mainly olive and almond, and livestock industry, mainly pork industry. During the XIX century, Almatret had a lot of coal mines, when the population was as high as 2000 inhabitants.

Almatret is classified within the Continental Mediterranean climate, Bsk in the Köppen-Geiger climate classification [22]. Therefore, the town climate is characterized by having a very marked seasonality (in distribution of temperature and rainfall). In other words, summers are hot and dry, and winters are very cold [23]. Its agriculture is dry land where the almond tree occupies 41% of the cultivated land (43% of the municipal total). In addition, olive and barley are grown. Industrial activity is concentrated in the production and elaboration of olive oil. During the 20th century, there was great economic activity promoted by the exploitation of small deposits of lignite, although, at the end of the century, it went into decline.

During this time of mining splendour, there was an expansion of Almatret with the construction of a large part of the town houses, as also happened in the Apuseni (Romania) [24]. There was an increase in the population that rapidly decreased due to the abandonment of this activity with the emigration of its inhabitants to nearby centres with more services. At present, Almatret is a municipality of less than 350 inhabitants.

Its geographical location makes it ideal for the installation of wind farms. In 2003, more than 25 turbines with a nominal power of more than 50,000 kW [25] were installed, to which the last project, the Vencills wind farm, consisting of only four mills, but of a size and a higher power (24 MW), was added. Almatret is located in a geographic pool of energy production, a few kilometers from the Ribaroja reservoir and the Maquinenza thermal power plant. Moreover, from the Asco nuclear power plant and also the settlement of various solar parks, a landscape conformed oriented to the production of energy, from a past with lignite mines to the present with the different mentioned alternatives sources.

3.2. The Project ALMIA

Around 2010, the municipality of Almatret decided to initiate a new project with the aim of reversing the loss of population, jobs, and quality of life in the municipality. This led to the creation of the ALMIA project, which develops a series of actions based on three basic principles:

1. The knowledge and involvement of the inhabitants of the municipality from the community empowerment [26]. To ensure the social dimension of sustainable development, it is important to enhance social participation, the capacity for initiative and co-responsibility. Only by building a local project, such as ALMIA, with the active participation of the population, will it be possible to generate social transformation.
2. Involvement and leadership of the local administration. It is important to develop a transformational leadership [27] that motivates and encourages the community to generate changes towards sustainable development.
3. Importance of creating collaborative networks to support transformation. Networks help to eliminate isolation in rural areas and to bring innovative proposals closer.

Therefore, the actions proposed in the ALMIA Project are developed as follows:

- Production and access to sustainable, reliable, and modern energy.
- Almatret today is a net positive energy village building because it produces more energy than it uses. As mentioned above, Almatret currently produces almost 150,000 MWh per year, and its energy consumption is very low. However, the homes use wood from the area around the village as a source for space heating and domestic hot water. Moreover, most homes do not have air conditioning due to the climate conditions and the building envelope (low cooling demand). Seeking not only a production, but also a sustainable consumption, the aim is to achieve a zero-carbon village, avoiding the use of fossil fuels. For this purpose, there is a first action with the future installation of electricity chargers and providing free electric vehicles within the village to go to the nearby city of Lleida, where most services are located (hospital, university, governmental services, etc.). The other big action is the production of renewable heat both from solar energy (see below the demonstration pilots from H2020 ongoing projects) and from biodigestion of livestock wastes. This action also requires installing a district heating system to distribute this energy to the municipality households.
- Sustainable development from research and innovation and improvement of infrastructures.
- Recognizing that only through innovation such changes are possible, the municipality of Almatret involved the University of Lleida in this project from the very beginning. Very soon, the possibility to involve Almatret in the European funded project Innova MicroSolar [28], installing the demonstration site in Almatret. The aim of the project is to develop a system to generate electricity and heat based on solar energy. A linear Fresnel solar field (Figure 2) is able to heat up oil up to 280 °C to expand an Organic Rankine Cycle (ORC) to generate 2.3 kW of electricity. The condensation process inside the ORC delivers 25 kW of heat to the youth hostel building for space heating

and domestic hot water applications. Moreover, a latent heat thermal energy storage system of 100 kWh of capacity is used to expand the ORC 4 h after solar availability. The system relies on the use of heat pipes to transfer the heat from oil to PCM and the way back.

Figure 2. Fresnel solar field and youth hostel building.

Shortly after, Almatret started its participation as full partner in the European funded project HYBUILD [29], again hosting one of the three demonstration sites. Within HYBUILD, two innovative hybrid storage concepts are developed and tested: one for Mediterranean climate, focused mainly on meeting the cooling demand in residential buildings, and one for Continental climate, for which the main focus is to provide heating and domestic hot water in residential buildings. Innovative components are integrated in both systems, such as a high-density latent storage, a reversible vapor compression heat pump, a DC bus connected with PV panels to run the heat pump in DC. Moreover, the Mediterranean climate system also contains a compact sorption module and a field of Fresnel solar collectors. Other than the thermal energy storage components, an electric storage is also used to store the surplus of electricity production of the PV system. An advanced control and building energy management system (BEMS) is also implemented in the systems to ensure a proper end efficient operation. The demonstration site in Almatret is a single-family residential building (Figure 3) where the Mediterranean climate system is implemented and tested for future assessment and validation.

Figure 3. Residential building where the HYBUILD Mediterranean system is implemented.

- Community empowerment and dimensioning of the ALMIA project from an educational perspective.

The vision of renewable energies as an integrative alternative to socio-economic and environmental challenges, together with the commitment to open innovation where citizens are participants in the advances in research in energy systems, is the commitment of the proposal of the Energy Interpretation Center (EIC) and the ALMIA museum that is being developed in Almatret. This current of openness about technological scientific advances coincides with the establishment of participation mechanisms in science and innovation that allow citizens to have information and be part of the innovation system. In addition, these initiatives seek to influence the economic and social revitalization of a territory that in recent decades has been immersed in unprecedented depopulation.

The objectives of EIC are, among others:

- To promote education in renewable energy and energy efficiency.
- To contribute to formal and non-formal participatory scientific education and open to all public.
- To preserve and value the scientific, industrial, and technological heritage and its integration with the territory.
- To disseminate, communicate and enhance the research and innovation activity in the field of renewable energies.
- To investigate energy efficiency technology solutions in a participatory way.
- To act as a benchmark for social cohesion, integration, and sustainability for the municipality of Almatret.

This interpretation center is based on the inclusion of contextualization elements of the existing active energy facilities such as the wind farm and photovoltaic solar installation, as well as the recovery of the territory heritage. It will also have a museum area with exhibition, educational, and research programs that allow the carrying out of education, communication, and heritage preservation activities. The creation of a research program for the generation of integrated solutions for renewable energy and energy efficiency and the possible application of these solutions in the municipality will allow the transfer of knowledge in this area and the participation of citizens in the innovation process. The technological facilities that currently exist in Almatret are the best starting point for the visitor to find inspiration in the curiosity for science and technology, in formulating questions about the impact on their past and present life, thus stimulating reflection and knowledge. EIC, in short, is an open and public space dedicated to scientific and technical knowledge, based on rigor, credibility, accessibility and understandability.

The pedagogical paradigm that frames the ALMIA proposal focuses on the empowerment of citizens [30,31] as active actors of the project, generating a committed implication and running with responsibility. This citizen empowerment begins with the generation of a promoter group that helps to start this community process. The promoting group is made up of specialists and experts in the field (teachers and researchers) as well as technicians from the local administration. This promoter group will guide the community diagnosis phase, prioritizing the intervention strategies, planning the actions that will be put into practice to finally evaluate the process and the impact. It is very important that the inhabitants of Almatret know about the project and are involved in all phases of development. The promoter group encourages this participation, but at the same time it must establish a network of entities and organizations that help to disseminate, implement and consolidate the project. ALMIA is planted from its socio-educational aspect as an example of community development with the involvement of citizens. At present, educational activities for schoolchildren have already been generated, educational material has been generated and different practical workshops have been piloted. All this begins an educational and scientific tourism [32] oriented to the interpretation of the different sources of energy that exist in the municipality.

All these activities created the first quality jobs, one for an engineer or similar, to be able to manage all the new systems installed in the village, and a second one to lead the educational activities and the relation with schools.

4. Consistency between the SDGs and ALMIA

The Action Plan for the Implementation of the 2030 Agenda in Spain [33] clearly states that "it is not possible to achieve the SDGs by leaving rural areas and their inhabitants behind".

ALMIA is presented as a strategic project which develops a series of specific actions within the framework of SDG deployment, which are specified in Table 1.

Table 1. Summary of the relationship between the ALMIA project and the SDGs.

SDGs	Description	Actions in Almatret (ALMIA)	First Results
SDG #1	Eradication of poverty.	Creation of jobs to fix the population to the territory.	Jobs to develop the EIC and attend visitors: one contract.
SDG #2	Attention to rural areas, investment in rural infrastructure, agricultural research, and technological development.	Generation of infrastructure to support technological development and modernization of the municipality.	Innovation infrastructure linked to the Innova MicroSolar and HYBUILD projects.
SDG #4	Education as an integral element of sustainable development ALMIA-Educational.	To generate educational and training activities for and in favor of sustainable development.	Group promoting the ALMIA-Educativa educational program to be developed with educational centers and the visiting public in general.
SDG #7	Clean and affordable energy ALMIA-Energia.	Transformation towards a positive energy village towards a zero-carbon village.	Positive village: produces more energy than it consumes: produces 150,000 MWh per year.
SDG #8	Sustained, inclusive and sustainable economic growth.	Sustainable tourism. Creation of EIC as a pole of attraction for sustainable, educational and research tourism.	Visits by educational centers during 2021 (approximately 250 people). Expansion of the visit program for the 2021–22 academic year to 1500 visitors.
SDG #9	Inclusive and sustainable industrial development.	Promotion of R&D&I with the involvement of the municipality in international projects in collaboration with university research groups.	European projects Innova MicroSolar and HYBUILD.
SDG #11	Sustainable transport.	Incentivize clean and sustainable transport related to all activities in Almatret (i.e., agriculture, mobility, tourism). Electric vehicles.	Charging points in the municipality and electric transport.
SDG #12	Ensure sustainable consumption and production patterns.	Energy transition of the municipality: transformation of energy sustainable buildings.	Refurbishment of municipal buildings for energy transition such as schools. Passive Haus construction.
SDG #13	Action against climate change.	Commitment to clean and sustainable energy generation with the creation of solar fields and wind turbines.	Solar fields, wind turbine fields, production of renewable heat from solar energy and from the biodigestion of livestock waste.

Rural development involves empowering the rural population and formulating policies that generate synergies between institutions, seeking common objectives to advance sustainable development, ensure quality of life and a better balance in population distribution [34].

When rural people are empowered, they are able to participate, decide, negotiate, influence and control so that they can strengthen their family and productive environments in the face of the invisibility imposed by the current socio-economic evolution.

Distributed and committed leadership is important, as in the case of the ALMIA project, where the city council assumed this role. It also seeks complicity with external agents such as the University of Lleida and the Provincial Council. In short, the role of the local administration is fundamental as an agent to lead the change and transition in the municipality towards a type of circular and sustainable economy.

In line with SDG #7 and SDG #13, a clear commitment has been made to the production of renewable energy, taking advantage of available local resources, in this case its privileged orographic situation. In addition, ALMIA represents a step forward in the research, development and application of new energy techniques that lead us to zero consumption of finite resources that are not very respectable with the environment. From this scientific, engineering and research dimension, ALMIA responds to SDG two and SDG nine with three projects funded by the EU:

- Innova Microsolar (2016): http://www.innova-microsolar.eu/ (Accessed on 31 July 2021)
- HYBUILD (2017): http://www.hybuild.eu/ (Accessed on 31 July 2021)
- Passive Haus construction: https://passivehouse.com/ (Accessed on 31 July 2021)

Aligned with SDG 11 and from this scientific and technological dimension, different proposals are being projected in the municipality to transform buildings into sustainable buildings with an important integration of renewable energies, promoting the consumption of this type of energy: rural buildings under self-consumption, means of transport, systems heating of homes, etc. It would be a balanced use of different renewable energies that are generated by the municipality, both those that have a greater technological maturity (wind and photovoltaic) and those that are more manageable (solar, thermoelectric, biomass, among others), always taking advantage of local resources and minimizing the environmental impact and in the territory.

It is also necessary to highlight the socio-educational dimension of this project. The aim is to turn the EIC into a pole of attraction for educational centres to influence the education of young people. To this end, activities are planned that involve the municipality, generate jobs and help disseminate scientific knowledge, laying the foundations for raising awareness and involving young people in the implementation of the 2030 Agenda and the ESD 2030.

The EIC, aligned with SDG four, becomes a Learning Camp open to students from different educational stages and levels, and to the general public. The added value is to learn in situ about energy production, transport, storage, energy transition, and responsible energy consumption.

The museum part of the project adds value to the scientific and technological dimension of ALMIA. The same inhabitants will help to create a local inventory of heritage resources (natural and cultural, tangible and intangible).

This is aligned with SDG eight where the rural nucleus is transformed into a good to know and to protect. To all this, other additions are:

- To contribute to a formal and non-formal science education that is both participatory and open to the public.
- To preserve and value the scientific, industrial, and technological heritage and its integration with the territory. It would try to revalue the heritage of the municipality of Almatret.
- To disseminate, communicate, and value research and innovation activity in renewable energy matters.
- To investigate energy efficiency technology solutions in a participatory way.
- To act as a benchmark for social cohesion, integration, and sustainability for the municipality of Almatret.

5. Discussion and Conclusions

The UN objectives cannot be achieved only with top-down actions. Bottom-up initiatives, such as the one presented and analyzed in the paper, are key and necessary to reach these challenges. Almatret is a clear example of a small village fighting to overcome depopulation and doing a good use of local resources in a sustainable way [35]. The ALMIA project is a clear example of a strategic organization based on community empowerment that promotes participation by fostering a sense of community and belonging to the group [36]. It is based on available resources, with a broad vision of the concept of resources beyond the economic, also taking into account personal capital and the possibilities of the context/environment. It is necessary to mobilize all the opportunities and strengths of the rural context to promote transformations.

AlMIA promotes networking, not only with surrounding municipalities and institutions, but also with scientific experts from the University of Lleida. This networking helps to share information, resources, processes, consolidates the commitment between actors, etc., increasing the chances of success, and all this is from the distributed leadership exercised by the municipal council, in decision making and in the development of the work, which provides greater flexibility and agility in all processes and much more efficient results [26].

The commitment to clean energy in Almatret is not limited to an economic issue but goes beyond that and is committed to the production and energy transition of the municipality itself. The promotion of the rehabilitation of buildings towards the concept of passive houses, the installation of charging points to encourage electric transport and facilitating access to sustainable, reliable and modern energy are key aspects to achieve the goal of turning this town into a positive-energy and zero-carbon village [37]. This is an approximation to what some authors propose as a leap towards a smart village based on sustainable energy with the possibility of disconnecting from the grid and producing more energy than it consumes.

Sustainable development has also been promoted through research and innovation (through European projects such as Innova MicroSolar and HYBUILD) and the improvement of infrastructures. Networking with experts who provide technical advice helps to eliminate the isolation of rural areas and to bring innovative proposals closer together. Citizens are involved in scientific initiatives and raise the problems of the municipality to find solutions. This is an effective way of implementing the SDGs, with the collaboration of citizen science, which helps to strengthen research designs in the field of renewable energies, taking into account the interests and needs of citizens [38–40].

Finally, in this effort to effectively implement the SDGs, focusing on energy and linking research, innovation and education, ALMIA proposes the creation of the EIC as a strategy to address education in renewable energy and energy efficiency, contributing to formal and non-formal participatory science education and open to all audiences, although preferably to the student population [41–45].

The idea is to influence social transformation by generating knowledge among the population that will help to involve society in the promotion of Agenda 2030 and ESD 2030. This is part of the strategy known as education for global citizenship, which seeks to involve civil society in the transformation, and this is only possible if it is educated in the commitment to the wellbeing and quality of universal life.

Influencing the education of future generations to participate and take active roles, locally but also globally, makes them proactive contributors in responding to the challenges of today's society [46,47].

Currently, the ALMIA project is beginning to bear fruit: there is increased tourism to the village with organized academic visits (more than 250 in the last year and 1500 are expected for next year). Direct jobs have been created to revitalize the EIC and indirect jobs (opening of restaurants). The production, transport, efficient consumption and storage of clean energy has become a catalytic issue, contributing to the effective implementation of the SDGs in this rural village.

The ALMIA project presents specific limitations and barriers. It is difficult to reverse the isolation and depopulation of the municipality when it is a marked trend of the last decades. Depopulation has led to the suppression of services: schools, health services, commercial infrastructures.

The low public investment by the regional and state administration hinders the consolidation of the project, which is based on local funding and the voluntary work of the project's generators (municipal managers, university professors and researchers). The leadership of the local administration is important, but changes of government after elections can slow down the implementation of ALMIA.

The environment, the investment in renewable energies, and the enthusiasm of the population of the municipality support the project, although the lack of young people involved, because of an aging population, can endanger the continuity of ALMIA over time.

Author Contributions: Conceptualization, L.F.C. and I.d.A.; methodology, L.F.C. and I.d.A.; investigation, L.F.C., I.d.A., A.R.-P., A.d.G. and G.Z.; resources, L.F.C.; data curation, L.F.C.; writing—original draft preparation, L.F.C. and A.R.-P.; writing—review and editing, I.d.A., A.d.G. and G.Z.; visualization, A.d.G. and G.Z.; supervision, L.F.C. and I.d.A.; project administration, L.F.C. and I.d.A.; funding acquisition, L.F.C. and I.d.A. All authors have read and agreed to the published version of the manuscript.

Funding: This project has received funding from the European Union's Horizon 2020 research and innovation programme under grant agreement No 768824 and under the grant agreement No 723596. This work is partially supported by ICREA under the ICREA Academia programme.

Institutional Review Board Statement: Not applicable.

Informed Consent Statement: Not applicable..

Data Availability Statement: Data available under request to the correspondence authors.

Acknowledgments: The authors would like to thank the Catalan Government for the quality accreditation given to their research group (2017 SGR 1537). GREiA is certified agent TECNIO in the category of technology developers from the Government of Catalonia. The authors would like to thank the Almatret municipality for their involvement in this challenging project.

Conflicts of Interest: The authors declare no conflict of interest.

Abbreviations

SDGs	Sustainable Development Goals. There are 17 Sustainable Development Goals (SDGs), which are an urgent call for action by all countries (developed and developing) in a global partnership
ESD	Education for Sustainable Development. This action empowers learners of all ages with the knowledge, skills, values and attitudes to address the interconnected global, environmental degradation, loss of biodiversity, poverty and inequality
BEMS	Building Energy Management System. It offers monitoring, metering, as well as submetering, functions which help collect energy data, giving property managers and owners a comprehensive insight on building's energy usage
ALMIA	Name of the project that aims at increasing the sustainable development of the village of Almatret by means of becoming a positive energy village. The acronym comes from joining the name of the village (Almatret) with the word energy in Catalan (energia), i.e., ALMatret+energIA
EIC	Energy Interpretation Center
DC	Direct Current

References

1. United Nations. *The 2030 Agenda for Sustainable Development*; United Nations: New York, NY, USA, 2016.
2. Guerra, P. Comunitarismo y solidaridad, ¿apuestas para llegar a un desarrollo a escala humana. In *Desarrollo, Ciudadanía Y Cambio Social. Desafíos Para Las Políticas Públicas En Colombia*; Uniminuto, Minuto de Dios, DanSocial, Fundación Carolina y Fodesep: Bogotá, Columbia, 2007; pp. 41–58.
3. Domínguez, R. Agenda Internacional 2030: Perspectivas De La Cooperación Para El Desarrollo. Ph.D. Thesis, Universidad de San Buenaventura, Bogotá, Colombia, 2016.
4. Del Hidalgo, M.M. Las ciudades como objetivo de desarrollo sostenible. *Bie3 Boletín IEEE* **2017**, *5*, 22–32.
5. Frick, S.A.; Rodríguez-Pose, A. Big or small cities? On city size and economic growth. *Growth Chang.* **2018**, *49*, 4–32. [CrossRef]

6. Willett, W.; Rockström, J.; Loken, B.; Springmann, M.; Lang, T.; Vermeulen, S.; Garnett, T.; Tilman, D.; DeClerck, F.; Wood, A.; et al. Food in the anthropocene: The EAT-lancet commission on healthy diets from sustainable food systems. *Lancet* **2019**, *393*, 447–492. [CrossRef]
7. Watts, N.; Amann, M.; Arnell, N.; Ayeb-Karlsson, S.; Belesova, K.; Berry, H.; Bouley, T.; Boykoff, M.; Byass, P.; Cai, W.; et al. The 2018 report of the Lancet Countdown on health and climate change: Shaping the health of nations for centuries to come. *Lancet* **2018**, *392*, 2479–2514. [CrossRef]
8. FAO. 2018 Panorama De La Pobreza Rural En América Latina Y El Caribe. Available online: https://www.fao.org/3/CA2275ES/ca2275es.pdf (accessed on 15 May 2021).
9. FAO. 2017 Marco De Protección Social De La FAO. Promoviendo El Desarrollo Rural Para Todos. Available online: https://www.fao.org/3/i7016s/i7016s.pdf (accessed on 15 May 2021).
10. Thacker, S.; Adshead, D.; Fay, M.; Hallegatte, S.; Harvey, M.; Meller, H.; O'Regan, N.; Rozenberg, J.; Watkins, G.; Hall, J.W. Infrastructure for sustainable development. *Nat. Sustain.* **2019**, *2*, 324–331. [CrossRef]
11. Sodiq, A.; Baloch, A.A.B.; Khan, S.A.; Sezer, N.; Mahmoud, S.; Jama, M.; Abdelaal, A. Towards modern sustainable cities: Review of sustainability principles and trends. *J. Clean. Prod.* **2019**, *227*, 972–1001. [CrossRef]
12. Wachsmuth, D.; Angelo, H. Green and gray: New ideologies of nature in urban sustainability policy. *Ann. Am. Assoc. Geogr.* **2018**, *108*, 1038–1056. [CrossRef]
13. Ehnert, F.; Frantzeskaki, N.; Barnes, J.; Borgström, S.; Gorissen, L.; Kern, F.; Strenchock, L.; Egermann, M. The acceleration of urban sustainability transitions: A comparison of brighton, budapest, dresden, genk, and stockholm. *Sustainability* **2018**, *10*, 612. [CrossRef]
14. Fastenrath, S.; Braun, B. Sustainability transition pathways in the building sector: Energy-efficient building in Freiburg (Germany). *Appl. Geogr.* **2018**, *90*, 339–349. [CrossRef]
15. Zawadzka, A.K. Making small towns visible in Europe: The case of cittaslow network—The strategy based on sustainable development. *Transylv. Rev. Adm. Sci.* **2017**, 90–106. [CrossRef]
16. Ramírez, M. El hierro, en busca de la isla 100% renovable. *Técnica. Ind.* **2019**, *323*, 4–6.
17. Dávalos, J. Ciudades sostenibles, inclusivas y resilientes: Gobiernos locales y participación ciudadana en la implementación de las agendas globales para el desarrollo. *INNOVA Res. J.* **2017**, *2*, 116–131. [CrossRef]
18. Gifreu, J. Ciudades adaptativas y resilientes ante el cambio climático: Estrategias locales para contribuir a la sostenibilidad urbana. *Rev. Aragon. Adm. Pública.* **2018**, *52*, 102–158.
19. Byrne, J.; Mouritz, M.; Taylor, M.; Breadsell, J.K. East village at knutsford: A case study in sustainable urbanism. *Sustainability* **2020**, *12*, 6296. [CrossRef]
20. Croese, S.; Green, C.; Morgan, G. Localizing the sustainable development goals through the lens of urban resilience: Lessons and learnings from 100 resilient cities and cape town. *Sustainability* **2020**, *12*, 550. [CrossRef]
21. Mapiberia f&b. 443-2 ALMATRET 1:25,000. Available online: https://www.mapiberia.com/producto/443-2-almatret-1-25-000/6169 (accessed on 15 May 2021).
22. Kottek, M.; Grieser, J.; Beck, C.; Rudolf, B.; Rubel, F. World map of the köppen-geiger climate classification updated. *Meteorol. Zeitschrift* **2006**, *15*, 259–263. [CrossRef]
23. Valladares, F. El hábitat mediterráneo continental: Un sistema humanizado, cambiante y vulnerable. In *Ambientes Mediterráneos. Funcionamiento, Biodiversidad Y Conservación De Los Ecosistemas Mediterráneos*; Paracuellos, M., Ed.; Instituto de Estudios Almerienses: Almeria, Spain, 2007; pp. 2019–2040.
24. Botezan, C.; Constantin, V.; Meltzer, M.; Radovici, A.; Pop, A.; Alexandrescu, F.; Stefanescu, L. Is there sustainable development after mining? a case study of three mining areas in the apuseni region (Romania). *Sustainability* **2020**, *12*, 9791. [CrossRef]
25. BOE.es. *Resolución De La Generalitat De Cataluña, Dirección General De Energía Y Minas, De Fecha 5 De Noviembre De 2003, De Autorización Administrativa, Declaración De Utilidad Pública Y Aprobación Del Proyecto De Ejecución Sobre La Instalación De Producción Eléc*; Generalitat de Catalunya: Barcelona, Spain, 2003.
26. Contreras, R. Empoderamiento campesino y desarrollo local. *Rev. Austral De Cienc. Soc.* **2000**, *4*, 55–68. [CrossRef]
27. Santacruz Espinoza, A.; Montenegro Muguerza, H.; Pizarro Alejandro, A.; Estacio Flores, H. Liderazgo transformacional y desarrollo sostenible ambiental verde en docentes de la Universidad Nacional Herminio Valdizán. *Rev. De La SEECI* **2020**, *24*, 135–151. [CrossRef]
28. Innova-Microsolar. 2016. Available online: http://innova-microsolar.eu/ (accessed on 15 May 2021).
29. HYBUILD. 2017. Available online: http://www.hybuild.eu/ (accessed on 4 December 2020).
30. Adamson, D.; Bromiley, R. Community empowerment: Learning from practice in community regeneration. *Int. J. Public Sect. Manag.* **2013**, *26*, 190–202. [CrossRef]
31. Surya, B.; Suriani, S.; Menne, F.; Abubakar, H.; Idris, M.; Rasyidi, E.S.; Remmang, H. Community empowerment and utilization of renewable energy: Entrepreneurial perspective for community resilience based on sustainable management of slum settlements in Makassar City, Indonesia. *Sustainability* **2021**, *13*, 3178. [CrossRef]
32. Jeong, E.; Shim, C.; Brown, A.D.; Lee, S. Development of a scale to measure intrapersonal psychological empowerment to participate in local tourism development: Applying the sociopolitical control scale construct to tourism (SPCS-T). *Sustainability* **2021**, *13*, 4057. [CrossRef]

33. De España, G. *Plan De Acción Para La Implementación De La Agenda 2030: Hacia Una Estrategia Española De Desarrollo Sostenible*; Centre for Global Education: Madrid, Spain, 2018.
34. Leguizamon, N.Z.P.; Garcia-Parra, M. Empoderamiento de las comunidades rurales a través de la proyección social del conocimiento científico. *Cult. Científica* **2017**, *15*, 124–133.
35. United Nations. *Informe Sobre Desarrollo Humano 2009: Superando Barreras: Movilidad Y Desarrollo Humanos*; United Nations: New York, NY, USA, 2009.
36. Zimmerman, M.A. Empowerment theory. In *Handbook of Community Psychology*; Springer: Boston, MA, USA, 2000; pp. 43–63. [CrossRef]
37. Smart Villages. New Thinking for Off-Grid Communities Worldwide. Available online: https://rael.berkeley.edu/wp-content/uploads/2015/07/Smart_Villages_New_Thinking_for_Off_grid_Comunities_Worldwide.pdf. (accessed on 15 May 2021).
38. Santana, M.; Cabello, J.; Cubas, R.; Medina, V. *Redes Sociales Como Soporte A La Gestión Del Conocimiento*; Esan ediciones: Lima, Perú, 2011.
39. Senabre, E.; Ferran-Ferrer, N.; Perelló, J. Participatory design of citizen science experiments. *Comunicar* **2018**, *26*, 29–38. [CrossRef]
40. Anglada, L.; Abadal, E. ¿Qué es la ciencia abierta? *Anu. ThinkEPI* **2018**, *12*, 292. [CrossRef]
41. Murga-Menoyo, M.Á. El camino hacia los ODS: Conformar una ciudadanía planetaria mediante la educación. *Comillas J. Int. Relat.* **2020**, *19*, 1–11. [CrossRef]
42. Murga-Menoyo, M.A. La Formación de la Ciudadanía en el Marco de la Agenda 2030 y la Justicia Ambiental. *Rev. Int. Educ. Para La Justicia Soc.* **2018**, *7*, 37–52. [CrossRef]
43. Murga-Menoyo, M.Á.; Novo, M. Sostenibilidad, desarrollo glocal y ciudadanía planetaria. Referentes de una Pedagogía para el desarrollo sostenible. *Teoría La Educ. Rev. Interuniv.* **2017**, *29*, 55–78. [CrossRef]
44. UNESCO. *Global Citizenship Education: Preparing Learners for The Challenges of The 21st Century*; UNESCO: Paris, France, 2014.
45. UNESCO. *Education 2030: Incheon Declaration and Framework for Action for The Implementation of Sustainable Development Goal 4: Ensure Inclusive and Equitable Quality Education and Promote Lifelong Learning Opportunities for All*; UNESCO: Paris, France, 2016.
46. da Silva, M.C. Global education guidelines. In *Concepts and Methodologies on Global Education for Educators and Policymakers*; Gobierno De España: Lisbon, Portugal, 2019.
47. O'Donoghue, R. Transformative pedagogy for global citizenship education. *UNESCO Forum Glob. Citizsh. Educ.* **2013**.

sustainability

Article

Institutional Quality, Governance and Progress towards the SDGs

Edward B. Barbier * **and Joanne C. Burgess**

Department of Economics, Colorado State University, Fort Collins, CO 80523-1771, USA; jo.barbier@colostate.edu
* Correspondence: edward.barbier@colostate.edu; Tel.: +1-9-70-491-6324

Abstract: Whether institutional quality and governance help or hinder progress towards the 17 Sustainable Development Goals (SDGs) of UN Agenda 2030 is an important issue to consider. These fundamental social structures are generally under-represented among the SDGs, but institutional quality and governance often have an important role in supporting or constraining efforts to achieve sustainable development. We compare estimates of the changes in net welfare that reflect progress towards the 17 SDGs over 2000–2018 with two institutional quality and governance indicators over the same period. We do this at the world level, for the group of low-income countries and for nine representative developing countries. We utilize the Worldwide Governance Indicators and OECD's Country Risk Classification as our two institutional quality and governance measures. We find that SDG welfare gains are somewhat correlated with institutional quality and highly correlated with lower country risk. These results suggest that good governance and institutional effectiveness are associated with long-run development and sustainability success. Long-term progress towards the SDGs may hinge on improved institutional quality and reduced country risk.

Keywords: country risk; developing economies; governance; institutional quality; low-income countries; SDGs; sustainable development

Citation: Barbier, E.B.; Burgess, J.C. Institutional Quality, Governance and Progress towards the SDGs. *Sustainability* **2021**, *13*, 11798. https://doi.org/10.3390/su132111798

Academic Editors: Margarita Martinez-Nuñez, Mª Pilar Latorre-Martínez and Antonio Boggia

Received: 1 September 2021
Accepted: 21 October 2021
Published: 26 October 2021

1. Introduction

Since the early 1990s, a large empirical literature in economics has established a persuasive link between institutions and long-run economic development [1–12]. Institutional quality is a broad concept that refers to law, individual rights and the provision of government regulation and services. A breakdown in these attributes undermines and weakens the institutional framework supporting economic development. Although the causality of the relationship between strong institutions and long-run economic development may be in doubt [2], the possibility that economic progress, institutional quality and good governance go hand-in-hand is rarely questioned. Recent evidence also shows that more effective institutions and governance are essential to reducing extreme poverty and achieving other development goals that are important to low and middle-income economies [13].

This raises an important question. As the world shifts its focus towards making development more "sustainable", as reflected in the UN's Agenda 2030, are weak institutions and governance inhibiting long-term progress towards sustainability, especially in poorer economies? For example, does a break down in factors such as the effective rule of law, an uncertain business climate, insecure property rights and the presence of social norms that are not conducive to market-based trade and transactions, constrain sustainable development? [10,11].

In 2015, the General Assembly of the United Nations (UN) formally adopted "The 2030 Agenda for Sustainable Development". This provides a framework for "peace and prosperity for people and the planet, now and into the future" [14]. The centerpiece of Agenda 2030 are the 17 Sustainable Development Goals (SDGs). As Jeffrey Sachs emphasized, the SDGs "aim for a combination of economic development, environmental sustainability and social inclusion" [15], p. 2206. Attaining the SDGs can be viewed as sustainable development in its broadest sense, through achieving progress across economic, social and environmental systems simultaneously [16–18]. This approach has been called the systems approach to sustainability, and is attributed to Barbier (1987) [19]. The 17 SDGs can be attributed to the economic, social and environmental systems. For example SDG 1: No Poverty, and SDG 8 Good Jobs and Economic Growth are goals within the economic system. SDG 13 Climate Action and SDG 14 Life below Water are environmental system goals. Within the social system goals, SDG 10 Reduced Inequality and SDG 16 Peace, Justice and Institutions reside [16–18].

The systems approach to sustainability contrasts with the economists' capital approach to sustainability [20]. The capital approach to sustainability treats nature as a form of capital. In order to ensure that future generations have at least the same economic opportunities, and thus the same economic welfare, as the present generation, the capital approach to sustainability entails managing and enhancing a portfolio of assets. This portfolio of assets comprise the total capital stock, which consists of physical, human and natural capital. In addition to maintaining or enhancing the total stock of capital, any essential natural capital needs to be kept "intact" due to imperfect substitution, irreversible losses and uncertainty over values [20].

Within the SDGs systems approach to sustainability, institutional attributes seem to be under-represented in the 17 SDGs of Agenda 2030. Only one of the goals, SDG 16 Peace, Justice and Strong Institutions, includes these attributes, and they are narrowly defined. For example, the 23 indicators currently proposed by the UN to track progress towards SDG 16 focus mainly on reducing violence, conflict and political instability rather than on broader measures of institutional quality and governance [21]. Although peace and political stability are essential to sustainable development, strong institutions also imply creating the economy-wide business and market conditions to sustain economic progress and development through promoting effective government, the rule of law, reducing unnecessary regulation, greater accountability and minimizing corruption and risk. As summarized by the World Bank: "Strong institutions and conducive business climates can set the stage for vigorous growth. Institutions can promote forms of economic activity that are associated with greater economic complexity and higher productivity growth by encouraging human capital accumulation and innovative activities." [22], p. 40.

In this article, we examine further whether better institutional quality and a more conducive business climate can help or hinder advancement towards the 17 Sustainable Development Goals (SDGs) of UN Agenda 2030. To assess attainment of these goals, we make use of our existing analysis of the net welfare changes of advancement towards the 17 SDGs over the period 2000–2018. We do this at the world level, for the group of low-income countries and for nine representative developing countries [17]. For the same sets of countries, we contrast and compare our estimates of net welfare gains or losses with two institutional quality and governance indicators over the same period.

Given that institutional quality is a broad concept that reflects the state of the law, individual rights and the provision of government regulation and services within a country, it is difficult to obtain precise indicators and reliable data. We use the Worldwide Governance Indicators (WGI) [23] and Country Risk Classification (CRC) of the Organization for Economic Cooperation and Development (OECD) [24] as our two broad institutional quality and governance measures. Although such indicators have attracted some criticisms, they have become widely accepted and used by policy makers and academics [11]. The WGI include not only an indicator representing political stability and absence of violence, but also indicators for control of corruption, government effectiveness, regulatory quality, rule of law, voice and accountability. The CRC measures broadly the risk of "doing business" in a country. It includes a measure of credit risk and a qualitative assessment of other factors that contribute to country level risk, such as natural disasters including floods, earthquakes and other natural hazards, as well as man-made crises such as war, expropriation, revolution, civil disturbance. We use the WGI to construct an overall indicator of institutional quality and the CRC to represent country risk.

We find that SDG welfare gains are somewhat correlated with institutional quality and highly correlated with lower country risk. Ineffective institutions and country risk seem to be especially associated with a lack of SDG progress in poorer economies. The implication is that the inability to improve governance may be constraining the long-term welfare and development of many poorer economies. On the positive side, countries with better quality institutions and lower risk appear to have made overall gains towards fulfilling the SDGs. These results suggest that effective institutions and good governance are associated with long-run development and greater sustainability. Long-term progress towards the SDGs may hinge on improved institutional quality and reduced country risk.

2. Materials and Methods

We have previously developed a methodological approach in order to assess progress towards the 17 SDGs, based on welfare economics [16,17]. Our approach to estimating the possible net benefits of making progress towards one SDG goal, while accounting for simultaneous declines or improvements in achieving other goals, is based on standard economic methods for measuring the welfare effects arising from changes in imposed quantities [25,26]. Here, we use the approach previously developed by Barbier and Burgess [16] to estimate the welfare effects of progress in attaining one SDG while accounting for interactions in achieving other SDGs. In essence, this analytical framework allows us to estimate the "willingness to pay" (WTP) in dollar terms by a representative individual for an improvement in one SDG indicator, whilst taking into account possible simultaneous changes–positive or negative–in other SDG indicators. The reason it is important to do this is that many assessments have shown that, since 2000, there has been considerable variation from country to country in the progress towards attaining the SDGs, as well as between the world and low-income economies. Furthermore, that progress in attaining any individual goal may have led to the reduction (or increase) of achievement in other goals [16,17,27–30].

We have applied our method at the world level, for the group of low-income countries and for nine representative developing countries over the period 2000–2018 [17]. We use the World Bank's classification of countries by income [31]. Low-income economies are defined by the World Bank as those with a Gross National Income (GNI) per capita of 1045 USD or less in 2020; lower middle-income economies are those with a GNI per capita between 1046 USD and 4095 USD; and upper middle-income economies are those with a GNI per capita between 4096 USD and 12,695 USD. The remaining countries of the world are classified as high income, with a GNI per capita of 12,536 USD or more. The nine representative countries we chose are Malawi, Rwanda and Uganda (three low-income economies), Bangladesh, Bolivia and the Kyrgyz Republic (three lower middle-income economies) and Colombia, Dominican Republic and Indonesia (three upper middle-income economies). In selecting these countries, we also took into account the extent to which a country has made progress since 2000 towards achieving the main goal used in our analysis, which is SDG 1 No Poverty. For each income group we chose three countries that showed long-term poverty reduction since 2000, and our nine countries vary between those with small to very large declines in poverty. Finally, in choosing countries, we also considered their geographic distribution. Consequently, we selected three countries each from Asia, Latin America and the Caribbean, and Sub-Saharan Africa [17].

Table 1 summarizes the 17 SDGs and the main indicators that we used to measure progress towards each goal. In our analysis, we chose SDG 1 No Poverty as the benchmark indicator. We estimate the change in per capita welfare from any reduction in 2000–2018 poverty rates, and adjust this to take account of any gains or losses that may occur when simultaneously achieving each of the other 16 SDGs. Further details on how we chose these indicators and the methods we use to measure these welfare gains or losses can be found in [11]. For SDG 16, Peace, Justice and Strong Institutions, we use as a representative indicator *political stability and absence of violence/terrorism* from the Worldwide Governance Indicators (WGI) [23] (Table 1).

Table 1. The 17 SDGs and their indicators for assessing progress.

	Sustainable Development Goal	Indicator [1]
1.	No Poverty	Poverty headcount ratio at $1.90 a day (2011 PPP) (% of population)
2.	Zero Hunger	Prevalence of undernourishment (% of population)
3.	Good Health and Well Being	Maternal mortality ratio (per 100,000 live births)
4.	Quality Education	Adolescents out of school (% of lower secondary school age)
5.	Gender Equality	Lower secondary completion rate, female (% of relevant age group)
6.	Clean Water and Sanitation	People using at least basic drinking water services (% of population)
7.	Affordable and Clean Energy	Access to clean fuels and technologies for cooking (% of population)
8.	Good Jobs and Economic Growth	Adjusted net national income per capita (annual % growth)
9.	Industry, Innovation and Infrastructure	Manufacturing, value added (% of GDP)
10.	Reduced Inequalities	GINI index
11.	Sustainable Cities and Communities	PM2.5 air pollution, population exposed to levels exceeding WHO guideline value (% of total)
12.	Responsible Consumption and Production	Adjusted net savings, excluding particulate emission damage (% of GNI)
13.	Climate Action	CO_2 emissions (metric tons per capita)
14.	Life Below Water	Total fisheries production (metric tons)
15.	Life on Land	Forest area (sq. km)
16.	Peace, Justice and Strong Institutions	Political stability and absence of violence/terrorism (−2.5 to 2.5)
17.	Partnerships for the Goals	Debt service (% of exports)

[1] All indicators are available from [31], except for political stability and absence of violence/terrorism, which is from [23]. GDP = gross domestic product. GNI = gross national income. Table 1 was created by the authors and adapted from [17].

In this article, we contrast and compare our estimates from our previous study of the net welfare changes of progress towards the 17 SDGs over 2000–2018 [17], with two institutional quality and governance measures over the same period at the world level, for the group of low-income countries and for nine representative developing countries. Here, we use the Worldwide Governance Indicators (WGI) [23] and Country Risk Classification (CRC) of the Organization for Economic Cooperation and Development (OECD) [24] for our two broad institutional quality and governance measures, respectively.

The World Governance Indicators consists of six measures of institutional quality and governance. These include *political stability and absence of violence/terrorism*, which measures perceptions of the likelihood of political instability and/or politically-motivated violence, including terrorism. The second indicator, *control of corruption*, reflects perceptions of the extent to which public power is exercised for private gain, including both petty and grand forms of corruption, as well as "capture" of the state by elites and private interests. The third indicator, *government effectiveness*, indicates perceptions of the quality of public services, the quality of the civil service and the degree of its independence from political pressures, the quality of policy formulation and implementation, and the credibility of the government's commitment to such policies. The fourth indicator, *regulatory quality*, reflects perceptions of the ability of the government to formulate and implement sound policies and regulations that permit and promote private sector development. The fifth indicator, *rule of law*, captures perceptions of the extent to which agents have confidence in and abide by the rules of society, and in particular the quality of contract enforcement, property rights, the police, and the courts, as well as the likelihood of crime and violence. And finally the sixth indicator, *voice and accountability*, measures perceptions of the extent to which a country's citizens are able to participate in selecting their government, as well as freedom of expression, freedom of association, and a free media [23].

These six measures of Worldwide Governance Indicators are scaled, with the lowest value at −2.5 and the highest value at 2.5. The current database covers the period from 1996 to 2019, and includes over 200 countries. Using this information, we are able to derive a measure of institutional quality based on an average of the six measures in the WGI from 2000 to 2018 at the world level, for the group of low-income countries and for nine representative developing countries. Finally, we rescale this average institutional quality indicator over 2000 to 2018 from 0 (lowest value) to 5 (highest value).

According to the OECD, its Country Risk Classification (CRC) is "one of the most fundamental building blocks of the Arrangement rules on minimum premium rates for credit risk" [24]. Consequently, the CRC is a broad measure of country risk, which "encompasses transfer and convertibility risk (i.e., the risk a government imposes capital or exchange controls that prevent an entity from converting local currency into foreign currency and/or transferring funds to creditors located outside the country) and cases of force majeure (e.g., war, expropriation, revolution, civil disturbance, floods, earthquakes)" [24].

The CRC is scaled from 0 (lowest) to 7 (highest). The current classification covers 201 countries from 1999 to 2021. Consequently, for each country we take the average of its final year score over 2000 to 2018, and thus construct an average measure of this period at the world level, for the group of low-income countries and for nine representative developing countries.

3. Results

Table 2 summarizes the comparison of our estimates of the changes in net welfare from progress towards the 17 SDGs over the period from 2000 to 2018 with average institutional quality and average country risk over the same period at the world level, for the group of low-income countries and for nine representative developing countries.

Table 2. Progress towards the SDGs, institutional quality and country risk, 2000–2018.

Countries	Net Welfare Change ($ per Capita) 2000–2018 [1]	Institutional Quality 2000–2018 Average [2]	Country Risk 2000–2018 Average [3]
World	$3633	2.5	4.5
Low Income Countries	−$29	1.4	6.9
Malawi	$784	2.1	7.0
Rwanda	−$218	2.1	6.8
Uganda	−$520	1.9	6.4
Bangladesh	$1115	1.6	5.7
Bolivia	−$2076	2.0	6.3
Kyrgyz Republic	−$5287	1.7	7.0
Colombia	$10,068	2.1	4.4
Dominican Republic	$14,968	2.2	5.1
Indonesia	$4363	2.0	4.4
Average 9 countries	$2578	2.0	3.0

[1] From [17]. [2] Institutional quality is the average of the 2000–2018 indicators for control of corruption, government effectiveness, political stability and absence of violence/terrorism, regulatory quality, rule of law, and voice and accountability of the Worldwide Governance Indicators [23]. Institutional quality has been rescaled from 0 (lowest quality) to 5 (highest quality). [3] Country risk is the average of the OECD Country Risk Classification [24]. Indicator ranges from 0 (lowest risk) to 7 (highest risk). Table 2 was created by the authors.

Our analysis of changes in net welfare from progress towards the 17 sustainable development goals shows that there are substantial differences in the level of interactions among the SDGs and the corresponding net welfare effects at the global level compared to that experienced by low-income countries [17]. At the global level, there are some welfare losses through tradeoffs with declining SDG indictors over 2000–2018, but these losses are mostly compensated by increases in other SDG indictors. Once such interactions are taken into account, there is an overall net gain in welfare of 3633 USD per person on average at the world level from 2000 to 2018 (Table 2). Therefore, our welfare analysis suggests that there has been an overall enhancement in sustainability at the global level from 2000 to 2018.

In comparison, for poor economies, the tradeoffs from declining SDG indicators surpass the gains in welfare from improving indicators over the period from 2000 to 2018. As a result, for low-income countries, these interactions imply that countries in this group have experience a net loss in welfare over the period from 2000 to 2018 of 29 USD per person on average. This means that in contrast to the world as a whole, low-income economies experienced an overall reduction rather than an improvement in sustainability from 2000 to 2018.

This difference in the sustainability performance of poorer economies is also demonstrated across our nine representative countries. Here, we see that all nine countries benefit from progress towards SDG 1 No Poverty. However, when we take into account interactions with other SDGs, the less well-off countries tend to perform less well. For example, two of the three low-income economies–Rwanda and Uganda–and two of the three lower middle-income economies–Bolivia and the Kyrgyz Republic–all experience an overall loss in sustainability over the period from 2000 to 2018. In contrast, over this same period, the three upper middle-income economies demonstrate substantial gains from overall increase in sustainability over the period from 2000 to 2018 (Table 2).

As Table 2 shows, a similar pattern emerges for average institutional quality over 2000–2018, which ranges from 0 (low quality) to 5 (high quality). The world displays a much higher level of effective institutions and governance (2.5) than low-income countries (1.4). However, institutional quality is more mixed among our nine representative countries. For example, Colombia, Dominican Republic and Indonesia (our three upper middle-income economies) generally have the highest institutional quality, as well as Bolivia (which is a lower middle-income economy), and also Malawi and Rwanda (which are low-income countries).

Country risk appears to be much more closely associated with net welfare gains and losses (Table 2). This indicator measure ranges from 0 (low risk) to 7 (high risk). Once again, low-income countries display much higher risk (6.9) compared to the world on average (4.5). This is also the case at the country level. For example, Malawi, Rwanda and Uganda (our three low-income countries) show country risk levels of 7.0, 6.8 and 6.4 respectively. Very high country risk (7.0) is also displayed by the Kyrgyz Republic, which is a lower middle-income economy, and Bangladesh and Bolivia (also middle-income economies), have slightly lower risk at 5.7 and 6.3 respectively. In contrast, the country risk for the three countries in the upper middle-income category are close to the world average.

4. Discussion

Figures 1 and 2 further aid the interpretation of these results, especially with respect to understanding whether institutional quality and governance help or hinder progress towards the 17 Sustainable Development Goals (SDGs) of UN Agenda 2030.

Figure 1 depicts the pairwise comparison between net welfare gains or losses from attaining the SDGs over 2000–2018 with average institutional quality over this period for the world level, for the group of low-income countries and for nine representative developing countries. Across these estimates, institutional quality displays modest correlation (0.49) with net welfare change in progress towards sustainability over 2000–2018.

As Figure 1 shows, the association between more effective institutions, good governance and progress towards the SDGs over 2000–2018 cannot be ruled out. On the whole, countries with better institutional quality achieved more progress compared to those with weaker institutions. Especially notable is that Colombia, Dominican Republic and Indonesia (the three upper middle-income countries), appear to show reasonable institutional quality for developing countries and achieved significant net welfare gains through SDG progress over 2000–2018.

Overall, by comparing and contrasting our welfare analysis [17] to average institutional quality over 2000–2018 we are able to provide some evidence to support the view that institutional effectiveness and good governance are essential for successful long-run sustainable development [1–13,22]. As our results show for our three upper middle-income countries, there appears to be a synergistic relationship between economic progress, sustainability and enhancements in institutional quality. Unfortunately, several countries appear to be experiencing a trade-off between institutional quality and making progress towards the 17 SDGs. In particular, for poorer countries, the lack of progress towards sustainability and strengthening governance may create a chronic problem that undermines progress on long-term development and improvements in welfare.

Figure 1. This figure plots the pairwise estimates from Table 2 of net welfare change per capita in achieving the 17 SDGs over 2000–2018 and the average institutional quality over 2000–2018 for the world level, for the group of low-income countries and for nine representative developing countries. These estimates have a positive correlation of 0.49.

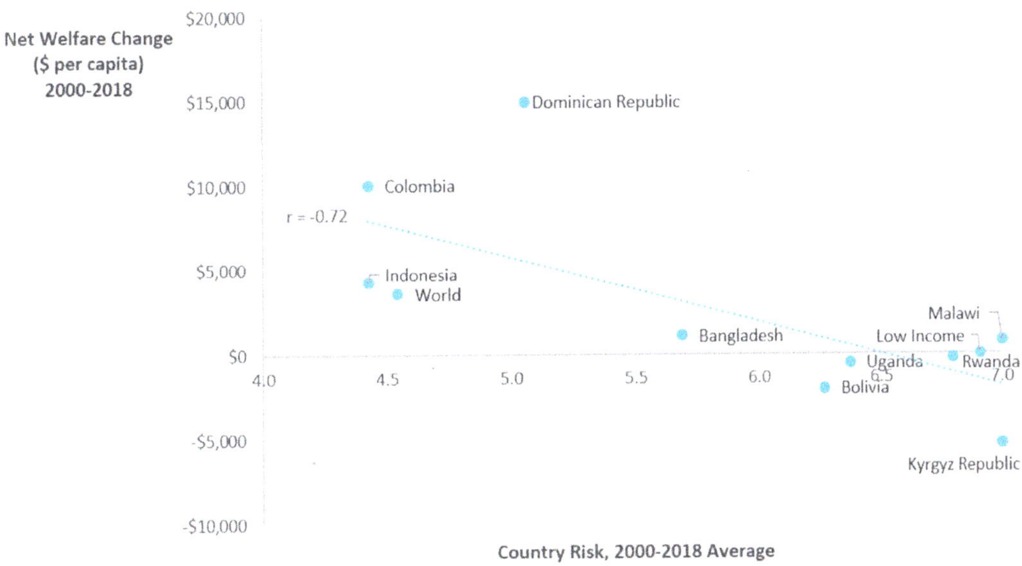

Figure 2. This figure plots the pairwise estimates from Table 2 of net welfare change per capita in achieving the 17 SDGs over 2000–2018 and the average country risk over 2000–2018 for the world level, for the group of low-income countries and for nine representative developing countries. These estimates have a negative correlation of −0.72.

Figure 2 displays the pairwise comparison between net welfare gains or losses from attaining the SDGs over 2000–2018 with average country risk over this period for the world level, for the group of low-income countries and for nine representative developing countries. Across these estimates, country risk displays high negative correlation (−0.72) with net welfare change in achieving the SDGs over 2000–2018.

Figure 2 shows that the association between lower country risk and improvements towards the SDGs over the period from 2000 to 2018 is very strong. Countries with lower risk achieved more progress compared to those with higher risk. Especially striking is the results for Colombia, Dominican Republic and Indonesia (our three upper middle-income countries), which generally have much lower risk and higher net welfare gains compared to less well-off countries and the group of low-income economies. The group of low-income economies tend to have much higher country risk and display only modest gains–and often net losses–in attaining the SDGs over 2000–2018.

The strong association between country risk and sustainability performance may seem surprising, but it fits with other analyses of the relationship between business climate and long-run economic performance and attaining development goals, such as poverty reduction [13,17,22]. It also accords with other analyses that show, as a whole, poorer economies have not been faring well in overall progress towards the SDGs, and especially have performed badly in terms of the environmental goals SDGs 11–15 [16,17,27,28,30]. As explained by Jeffrey Sachs and colleagues, ineffective institutions may be a factor: poorer economies not only have lower overall SDG Index scores but also "they tend to lack adequate infrastructure and mechanisms to manage key environmental challenges covered under SDGs 12–15" [30], p. 25.

These results provide further support to the substantial empirical literature in economics that has established a persuasive link between institutions and long-run economic development [1–13,22]. These results also confirm recent studies that have shown that more effective institutions and governance are essential to reducing extreme poverty and achieving other development goals that are important to low and middle-income economies [12,13].

5. Conclusions

This article compares and contrasts estimates of the changes in net welfare from progress towards the 17 SDGs over the period from 2000 to 2018 with measures of institutional quality and country risk over the same period for the world level, for the group of low-income countries and for nine representative developing countries. These comparisons shed some light on whether institutional quality and governance help or hinder progress towards the 17 Sustainable Development Goals (SDGs) of UN Agenda 2030.

Although it is difficult to obtain precise indicators and reliable data on the broad concept of institutional quality and governance, and such indicators have attracted notable criticisms in the past, the WGI and CRC measures of institutional quality and governance are now widely accepted by policy makers and academics [11]. Overall, we find that SDG welfare gains are somewhat correlated with institutional quality and highly correlated with lower country risk. Countries with better quality institutions and lower risk appear to have made overall gains towards fulfilling the SDGs. These results suggest that good governance and institutional effectiveness are associated with success in achieving long-run sustainable development objectives. Therefore, long-term progress towards the SDGs may hinge on improved institutional quality and reduced country risk.

Unfortunately, ineffective institutions and country risk seems especially associated with lack of SDG progress in poorer economies. The implication is that, for many low-income countries, the lack of progress towards sustainability and improving governance may be a fundamental problem that is undermining their long-term development and welfare. This does not bode well for poorer economies, who generally display poorer institutional quality and much higher country risk. In addition, these economies have struggled to achieve progress towards attaining the 17 Sustainable Development Goals, and especially SDGs 11–15 [16,17,27,28,30].

A major concern is that poorer economies are facing even greater development burdens with the ongoing COVID-19 pandemic. They are particularly affected by mounting debt, inequality and poverty. These challenges will further constrain their ability to build strong institutions, improve governance and reduce credit risk.

Due to the pandemic, global debt reached 289 USD by the end of the first quarter of 2021, and accounts for just over 369% of global GDP [32]. Around 86 trillion USD of this debt is in emerging market economies [33]. Research has shown that mounting debt can severely exacerbate the duration and intensity of recessions in emerging market economies, in part due to less supportive fiscal policies in these countries during times of crises [34]. Emergency measures established during the COVID-19 pandemic, including the Debt Service Suspension Initiative that was established by the International Monetary Fund and the World Bank, has provided poorer countries a short-term respite from payments on debt. However, there is not yet any sign of a longer-term comprehensive debt relief program for the world's poorest countries [35]. Growing indebtedness in developing countries will further undermine their credit worthiness and make it extremely difficult to ameliorate their high levels of country risk.

Furthermore, inequality has increased during the pandemic as the world's richest have become wealthier and poverty reduction has been setback substantially [26–28]. Worldwide in 2020, there was an increase of 3.9 trillion USD in the wealth of billionaires. In contrast, the total number of people living in poverty may have increased by 200 to 500 million during the pandemic [36]. As many as 70 to 100 million people across the world could fall into extreme poverty, which is the first rise in extreme poverty over two decades [37,38]. Shared prosperity–the relative increase in the incomes of the bottom 40% of the population compared to that of the entire population–is anticipated to decrease sharply in nearly all countries in 2020–2021. This decline in shared prosperity will be even more significant if the pandemic's economic impacts continue to fall disproportionately on poor people [38].

The pandemic could be especially devastating for the inclusivity of global development seriously in terms of extreme poverty and shared prosperity. Even before the COVID-19 pandemic, the global community was still a long way from achieving critical sustainability and development objectives for the most vulnerable people and countries. For example, in 2019, as many as 736 million people lived in extreme poverty, 821 million were undernourished, 785 million people lacked basic drinking water services, and 673 million people across the globe were without sanitation [39]. About 3 billion people did not have access to clean cooking fuels and technology, and on top of this, of the 840 million people without electricity, 87% lived in rural areas. It has been projected that as many as 28 poor countries could fall short of attaining SDGs 1–4, 6 and 7 by 2030 [18].

As a final observation, our article provides support for the view that long-term progress towards the SDGs may be associated with improved institutional quality and reduced country risk. However, this association does not necessarily imply that "the causality runs from institutions to economic development, ignoring the important possibility that economic development changes institutions" [2], p. 476. Clearly, further research and more country-level data are required to statistically analyze the relationship between net welfare changes, institutional qualities and country risks to determine conclusively whether improved institutions and governance will necessarily lead to better progress towards sustainability, as reflected in the 17 Sustainable Development Goals. As our article suggests, this is a rich and important area for further research.

Author Contributions: Conceptualization, E.B.B. and J.C.B.; methodology, E.B.B.; formal analysis, E.B.B.; writing—original draft preparation, E.B.B. and J.C.B.; writing—review and editing, E.B.B. and J.C.B. All authors have read and agreed to the published version of the manuscript.

Funding: This research received no external funding.

Institutional Review Board Statement: Not applicable.

Informed Consent Statement: Not applicable.

Data Availability Statement: Not applicable.

Conflicts of Interest: The authors declare no conflict of interest.

References

1. Acemoglu, D.; Johnson, S.; Robinson, J.A. Institutions as the fundamental cause of long-run growth. In *Handbook of Economic Growth*; Aghion, P., Durlauf, S., Eds.; North-Holland: Amsterdam, The Netherlands, 2005; pp. 385–472.
2. Chang, H.-J. Institutions and economic development: Theory, policy and history. *J. Inst. Econ.* **2010**, *7*, 473–498. [CrossRef]
3. Easterly, W. Institutions: Top down or bottom up? *Am. Econ. Rev.* **2008**, *98*, 95–99. [CrossRef]
4. Glaeser, E.; La Porta, R. Do institutions cause growth? *J. Econ. Growth* **2004**, *9*, 271–303. [CrossRef]
5. Gradstein, M. Governance and growth. *J. Dev. Econ.* **2004**, *73*, 505–518. [CrossRef]
6. Rodrik, D.; Subramanian, A.; Trebbi, F. Institutions rule: The primacy of institutions over geography and economic integration in economic development. *J. Econ. Growth* **2004**, *9*, 131–165. [CrossRef]
7. Acemoglu, D.; Robinson, J.A. *The Role of Institutions in Growth and Development*; World Bank: Washington, DC, USA, 2008; Volume 10.
8. Acemoglu, D.; Robinson, J.A. *Why Nations Fail: The Origins of Power, Prosperity, and Poverty*; Crown Business: New York, NY, USA, 2012.
9. Lehne, J.; Mo, J.; Plekhanov, A. *What Determines the Quality of Economic Institutions? Cross-Country Evidence*; Working Paper 171; European Bank for Reconstruction and Development: London, UK, 2014.
10. Kaufmann, D.; Kraay, A. *Growth without Governance*; Policy Research Working Paper 2002. No. 2928; World Bank: Washington, DC, USA, 2002.
11. Kaufmann, D.; Kraay, A.; Mastruzzi, M. *Worldwide Governance Indicators Project: Answering the Critics*; Policy Research Working Paper 2007. No. 4149; World Bank: Washington, DC, USA, 2002.
12. Wilson, R. Does governance cause growth? Evidence from China. *World Dev.* **2016**, *79*, 138–151. [CrossRef]
13. Asadullah, M.N.; Savoia, A. Poverty reduction during 1990–2013: Did millennium development goals adoption and state capacity matter? *World Dev.* **2018**, *105*, 70–82. [CrossRef]
14. United Nations. *Transforming Our World: The 2030 Agenda for Sustainable Development*; United Nations: New York, NY, USA, 2015.
15. Sachs, J.D. From millennium development goals to sustainable development goals. *Lancet* **2012**, *379*, 2206–2211. [CrossRef]
16. Barbier, E.B.; Burgess, J.C. Sustainable development goal indicators: Analyzing trade-offs and complementarities. *World Dev.* **2019**, *122*, 295–305. [CrossRef]
17. Barbier, E.B.; Burgess, J.C. *Economics of the SDGs: Putting the Sustainable Development Goals into Practice*; Palgrave Macmillan: London, UK; New York, NY, USA, 2021.
18. Barbier, E.B.; Burgess, J.C. Sustainability, the systems approach and the sustainable development goals. *Cah. D'économie Polit. Pap. Political Econ.* **2021**, *79*, 31–59. [CrossRef]
19. Barbier, E.B. The concept of sustainable economic development. *Environ. Conserv.* **1987**, *4*, 101–110. [CrossRef]
20. Barbier, E.B. *Capitalizing on Nature: Ecosystems as Natural Assets*; Cambridge University Press: Cambridge, UK; New York, NY, USA, 2011; p. 321.
21. United Nations. *Global Indicator Framework for the Sustainable Development Goals and Targets of the 2030 Agenda for Sustainable Development*; United Nations: New York, NY, USA, 2020. Available online: https://unstats.un.org/sdgs/indicators/Global%20Indicator%20Framework%20after%202020%20review_Eng.pdf (accessed on 23 August 2021).
22. World Bank. *Data, Digitalization, and Governance*; Europe and Central Asia Economic Update, Office of the Chief Economist, Spring 2021; World Bank: Washington, DC, USA, 2021.
23. World Bank. *Worldwide Governance Indicators*; World Bank: Washington, DC, USA. Available online: https://databank.worldbank.org/source/worldwide-governance-indicators (accessed on 23 August 2021).
24. OECD. *Country Risk Classification*; Organization for Economic Cooperation and Development (OECD): Paris, France. Available online: http://www.oecd.org/tad/xcred/crc.htm (accessed on 18 August 2021).
25. Freeman, A.M., III. *The Measurement of Environmental Values: Theory and Methods*, 2nd ed.; Resources for the Future: Washington, DC, USA, 2003.
26. Lankford, R.H. Measuring welfare changes in settings with imposed quantities. *J. Environ. Econ. Manag.* **1988**, *15*, 45–63. [CrossRef]
27. Campagnolo, L.; Eboli, F.; Farnia, L.; Carraro, C. Supporting the UN SDGs transition: Methodology for sustainability assessment and current worldwide ranking. *Economics* **2018**, *12*, 1–31. [CrossRef]
28. Moyer, J.D.; Hedden, S. Are we on the right path to achieve the sustainable development goals? *World Dev.* **2020**, *127*, 104749. [CrossRef]
29. Pradhan, P.; Costa, L.; Rybski, D.; Lucht, W.; Kropp, J.P. A systematic study of sustainable development goal (SDG) interactions. *Earth's Future* **2017**, *5*, 1169–1179. [CrossRef]
30. Sachs, J.; Schmidt-Traub, G.; Kroll, C.; Lafortune, G.; Fuller, G.; Woelm, F. *The Sustainable Development Goals and COVID-19. Sustainable Development Report 2020*; Cambridge University Press: Cambridge, UK, 2020.
31. World Bank. *World Bank Country and Lending Groups*; World Bank: Washington, DC, USA. Available online: https://datahelpdesk.worldbank.org/knowledgebase/articles/906519-world-bank-country-and-lending-groups (accessed on 18 August 2021).
32. World Bank. *World Development Indicators*; World Bank: Washington, DC, USA. Available online: https://databank.worldbank.org/source/world-development-indicators (accessed on 28 October 2020).

33. Tiftik, E.; Mahmood, K. *Chipping Away at the Mountain? Global Debt Monitor*; Institute of International Finance: Washington, DC, USA, 2021. Available online: https://www.iif.com/Portals/0/Files/content/Global%20Debt%20Monitor_May2021_vf.pdf (accessed on 23 August 2021).
34. Bernardini, M.; Forni, L.; Pattillo, C.A. *Private and Public Debt*; IMF Working Papers, 2017(061); A001; International Monetary Fund: Washington, DC, USA, 2017. Available online: https://www.elibrary.imf.org/view/journals/001/2017/061/article-A001-en.xml (accessed on 30 August 2021).
35. Volz, U.; Akhtar, S.; Gallagher, K.P.; Griffith-Jones, S.; Haas, J.; Kraemer, M. *Debt Relief for a Green and Inclusive Recovery: A Proposal*; Heinrich-Böll-Stiftung: Berlin, Germany; SOAS, University of London: London, UK; Boston University: Boston, MA, USA, 2020. Available online: https://eprints.soas.ac.uk/34346/1/DRGR-report.pdf (accessed on 23 August 2021).
36. Oxfam. *The Inequality Virus: Bringing Together a World Torn Apart by Coronavirus through a Fair, Just and Sustainable Economy*; Oxfam: Oxford, UK, 2021. Available online: https://policy-practice.oxfam.org/resources/the-inequality-virus-bringing-together-a-world-torn-apart-by-coronavirus-throug-621149/ (accessed on 23 August 2021).
37. United Nations (UN). *Sustainable Development Goals Report 2020*; United Nations: New York, NY, USA, 2020. Available online: https://unstats.un.org/sdgs/report/2020/The-Sustainable-Development-Goals-Report-2020.pdf (accessed on 23 August 2021).
38. World Bank. *Poverty and Shared Prosperity 2020: Reversals of Fortune*; World Bank: Washington, DC, USA, 2020. Available online: https://www.worldbank.org/en/publication/poverty-and-shared-prosperity (accessed on 23 August 2021).
39. United Nations (UN). *The Sustainable Development Goals Report 2019*; United Nations: New York, NY, USA, 2019.Available online: https://unstats.un.org/sdgs/report/2019/The-Sustainable-Development-Goals-Report-2019.pdf (accessed on 23 August 2021).

sustainability

MDPI

Article

Observational Scale Matters for Ecosystem Services Interactions and Spatial Distributions: A Case Study of the Ussuri Watershed, China

Jian Zhang [1,*], Hengxing Xiang [2,3], Shizuka Hashimoto [1] and Toshiya Okuro [1]

1 Graduate School of Agricultural and Life Sciences, The University of Tokyo, 1-1-1, Yayoi, Bunkyo-ku, Tokyo 113-8657, Japan; ahash@g.ecc.u-tokyo.ac.jp (S.H.); aokuro@mail.ecc.u-tokyo.ac.jp (T.O.)
2 Key Laboratory of Wetland Ecology and Environment, Northeast Institute of Geography and Agroecology, Chinese Academy of Sciences, Changchun 130102, China; xianghengxing@iga.ac.cn
3 University of Chinese Academy of Sciences, Beijing 100049, China
* Correspondence: zhangjian628@g.ecc.u-tokyo.ac.jp

Abstract: Understanding how observational scale affects the interactions and spatial distributions of ecosystem services is important for effective ecosystem assessment and management. We conducted a case study in the Ussuri watershed, Northeast China, to explore how observational scale (1 km to 15 km grid resolution) influences the correlations and spatial distributions of ecosystem services. Four ecosystem services of particular importance for the sustainable development of the study area were examined: carbon sequestration, habitat provision, soil retention, and water retention. Across the observational scales examined, trade-offs and synergies of extensively distributed ecosystem services were more likely to be robust compared with those of sparsely distributed ecosystem services, and hot/cold-spots of ecosystem services were more likely to persist when associated with large rather than small land-cover patches. Our analysis suggests that a dual-purpose strategy is the most appropriate for the management of carbon sequestration and habitat provision, and cross-scale management strategies are the most appropriate for the management of soil retention and water retention in the study area. Further studies to deepen our understanding of local landscape patterns will help determine the most appropriate observational scale for analyzing the spatial distributions of these ecosystem services.

Keywords: ecosystem service; observational scale; trade-off; hot/cold-spot; Ussuri watershed

check for updates

Citation: Zhang, J.; Xiang, H.; Hashimoto, S.; Okuro, T. Observational Scale Matters for Ecosystem Services Interactions and Spatial Distributions: A Case Study of the Ussuri Watershed, China. *Sustainability* **2021**, *13*, 10649. https://doi.org/10.3390/su131910649

Academic Editors: Margarita Martinez-Nuñez and Mª Pilar Latorre-Martinez

Received: 18 August 2021
Accepted: 22 September 2021
Published: 25 September 2021

Publisher's Note: MDPI stays neutral with regard to jurisdictional claims in published maps and institutional affiliations.

1. Introduction

Understanding the impact of observational scale on the interactions and spatial distribution of ecosystem services is an integral part of mainstreaming the incorporation of ecosystem services knowledge into ecosystem management strategies at the science–policy interface [1–3]. In ecological studies, observational scale can be defined in several ways depending on the context. For example, in studies based on remote sensing and modeling, observational scale is usually described as having four components: (1) level of spatial detail, (2) numerical fraction, (3) spatial extent, and (4) process scale [4–6], which together indicate that ecological phenomena and objects each have their own distinct scale, or a range of scales, at which their characteristics and patterns are best observed [7,8]. An example to clarify the importance of observational scale selection is the competition between individual plants, which can be observed and discussed at the habitat scale but not at the regional or global scale [9,10]. At these larger scales, the break-out of pests or diseases (regional scale) and climate change (global scale) are more suitable topics for observation [11].

Ecosystem services, which, broadly speaking, are the benefits humans receive from the natural environment, are examples of ecological phenomena that have distinct observational scales at which their dynamics can be most efficiently observed and understood. In the context of ecosystem service assessment, observational scale is usually defined as the scale at which samples are collected from the ecosystem service [8]. However, if observational scale is assumed to comprise several hierarchical levels, with each level associated with a scale break [9,10], it can be expected that an ecosystem service could have different characteristics at different observational scales. Recently, there has been increased interest in the issue of appropriate observational scale selection, which has resulted in the terms "scale effect" and "scale dependence" being used to describe the differences in ecological patterns and processes when observed at different scales [12,13]. However, despite ecosystem services currently being observed at many different observational scales [2], the effects of observational scale on ecosystem service assessment remain under-explored [8].

Identifying strategies that favor the management of multiple ecosystem services is an issue that is impacted by observational scale selection because ecosystem service management practices are developed based on feedback obtained from direct observation of ecosystem services [14–16]. Some scholars have recommended that ecosystem service assessments be made at the spatial scale where decision making occurs (e.g., at the local, subnational, national, regional, continental, or global scale) to provide assessments that are relevant to pre-defined social concerns [17]. However, such social concerns often span multiple spatial scales, and addressing those concerns requires an in-depth understanding of the complexities associated with multi-scale assessments [18]. Thus, elucidating how to identify the most appropriate observational scales at which to conduct ecosystem service assessments remains an important concern in the area of ecosystem services management.

Recently, a study in northeast China has suggested the need to reduce the ecosystem and service losses resulting from inappropriate policymaking and to improve the effectiveness of ecosystem management in the area [19]. However, a simple tally of ecosystem services would not reveal which factors are most important in this regard. Sometimes, factors that hamper the effectiveness of ecosystem management could be the intrinsic trade-offs among ecosystem services that were not anticipated during the management design phase [1]. Typical trade-offs among ecosystem services could occur across observational scales [20]. For example, strengthened food production observed at the site scale could impact habitat quality at a broader observational scale and potentially threaten water quality when observing at the watershed scale [21]. Thus, comprehending the impact of observational scale on ecosystem services' interactions is crucial for promoting the efficiency of ecosystem management.

In previous ecological studies, the concept of hot/cold-spots has been used to delineate areas of ecological importance [22]. In the context of ecosystem service assessment, hotspots indicate areas with a high concentration of ecosystem service value, whereas coldspots indicate the contrary. Greater understanding of the hot/cold-spots of ecosystem services has led to the spatial distributions of ecosystem services becoming essential topics of research [8,21]. However, most previous studies have used administrative divisions or the original resolution at which the data were collected as the observational scale rather than taking a multi-scale approach, which may have distorted the study outcomes [8]. Therefore, improving our understanding of how observational scale affects the spatial distributions of ecosystem services is expected to contribute to the development of improved ecosystem service management strategies.

One approach to better understand observational scales' impact on ecosystem services is to examine the robustness of ecosystem service assessments across various observational scales and determine how the characteristics of each ecosystem service are influenced by observational scale. Here, taking the Ussuri watershed in Northeast China as the study area, we examined how observational scale affects four ecosystem services in the area. For our observations, instead of using a conventional administrative scale (e.g., county, township), we used grids with resolutions ranging from 1 to 15 km. We then used our data to examine the changes in correlations between pairs of ecosystem services at increasing observational scales and how ecosystem service distributions react to different observational scales. Based on our findings, we discuss policy-relevant implications of choosing a specific scale for ecosystem service observation and assessment.

2. Data and Methods

2.1. Study Site and Ecosystem Services

The Ussuri watershed in the Northern China Plain covers an area of approximately 61,460 km^2 and comprises a range of land covers and natural habitats, although it is predominantly plain in character (Figure 1A). The region is dominated by cropland (approx. 55% of the watershed in 2015) followed by green spaces (grassland, forest, and wetland; approx. 40% in 2015), but it also contains sparsely distributed rural and urban settlements (approx. 2.5% in 2015). The region is a typical peri-urban agricultural landscape that has been subjected to progressively industrialized farming, settlement development growth, and tourism exploitation (Figure 1B). A total of 15 cities/counties are entirely or partially located in the watershed, giving it a population of about 3,700,000. The Ussuri watershed is the largest component of the Sanjiang Plain (the largest marsh area in China and an important grain production base). However, the region has experienced tremendous wetland loss since the 1950s [23,24] and has been the recipient of strict management attention since 2000 [19,25,26].

Management for the rational and sustainable use of resources in the watershed is overseen by an assortment of government and public institutions. Alternative livelihood activities have been jointly implemented by the government of Heilongjiang, the Asian Development Bank, and the Global Environmental Facility to offer support and build consensus on common objectives for the management and use of forest/wetland resources, recognizing that the success of management strategies ultimately rests on the involvement of individuals [26]. The National Wetland Conservation Project of the National Forestry Administration is presently overseeing wetland resources management and is therefore responsible for the management framework [27].

Ecosystem services in the Ussuri watershed and adjacent regions are frequently mentioned in the literature, providing methodologies and data for diagnosing management problems [19,28,29]. In the present study, we examined four ecosystem services (i.e., carbon sequestration, habitat provision, water retention, and soil retention) that are related to the main concerns addressed by present ecosystem management strategies in the watershed (i.e., climate change mitigation, habitat rehabilitation, and headwater protection) [27].

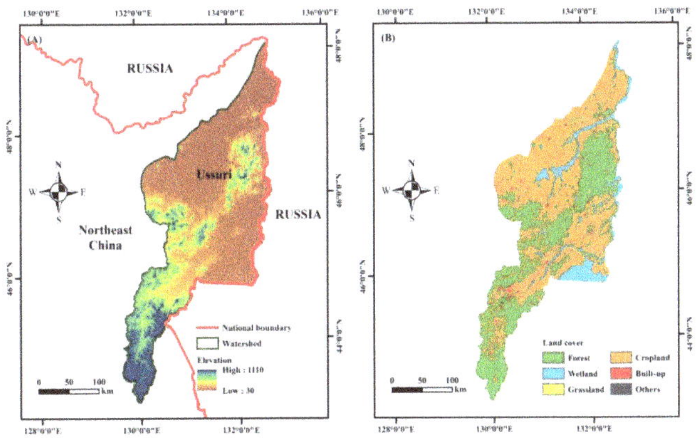

Figure 1. Maps of the study site and elevation (m) (**A**) and land cover in 2015 (**B**). Forest: mixed forest, deciduous broadleaf forest, deciduous conifer, deciduous shrub, evergreen conifer, evergreen shrub, and tree garden; wetland: lake, reservoir, river, tree wetland, shrub wetland, herbaceous wetland, and canal; grassland: temperate steppe, tussock, and lawn; cropland: paddy field and dry farmland; built-up area: mine, transportation network, and settlements; other: barren land and desert [28].

2.2. Methodological Steps

Figure 2 summarizes the methodological steps of estimating and analyzing ecosystem services across varied observational scales. The following sub-sections describe each step in detail.

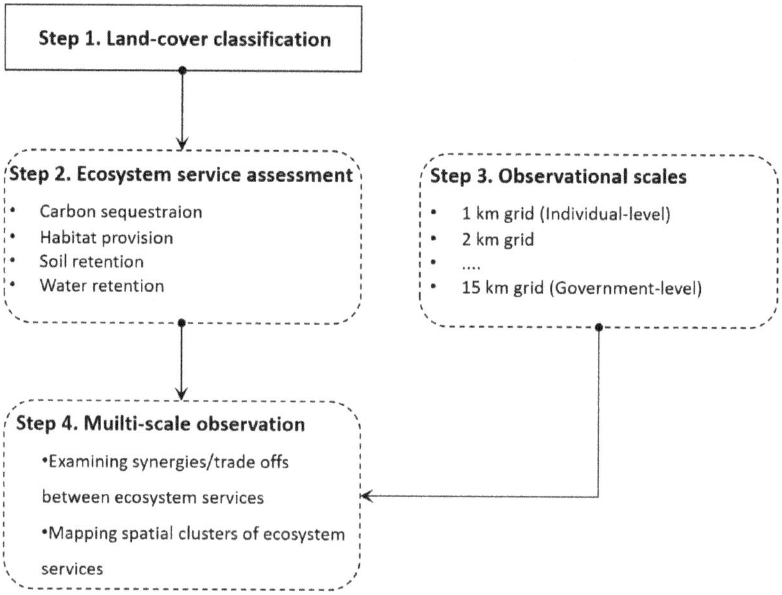

Figure 2. Diagram of methodologies used in this study.

2.3. Land-Cover Classification

Panchromatic images with 15/30 m spatial resolution collected in 2015 by the LANDSAT Enhanced Thematic Mapper Plus (NASA) and Operational Land Imager (NASA) were used as the source data for land-use classification (Table 1). Cloudless satellite images (cloud coverage < 8%) collected in July, August, and early September were used for object-based classification. A total of 25 land-cover types were identified by using the multi-resolution segmentation and object-based classification approach of Mao et al. [19]. The accuracy of classification was assessed based on a total of 2388 historical ground-truth samples, affording an overall accuracy of 94%. Classification results were further compared with the land-use maps presented in recent studies to guarantee the applicability of the results for ecosystem service calculation [28–30]. For the quantification of ecosystem services, we used the original land-cover map with 25 land-cover types; for clearer display, the land-cover map for 2015 was further re-classified into six major land-cover types (forest, wetland, grassland, cropland, built-up area, and other) using the classification system of Wang et al. [28] (Figure 1B).

Table 1. Descriptions of data used in land cover classification and ecosystem service assessment. All the websites were accessed on 16th April 2020).

Data	Resolution	Type	Data Sources
Satellite image	15/30 m	Raster	Geospatial Data Cloud (http://www.gscloud.cn).
Precipitation	1 km	Raster	National Meteorological Information Center (http://data.cma.cn/user/toLogin.html)
Daily minimum/ maximum temperature	1 km	Raster	National Meteorological Information Center (http://data.cma.cn/user/toLogin.html)
DEM	30 m	Raster	Geospatial Data Cloud (http://www.gscloud.cn).
Soil features	1 km	Raster	National Earth System Science Data Center
Carbon density	-	Text	Reference: Xiang et al., 2020.

2.4. *Quantification of Ecosystem Services*

2.4.1. Carbon Sequestration

Carbon sequestration service (CS) measures the capture and secure storage of carbon dioxide that would otherwise be emitted to, or remain, in the atmosphere and is of great importance to tackling global warming. We used the Carbon module of InVEST 3.2.0 to generate a distribution map of carbon sequestration in the study area [31] (Figure 3A). The amounts of carbon stored in four carbon pools (above-ground biomass, below-ground biomass, soil, and litter layer organic matter) for all land cover types was determined by using the formula:

$$C_{total} = \sum_{i}^{n} C_i \times S_i \tag{1}$$

where C_{total} is the total amount of carbon sequestration in the Ussuri watershed, C_i is the summed carbon density in the four carbon pools for land cover i, S_i is the area of land cover *i*, and *n* is the number of land cover types we detected in the land-cover classification phase (n = 25). Carbon density data were obtained from Xiang et al. (Table 1) [32].

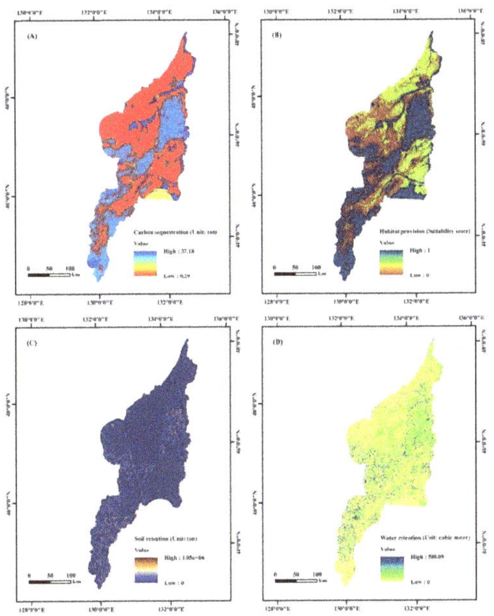

Figure 3. Distribution maps for the four ecosystem services examined in the present study. (**A**) Carbon sequestration; (**B**) habitat provision; (**C**) soil retention; (**D**) water retention.

2.4.2. Habitat Provision

Habitat provision services (HP) are relevant to both permanent and transient populations of wildlife [33] and are extremely important for maintaining biodiversity [34]. We used the Habitat Quality module of InVEST to rate habitat quality on a scale of 0 to 1 (higher value indicates higher habitat quality) (Figure 3B). This module uses the following formula to estimate the spatial extent, vegetation type across a landscape, and state of degradation by combining information on land cover and threats to biodiversity:

$$Q_{xj} = H_j\left(1 - \left(\frac{D_{xj}^z}{D_{xj}^z + k^z}\right)\right) \tag{2}$$

where Q_{xj} is the habitat quality of pixel x in land-cover type j, H_j is the expert knowledge-based habitat quality score obtained from the InVEST 3.2.0 User's Guide, D_{xj} is the state of degradation of pixel x in land-cover type j [31], and K is the half-saturation constant taken as 0.3 [29]. Note that the exponent Z assigned to D_{xj} and K refers to the scaling parameter, which was taken as 2.5 [29]. The data prepared for the estimation of Q_{xj} comprise a land-cover map, the sensitivity of each land cover to each threat, and a list of threats and their features. The sensitivity of each land-cover type to each threat and the weight of their impact were obtained from Xiang et al. [29].

2.4.3. Soil Retention

Soil retention (SR) (Figure 3C), here defined as the difference between the potential worst case soil erosion under bare soil conditions and the actual soil erosion calculated by using the Universal Soil Loss equation [35], was calculated as follows:

$$SR = R \times K \times LS \times (1 - C) \tag{3}$$

where R is rainfall erosivity (MJ mm, ha^{-2} ha^{-1} yr^{-1}) (calculated based on daily precipitation), K is soil erodibility (t h MJ^{-1} mm^{-1}), LS is slope length gradient factor (calculated based on DEM), and C is the percentage of vegetation coverage. Data sources of daily precipitation, soil erodibility, and DEM are from the National Earth System Science Data Center (http://www.geodata.c, accessed 16 April 2020.) (Table 1).

2.4.4. Water Retention

Water retention service (WR) measures soil's ability to retain water. A two-step process was used to estimate water retention (Figure 3D). First, water yield was calculated using the Annual Water Yield module of InVEST and the following formula [31]:

$$WY_{total} = \sum_{n}^{i}(P_i - AET_i) \times A_i \tag{4}$$

where WY_{total} is annual total water yield (t yr^{-1}), P_i is annual precipitation (mm), AET_i is evapotranspiration (mm), A_i is area (km^2), and i is the pixel of interest. The calculation of WY_{total} followed the detailed methods presented in InVEST 3.2.0 User's Guide. We then used the value of WY_{total} to calculate the water retention capacity using the method of Wang et al. [36] as follows:

$$WR_{capacity} = \text{MIN}\left(1, \frac{249}{V}\right) \times \text{MIN}\left(1, \frac{0.9 \times TI}{3}\right) \times \text{MIN}\left(1, \frac{K_{sat}}{300}\right) \times WY_{total} \tag{5}$$

$$\text{TI} = \text{Log}\left(\frac{Surface}{SoilDepth \times PercentSlope}\right) \tag{6}$$

where $WR_{capacity}$ is the annual average water retention capacity (mm); V is the velocity coefficient, which is a constant value (dimensionless); TI is the topographic index (dimensionless); and K_{sat} is the saturated hydraulic conductivity (mm/d). "$Surface$" in Formula (6) is the number of grids in the watershed, "$SoilDepth$" is the soil thickness (mm), and "$PercentSlope$" is the slope percentage (Table 1). The calculation methods of TI and K_{sat} were referenced from Li et al. [37], and the data sources and biophysical parameters used in Formulas (4)–(6) were obtained from Xiang et al. [29].

2.5. Method of Analysis

2.5.1. Observational Scales

The observational scales employed in this study were developed on the basis of management scales (scales at which ecosystem management is formally implemented in the Ussuri watershed) and scale hierarchies [10]. The management scales were determined by identifying the principal managers of ecosystem services. Usually, principal managers of ecosystem services refer to individuals who are formally incentivized to engineer the landscape to manage the ecosystem, and the institutions that develop rules to regulate access to ecosystem services [8]. In the Ussuri watershed, the principal managers of ecosystem services were (1) farmers who grant part of their croplands for wetland and forest restoration or manage their farmland for soil condition and grain yield, and (2) the government bureau that supervises national nature reserves therein [19,26,29]. The management decisions and implementations often occur at the individual level, government level, or some compromised level between them. Therefore, the 1 km grid, which was considered to approximate the spatial scale at which individual land-use management occurs, was designated as the finest observational scale. Meanwhile, the 15 km grid, which was considered to approximate areas of land similar in size to the smallest nature reserve (the Qixinghe wetland reserves that covers an area of 208 km^2) and the spatial scale at which national directives and local interventions are applied, was designated as the coarsest observational scale. In addition, 13 intermediate observation scales (scales between the 1 km and 15 km grids where sampling is conducted and measurements are taken) were added to clarify how features of ecosystem services react to changes in observational scale [8,38]. These intermediate observational scales were simulated with a 1 km grid interval (i.e., from 2 to 14 km grid resolution). All 15 grids were generated in ArcGIS 10.7 using the Fishnet tool.

2.5.2. Correlations between Ecosystem Service Pairs

A correlation analysis was conducted to examine the suitability of using a dual-purpose management strategy (a management strategy that regulates two ecosystem services at the same time) for each pair of ecosystem services. We hypothesized that (1) a dual-purpose management alignment will occur at an observational scale when there is synergy between a pair of ecosystem services, indicating that the two ecosystem services can be managed simultaneously, and (2) a dual-purpose management mismatch will occur at an observational scale when there is a trade-off or no correlation between a pair of ecosystem services, indicating that the two ecosystem services cannot be managed simultaneously.

Pearson's parametric correlation test was performed in IBM SPSS 22 to identify potential synergies and trade-offs among pairs of ecosystem services. Min max normalization was applied to nondimensionalize the data before analysis. The Shapiro–Wilk test was used to verify the normality of the data before the correlation analysis. If the coefficient between pairwise ecosystem services was significant ($p < 0.05$), the correlation was considered valid. A significant negative coefficient was considered to indicate the existence of a trade-off (one ecosystem service increases while the other decreases), and a significant positive correlation was considered to indicate the existence of a synergy (both ecosystem services increase) [39]. The correlation analysis was performed at each of the predetermined observational scales to examine the changes in ecosystem service interactions across observational scales. When using the coarsest observational scale (15 km grid), the maximum sample size of ecosystem services was 229 in this study. Therefore, the ecosystem service samples at each observational scale were all set to 229 to avoid the impact of varied sample sizes on the significance test. The sampling points at all observational scales except for the coarsest observational scale were randomly selected using ArcGIS 10.7 and manually edited to avoid an over-concentrated distribution.

2.5.3. Spatial Patterns of Ecosystem Services

To quantify the spatial patterns of the ecosystem services, we used Anselin's local Moran's indicator [40]. This indicator is used to decompose a global statistic into its constituent parts [40], and then each part is classified as a hot-spot (high–high clusters), cold-spot (low–low clusters), outlier (high–low or low–high clusters), or non-significant spot. We used this approach to identify areas of the watershed where different management approaches could be successful [22]. The hot-spots indicated areas where high-value ecosystem services are highly aggregated, suggesting that a reactive management approach would be appropriate; the cold-spots indicated areas where low-value ecosystem services are highly aggregated, suggesting that a proactive management approach would be appropriate.

A step-wise process was used to analyze the spatial clusters of the ecosystem services. First, the spatially explicit evaluation of the ecosystem services derived by InVEST simulation was re-calculated in ArcGIS for all observational scales. Then, Anselin's local Moran's indicator [40] with queen contiguity was calculated in ArcGIS and compared at all observational scales. Finally, the hot/cold-spots were screened out and counted. A local regression (LOESS) curve fitting was performed to visualize the general trends as the observational scale was changed.

3. Results

3.1. Correlations between Ecosystem Service Pairs at Different Observational Scales

Correlation analysis revealed only one ecosystem service pair, CS–HP, with significant, high synergy at all observational scales (r = 0.860–0.923; Table 2). A mix of synergies and trade-offs were found for the other ecosystem service pairs depending on the observational scale.

Table 2. Pearson correlation coefficients for pairs of ecosystem services at different observational scales. Synergies are shown in green, trade-offs are shown in red, and no correlations are shown in yellow. ** $p < 0.01$; * $p < 0.05$. CS, carbon sequestration; HP, habitat provision; SR, soil retention; WR, water retention.

	CS-HP	CS-SR	CS-WR	HP-SR	HP-WR	SR-WR
1 km grid	0.893 **	0.423 **	−0.127	0.373 **	−0.198	−0.191
2 km grid	0.894 **	0.386 **	−0.136	0.351 *	−0.211	0.026
3 km grid	0.891 **	0.331 *	−0.422 **	0.318 *	−0.555 **	−0.062
4 km grid	0.882 **	0.306 *	−0.398 **	0.296 *	−0.564 **	−0.031
5 km grid	0.877 **	0.253	−0.401 **	0.238	−0.580 **	0.004
6 km grid	0.875 **	0.244	−0.373 **	0.209	−0.551 **	0.120
7 km grid	0.874 **	0.253	−0.355 *	0.194	−0.540 **	0.173
8 km grid	0.873 **	0.254	−0.306 *	0.169	−0.499 **	0.187
9 km grid	0.869 **	0.291 *	−0.239	0.184	−0.459 **	0.253
10 km grid	0.868 **	0.302 *	−0.231	0.188	−0.448 **	0.304 *
11 km grid	0.867 **	0.322 *	−0.217	0.189	−0.442 **	0.335 *
12 km grid	0.860 **	0.301 *	−0.224	0.159	−0.471 **	0.380 **
13 km grid	0.862 **	0.309 *	−0.213	0.161	−0.452 **	0.375 **
14 km grid	0.864 **	0.323 *	−0.194	0.164	−0.440 **	0.427 **
15 km grid	0.923 **	0.674 **	0.288 *	0.630 **	0.181	0.613 **

Synergies were observed for CS–SR, HP–SR, and WR–SR. For CS–SR, low synergy was observed at 1–4 km grid resolution (r = 0.306–0.423) and 9–14 km grid resolution (r = 0.291–0.323), and high synergy was observed at 15 km grid resolution (r = 0.674). For HP–SR, low synergy was observed at 1–4 km grid resolution (r = 0.296–0.373) and high synergy was observed at 15 km grid resolution (r = 0.630). For WR–SR, low to high synergy was observed at 10–15 km grid resolution (r = 0.304–0.613).

Negative correlations, which indicate trade-offs between ecosystem services, were observed for CS–WR and HP–WR. For CS–WR, trade-offs were observed at 3–8 km grid resolution (r = −0.422 to −0.306) but synergy was observed at 15 km grid resolution (r = 0.288). For, HP–WR, trade-offs were observed at 3–14 km grid resolution (r = −0.580 to −0.440).

3.2. Spatial Distributions of the Ecosystem Services at Different Observational Scales

We used the local Moran's indicator to examine the spatial distribution of each of the four ecosystem services across the 15 observational scales. Figure 4 shows the distributions of hot-spots (high–high clusters), cold-spots (low–low clusters), outliers (high–low or low–high clusters), and non-significant spots. When the observational scale was increased, adjacent grid areas of any cluster type were merged and then smoothed before the new grid area was assigned a new cluster type. This resulted in the scaling behavior of the different ecosystem service clusters across the observational scales being characterized as either merge–shrink or merge–expand. For example, for WR, small hot-spots were scattered across the study area at the finest observational scale (1 km grid resolution); however, as the observational scale was increased, these areas merged together and increased in size, resulting in a merge–expand behavior.

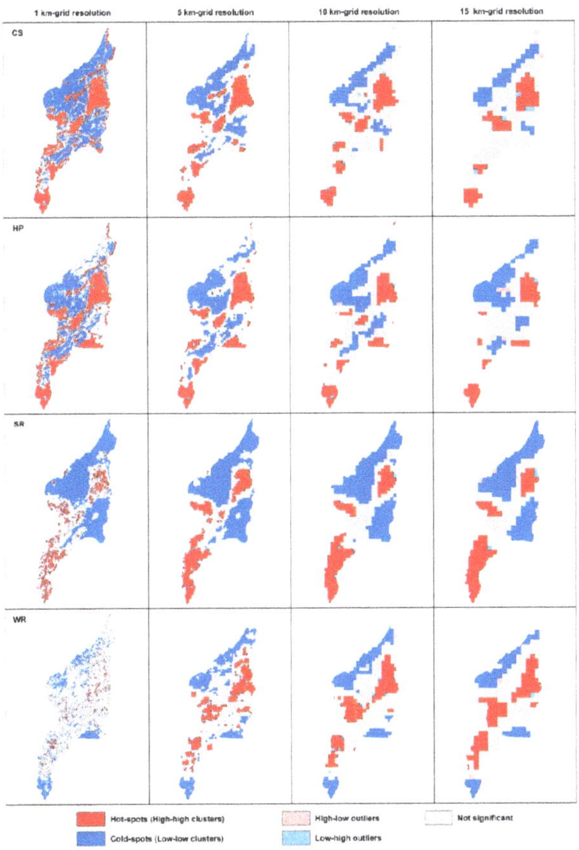

Figure 4. Spatial patterns of ecosystem service clusters (99% confidence interval) at four representative grid resolutions. CS, carbon sequestration; HP, habitat provision; SR, soil retention; WR, water retention.

Figure 5 shows the changes in the proportion of land classified as hot/cold-spots for each ecosystem service with increasing observational scale; LOESS curves are shown to clarify the general trends. The proportion changes of 1 km vs. 15 km grid resolution observation were calculated. For CS, HP, and SR, the proportion of cold-spots decreased with increasing observational scale, with decreases of 25.2% for CS, 3.3% for HP, and 7.3% for SR (Figure 5A–C). In contrast, for WR, the proportion of cold-spots increased by 4.5% (Figure 5D). For CS and HP, the proportion of hot-spots decreased by 8.6% and 15.1%, respectively (Figure 5A,B), whereas for SR and WR, the proportion of hot-spots increased by 13.0% and 21.8%, respectively (Figure 5C,D).

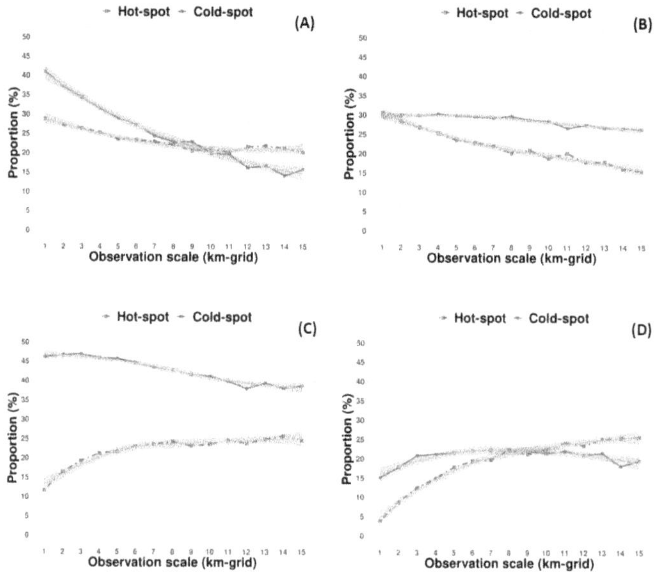

Figure 5. Changes in the proportion of land classified as hot/cold-spots with increasing observational scale for the four ecosystem services. (**A**) Carbon sequestration. (**B**) Habitat provision. (**C**) Soil retention. (**D**) Water retention. Gray curves are fitted curves that indicate the general trend (95% confidence interval).

Shifts in whether hot- or cold-spots were dominant were observed for CS, HP, and WR (Figure 5A,B,D). For CS and WR, there was a greater proportion of cold-spots compared with hot-spots at lower grid resolutions, but a greater proportion of hot-spots compared with cold-spots at higher resolutions; this reversal occurred at 10 km grid resolutions for CS and at 8 km grid resolution for WR. For HP, the proportion of hot-spots was greater than that of cold-spots at 1 km grid resolution, but this situation reversed from 2 km grid resolution.

4. Discussion

Understanding the spatial distributions and interactions of ecosystem services is crucial for the development of effective ecosystem service management strategies. The scales at which ecosystem services are observed or monitored fundamentally shape this understanding. Here, we conducted a case study of the area of the Ussuri watershed within the border of the People's Republic of China to examine how observational scale affects the mapping of ecosystem services and their pairwise synergies and trade-offs. Based on our findings, we discuss how best to select the most appropriate observational scale for ecosystem service assessment and management.

4.1. Synergies and Trade-Offs across Observational Scales

First, we determined pairwise correlations among the four ecosystem services to examine how their correlations change with increasing observational scale (Table 2). The authors of previous case studies conducted for Quebec, Canada [8], and the Ningxia Hui Autonomous Region, China [12], concluded that most pairwise correlations are robust across different observational scales and that significant correlations are more often observed at finer than at coarser observational scales. However, based on our present data, we do not agree with these previous conclusions, although it must be noted that ecosystem service selection and the biophysical context of the study area must be taken into consideration when comparing the present and previous data.

In our analysis, we found that the correlation between CS and HP was robustly synergistic across all observational scales. Our distribution maps for CS and HP revealed that these ecosystem services are distributed extensively throughout the study area and are frequently coincided with certain types of land cover (Figure 3). For example, areas of high CS value and HP suitability score were found to be areas classified as forest and wetland, which is natural given that these two land-cover types are characterized by their large carbon pools and abundance of wildlife [19,41,42]. Turner et al. noted that land cover types with extensive distribution tend to change evenly across observational scales because the local configuration does not influence scaling [43]. We consider that Turner's summing applies equally to correlations between ecosystem services that are extensively distributed.

In contrast to the distribution maps for CS and HP, those for SR and WR revealed that these ecosystem services were unevenly and sparsely distributed throughout the study area. Areas of high SR were primarily areas of forest at high elevation, and that of WR shared a similar distribution but appeared as more fragmented patches (Figure 3). In addition, we found that the correlations involving these ecosystem services were a mix of synergies, trade-offs, and no correlations (Table 2). For example, for CS–WR, trade-offs of various strengths were observed from 3 to 8 km grid resolution; no correlations were observed at 1, 2, and 9–14 km grid resolutions; and low synergy was observed at 15 km grid resolution.

Collectively, the results of our correlation analysis suggest that the distribution of an ecosystem service is an indicator of how robust its pairwise correlations are across observational scales; that is, correlations among extensively distributed ecosystem services are more likely to be robust, whereas those of ecosystem services with patchy distributions are more likely less robust, suggesting that more judicious selection of observational scale would be required when conducting assessments of these ecosystem services.

Assuming that observational scale equals the spatial scale at which future management occurs, our data suggest two implications with regard to the use of a dual-purpose ecosystem service management strategy. First, the robust synergy between CS–HP at all observational scales suggests that dual-purpose management of these two ecosystem services may be a cost-effective means of providing synergistic enhancements to both ecosystem services, and that such a management strategy could be applied at any scale. Evidence supporting our viewpoint could be found in several previous studies, where Zheng et al. [44] and Xiang et al. [29] have reported that widespread natural habitat restoration has increased biodiversity in the Sanjiang plain. The restoration of high-diversity ecosystems on degraded or abandoned land merits further implementation for its potential to provide increased CS [45].

Second, despite the present dual-purpose management of WR and SR in the Ussuri watershed by the Grain for Green Project [19], which is overseeing the reforestation of marginal cropland to reduce soil erosion and water loss, we only found synergies between these two ecosystem services at the coarser observational scales examined (10 to 15 km grid resolution; Table 2), suggesting that the current dual-purpose management approach is not suitable at all observational scales. Improving SR by improving erosion control in specific areas also improves WR [46,47]; however, WR often relies on large-scale landscape patterns and watershed dynamics [48–50]. These characteristics of the two ecosystem services are consistent with our findings of a dual-purpose management mismatch between WR–SR at the fine and intermediate scales. One means of resolving this mismatch could be to increase the number of soil erosion control sites in the Ussuri watershed to create a more extensive distribution pattern; however, soil erosion control is a costly investment when implemented over a sizeable spatial scale. Therefore, we suggest introducing a cross-scale strategy for the management of SR and WR [8]. That is, we suggest incentivizing the participation of individual managers or management institutions in improving WR via implementing erosion control, which will reduce the workload and financial burdens placed on government-level WR management. Meanwhile, the government-level managers need to carry out plans to specify the spatial extent in the Ussuri watershed to guide individual and institutional managers to put erosion control into effect. It is also noteworthy that the plans should be based on the in-depth knowledge of the local landscape and spatial pattern of ecosystem services.

Despite our observation of moderate synergies for CS–SR and HP–SR at the 15 km grid resolution, more low-level synergies and no-correlations occurred for them during the scaling-up of observation (Table 2). Further research is needed to explore the potential of dual-purpose management for such ecosystem service pairs in the Ussuri watershed.

4.2. Ecosystem Service Clusters at Different Observational Scales

Mapping hot/cold-spots provides straightforward information for determining where to implement different management options. Cold-spots are areas where there is a lower possibility of harvesting an ecosystem service, suggesting that the ecosystem service therein is better left undisturbed. In contrast, hot-spots are areas where there is a higher possibility of harvesting an ecosystem service, suggesting that the ecosystem service therein may potentially be harvested with high acceptance by management institutions and other stakeholders.

Here, we found that observational scale substantially influenced the spatial distributions of hot/cold-spots in two ways (Figure 4). First, increasing the observational scale altered the location and size of the hot/cold-spot clusters, such that the land-cover type associated with some of the clusters also changed. For example, whereas CS hot-spots associated with large patches of forest and cold-spots associated with continuous cropland were relatively preserved across the observational scales, those associated with wetland in the central part of the study area at fine observational scales (1–3 km grid resolution) had disappeared at the intermediate and coarse observational scales. Similar findings were also observed for HP, SR, and WR, although the clusters associated with continuous cropland, large patches of forest, and sizeable water bodies were relatively persistent across the observational scales.

Second, changing the observational scale altered the proportion of land covered by the cold/hot-spots, and for CS, HP, and WR, increasing the observational scale altered whether it was hot- or cold-spots that were the dominant cluster type (Figure 5). Both the trends and the spatial extent of the ecosystem service clusters varied at different observational scales. We speculate that regardless of whether or not an ecosystem service is extensively distributed throughout an area, observation at certain scales will fail to capture the real spatial pattern of hot/cold-spots because land-cover features shape the distribution of these clusters as well as how these clusters react at different observational scales. A widely used principle that emerged in the field of mapping ecosystem services is observing ecosystem services at the scale of administrative, policy, and management boundaries to facilitate the relevance between the assessment output and management decision making [8,19]. This principle implies that the mapping of ecosystem services should be based on the question being asked and the type of details required. We argue that, indeed, the assessment needs to match with the need of decision makers, but, more importantly, the assessment should capture the complexities of ecosystem services. Therefore, the selection of an appropriate observational scale for ecosystem service cold/hot-spots can only be interpreted by taking into consideration the social-ecological heterogeneity of the local landscape. The landscape pattern in the Ussuri watershed is a combined result of the gradual encroachment of cropland on wetland since the 1980s, the implementation of forest and wetland conservation measures around 2000 [19], and the watershed's distance from the urban agglomeration of Harbin. Therefore, understanding the underlying mechanisms that shape local social-ecological conditions can help define the appropriate scales for observing the spatial patterns of ecosystem service hot/cold-spots.

We argue that, indeed, the assessment needs to match with the need of decision makers, but, more importantly, the assessment should capture the complexities of ecosystem services.

4.3. Methodological Limitations and Future Study

Many previous studies have elucidated the factors that affect the correlations and distributions of ecosystem services (e.g., the accuracy of input data and aggregation method used), but observational scale has not yet been examined [51–53]. We found here that the effects of changing observational scale are similar to those produced by smoothing to create an approximate function that captures important patterns in a data set while leaving out noise and outliers. That is, we found that when the observational scale was changed, data points of value were modified so that individual points higher in value than the adjacent points were reduced, and points that were lower in value than the adjacent points were increased, leading to compromised values. Therefore, observational scale should be considered a double-edged sword [47,54]. On the one hand, because it clearly determines the fitness and coarseness of the data and patterns that are obtained, it brings convenience to decision makers by letting the assessment results feed the management goal; on the other hand, it brings challenges with respect to accuracy because increasing the observational scale may introduce redundant or inaccurate data.

Other limitations of the present study are the simulation method and quality of input data used to measure the ecosystem services. The InVEST model provides a straightforward approach to map and monitor habitat quality that can be used as an estimate of biodiversity [55]. However, when there are multiple definitions of natural habitats and threats (i.e., habitat patches would be defined differently by large mobile wildlife compared to rare species of plants) [53], different spatial distributions or other land cover-based data may be obtained. Although we conducted our analysis using the best available data to provide qualified results, our use of average inventory values for carbon density associated with specific land covers and default model parameters in the WR and SR simulations may have failed to capture the effects of management types, climate factors, and geographic traits. In addition, validation is often absent in ecosystem service mapping and monitoring, so a better understanding of the uncertainties involved in the models used is needed [56].

Regarding future study, we suggest including more aspects of ecosystem services, such as the societal values and consumption of ecosystem services, in multi-observational scale analysis to select suitable observational scales [8,15]. The societal values of ecosystem services influence the rules and actions that alter the provision and access to ecosystem services [15], and the consumption of ecosystem services implies appropriate management incentives [8]. Comprehending how these aspects of ecosystem services vary across different observational scales allows identifying potential conflicts in ecological/environmental management, particularly among different stakeholders and managers, thereby building an effective, accountable, and inclusive framework to guarantee the sustainability of ecosystems in the Ussuri watershed.

5. Conclusions

Here, we present a case study conducted in the Ussuri watershed, Northeast China, in which we examined how observational scale affects ecosystem service assessment. We examined four ecosystem services (regulating and supporting services) at 15 observational scales (1 to 15 km grid resolution), which included two approximate scales at which ecosystem management is formally implemented. Correlation analysis revealed that ecosystem service distribution may be an indicator of the robustness of the correlation between pairs of ecosystem services across various observational scales. That is, a correlation is likely to be robust when an ecosystem service is extensively distributed across the study area (i.e., CS and HP in the present study), but not robust when an ecosystem service is sparsely distributed (i.e., SR and WR). Based on our findings, we suggest that a dual-purpose management strategy is most appropriate for the management of CS–HP in the Ussuri watershed, and that cross-scale management strategies are the most appropriate for the management of WR and SR.

The pattern of ecosystem service clusters (cold/hot-spots) across the various observational scales was associated with the size of land cover patches. Indeed, complex networks of ecosystem service clusters were observed in association with dispersed land covers at finer observational scales, but these networks became less complex, homogenous clusters at coarser scales. In contrast, the ecosystem service clusters associated with continuous land covers were persistent across the various observational scales. Thus, selecting an appropriate observational scale for delineating ecosystem service clusters should take into account the local landscape patterns and an understanding of local social-ecological complexity.

Though our results have the potential to improve the management decision making in ecosystem services, there are still several limitations that need to be addressed in future study. Technically, enhancing the sensitivity of the models used results in better accuracy of the input data, and the use of local parameters could help achieve a better estimation of ecosystem services. Moreover, including societal values and consumption of ecosystem services in the multi-observational scale analysis will bring better negotiation and coordination between stakeholders and managers at different scales, thereby ensuring the sustainable development of the Ussuri watershed.

Author Contributions: J.Z. and H.X. conceived the study and generated the datasets. J.Z. performed the data analysis and prepared the manuscript. T.O., S.H. and H.X. provided critical feedback and edited the manuscript. All authors have read and agreed to the published version of the manuscript.

Funding: This research was funded by a scholarship from the Japan Ministry of Education, Culture, Sports, Science and Technology and the National Natural Science Foundation of China, grant number "41671219", the Scientific and Technological Development Program of Jilin Province, China (No. 20200301014RQ).

Institutional Review Board Statement: Not applicable.

Informed Consent Statement: Not applicable.

Acknowledgments: We thank the Northeast Institute of Geography and Agroecology, Chinese Academy of Sciences; the National Earth System Science Data Center (http://www.geodata.cn); Geospatial Data Cloud (http://www.gscloud.cn), and National Meteorological, Information Center (http://data.cma.cn/user/toLogin.html), for their data support. We are grateful to the all the editors and reviewers for their constructive feedback and edits.

Conflicts of Interest: The authors declare no conflict of interest.

References

1. Tallis, H.; Kareiva, P.; Marvier, M.; Chang, A. An ecosystem services framework to support both practical conservation and economic development. *Proc. Natl. Acad. Sci. USA* **2008**, *105*, 9457–9464. [CrossRef]
2. Malinga, R.; Gordon, L.J.; Jewitt, G.; Lindborg, R. Mapping ecosystem services across scales and continents–A review. *Ecosyst. Serv.* **2015**, *13*, 57–63. [CrossRef]
3. Costanza, R.; De Groot, R.; Braat, L.; Kubiszewski, I.; Fioramonti, L.; Sutton, P.; Farber, S.; Grasso, M. Twenty years of ecosystem services: How far have we come and how far do we still need to go? *Ecosyst. Serv.* **2017**, *28*, 1–16. [CrossRef]
4. Farina, A. *Principles and Methods in Landscape Ecology*, 1st ed.; Chapman & Hall: London, UK, 1998.
5. Goodchild, M.F. Models of scale and scales of modeling. In *Modeling Scale in Geographical Information*; Tate, N.J., Atkinson, P.M., Eds.; Willey & Sons: Chichester, UK, 2001; pp. 3–10.
6. Wu, H.; Li, Z.L. Scale issues in remote sensing: A review on analysis, processing and modeling. *Sensors* **2009**, *9*, 1768–1793. [CrossRef]

7. Emmett, B.A.; Cooper, D.; Smart, S.; Jackson, B.; Thomas, A.; Cosby, B.; Evans, C.; Glanville, H.; McDonald, J.E.; Malham, S.K.; et al. Spatial patterns and environmental constraints on ecosystem services at a catchment scale. *Sci. Total Environ.* **2016**, *572*, 1586–1600. [CrossRef]
8. Raudsepp-Hearne, C.; Peterson, G.D. Scale and ecosystem services: How do observation, management, and analysis shift with scale—Lessons from Québec. *Ecol. Soc.* **2016**, *21*, 16. [CrossRef]
9. O'Neill, R.V.; Gardner, R.H.; Milne, B.T.; Turner, M.G.; Jackson, B. *Heterogeneity and Spatial Hierarchies*; Kolasa, J., Pickett, S.T.A., Eds.; Ecological Heterogeneity; Springer: New York, NY, USA, 1991; pp. 85–96.
10. Wu, J. Hierarchy and scaling: Extrapolating information along a scaling ladder. *Can. J. Remote Sens.* **1999**, *25*, 367–380. [CrossRef]
11. Kremen, C. Managing ecosystem services: What do we need to know about their ecology? *Ecol. Lett.* **2005**, *8*, 468–479. [CrossRef]
12. Xu, S.N.; Liu, Y.F.; Wang, X.; Zhang, G.X. Scale effect on spatial patterns of ecosystem services and associations among them in semi-arid area: A case study in Ningxia Hui Autonomous Region, China. *Sci. Total Environ.* **2017**, *598*, 297–306. [CrossRef]
13. Zhang, Y.H.; He, N.P.; Loreau, M.; Pan, Q.M.; Han, X.G. Scale dependence of the diversity-stability relationship in a temperate grassland. *J. Ecol.* **2017**, *106*, 1227–1285. [CrossRef] [PubMed]
14. Fisher, B.; Turner, R.K.; Morling, P. Defining and classifying ecosystem services for decision making. *Ecol. Econ.* **2009**, *68*, 643–653. [CrossRef]
15. Reyers, B.; Biggs, R.; Cumming, G.S.; Elmqvist, T.; Hejnowicz, A.P.; Polasky, S. Getting the measure of ecosystem services: A social–ecological approach. *Front. Ecol. Environ.* **2013**, *11*, 268–273. [CrossRef]
16. Bennett, E.M.; Cramer, W.; Begossi, A.; Cundill, G.; Díaz, S.; Egoh, B.N.; Geijzendorffer, I.R.; Krug, C.B.; Lavorel, S.; Lazos, E.; et al. Linking biodiversity, ecosystem services, and human well-being: Three challenges for designing research for sustainability. *Curr. Opin. Env. Sust.* **2015**, *14*, 76–85. [CrossRef]
17. Gitay, H.; Raudsepp-Hearne, C.; Blanco, H.; Garcia, K.; Pereira, H. Assessment process. In *Ecosystems and Human Well–Being. Volume 4, Multiscale Assessments*; Capistrano, D., Samper, C., Lee, M.J., Raudsepp-Hearne, C., Eds.; Island: Washington, DC, USA, 2005; pp. 119–140.
18. Scholes, R.J.; Reyers, B.; Biggs, R.; Spierenburg, M.J.; Duriappah, A. Multi-scale and cross-scale assessments of social–ecological systems and their ecosystem services. *Curr. Opin. Environ. Sust.* **2013**, *5*, 6–25. [CrossRef]
19. Mao, D.H.; He, X.Y.; Wang, Z.M.; Tian, Y.L.; Xiang, H.X.; Yu, H.; Man, W.D.; Jia, M.M.; Ren, C.Y.; Zheng, H.F. Diverse policies leading to contrasting impacts on land cover and ecosystem services in Northeast China. *J. Clean Prod.* **2019**, *240*, 117961. [CrossRef]
20. Rodríguez, J.P.; Beard, T.D.; Bennett, E.M.; Cumming, G.S.; Cork, S.J.; Agard, J.; Dobson, A.P.; Peterson, G.D. Trade-offs across space, time, and ecosystem services. *Ecol. Soc.* **2006**, *11*, 28. [CrossRef]
21. Holland, R.A.; Eigenbrod, F.; Armsworth, P.R.; Anderson, B.J.; Thomas, C.D.; Heinemeyer, A.; Gillings, S.; Roy, D.B.; Gaston, K.J. Spatial covariation between freshwater and terrestrial ecosystem services. *Ecol. Appl.* **2011**, *21*, 2034–2048. [CrossRef] [PubMed]
22. Schröter, M.; Remme, R.P. Spatial prioritisation for conserving ecosystem services: Comparing hotspots with heuristic optimisation. *Landsc. Ecol.* **2016**, *31*, 431–450. [CrossRef] [PubMed]
23. Yan, F.; Zhang, S. Ecosystem service decline in response to wetland loss in the Sanjiang Plain, Northeast China. *Ecol. Eng.* **2019**, *130*, 117–121. [CrossRef]
24. Shi, S.X.; Chang, Y.; Wang, G.D.; Li, Z.; Hu, Y.M.; Liu, M.; Li, Y.H.; Li, B.L.; Zong, M.; Huang, W.T. Planning for the wetland restoration potential based on the viability of the seed bank and the land-use change trajectory in the Sanjiang Plain of China. *Sci. Total Environ.* **2020**, *733*, 139208. [CrossRef]
25. Day, K.A. *China's Environment and the Challenge of Sustainable Development*, 1st ed.; Routledge: Armonk, NY, USA, 2005.
26. Asian Development Bank (ADB). *Peoples' Republic of China: Sanjiang Plain Wetland Protection Project*; Asian Development Bank: Changchun, China, 2016.
27. Wang, Z.M.; Wu, J.G.; Madden, M.; Mao, D.H. China's Wetlands: Conservation Plans and Policy Impacts. *Ambio* **2012**, *41*, 782–786. [CrossRef]
28. Wang, Z.M.; Mao, D.H.; Li, L.; Jia, M.M.; Dong, Z.Y.; Miao, Z.H.; Ren, C.Y.; Song, C.C. Quantifying changes in multiple ecosystem services during 1992–2012 in the Sanjiang Plain of China. *Sci. Total Environ.* **2015**, *514*, 119–130. [CrossRef] [PubMed]
29. Xiang, H.M.; Wang, Z.M.; Mao, D.H.; Zhang, J.; Xi, Y.B.; Du, B.J.; Zhang, B. What did China's National Wetland Conservation Program Achieve? Observations of changes in land cover and ecosystem services in the Sanjiang Plain. *J. Environ. Manag.* **2020**, *267*, 110623. [CrossRef] [PubMed]
30. Song, K.S.; Wang, Z.M.; Du, J.; Liu, L.; Zeng, L.H.; Ren, C.Y. Wetland degradation: Its driving forces and environmental impacts in the Sanjiang Plain, China. *Environ. Manag.* **2014**, *54*, 255–271. [CrossRef]
31. Sharp, R.; Chaplin-Kramer, R.; Wood, S.; Guerry, A.; Tallis, H.; Ricketts, T. *InVEST 3.2.0 User's Guide. The Natural Capital Project. Stanford University, University of Minnesota*; The Nature Conservancy; World Wildlife Fund: Stanford, CA, USA, 2015.
32. Xiang, H.X.; Jia, M.M.; Wang, Z.M.; Li, L.; Mao, D.H.; Zhang, D.; Cui, G.S.; Zhu, W.H. Impacts of land cover changes on ecosystem carbon stocks over the transboundary Tumen River Basin in Northeast Asia. *Chin. Geogr. Sci.* **2018**, *28*, 973–985. [CrossRef]
33. Groot, R.D.; Wilson, M.; Boumans, R. A Typology for the Classification Description and Valuation of Ecosystem Functions, Goods and Services. *Ecol. Econ.* **2002**, *41*, 393–408. [CrossRef]

34. Krauss, J.; Bommarco, R.; Guardiola, M.; Heikkinen, R.; Helm, A.; Kuussaari, M.; Lindborg, R.; Öckinger, E.; Pärtel, M.; Pino, J.; et al. Habitat fragmentation causes immediate and time-delayed biodiversity loss at different trophic levels. *Ecol. Lett.* **2010**, *13*, 597–605. [CrossRef]
35. Jiang, C.; Zhang, H.; Zhang, Z. Spatially explicit assessment of ecosystem services in China's Loess Plateau: Patterns, interactions, drivers, and implications. *Glob. Planet. Chang.* **2018**, *161*, 41–52. [CrossRef]
36. Wang, Y.C.; Zhao, J.; Fu, J.W.; Wei, W. Effects of the grain for green program on the water ecosystem services in an arid area of China—Using the shiyang river basin as an example. *Ecol. Indic.* **2019**, *104*, 659–668. [CrossRef]
37. Li, M.Y.; Liu, T.X.; Luo, Y.Y.; Duan, L.M.; Zhang, J.Y.; Zhou, Y.J.; Scharaw, B. Pedo-transfer function and remote-sensing-based inversion saturated hydraulic conductivity of surface soil layer in Xilin river basin. *Acta Pedol. Sin.* **2019**, *56*, 90–100.
38. Wu, J.; Jones, K.B.; Li, H.; Loucks, O.L. *Scaling and Uncertainty Analysis in Ecology: Methods and Applications*; Springer: Berlin/Heidelberg, Germany, 2006.
39. Bennett, E.M.; Peterson, G.D.; Gordon, L.J. Understanding relationships among multiple ecosystem services. *Ecol. Lett.* **2009**, *12*, 1394–1404. [CrossRef]
40. Anselin, L. Local indicators of spatial association-LISA. *Geogr. Anal.* **1995**, *27*, 93–115. [CrossRef]
41. Chiang, L.C.; Lin, Y.P.; Huang, T.; Schmeller, D.S.; Verburg, P.H.; Liu, Y.L.; Ding, T.S. Simulation of ecosystem service responses to multiple disturbances from an earthquake and several typhoons. *Landsc. Urban Plan.* **2014**, *122*, 41–55. [CrossRef]
42. Liu, X.; Meng, M.; Wang, Q.; Zhou, X.H.; Zhao, Y.N. Economic mechanisms for oriental white stork conservation. *Chin. J. Wild. life.* **2019**, *40*, 240–246.
43. Turner, M.G. Disturbance and landscape dynamics in a changing world. *Ecology* **2010**, *91*, 2833–2849. [CrossRef] [PubMed]
44. Zheng, H.F.; Shen, G.Q.; Shang, L.Y.; Lv, X.G.; Wang, Q.; McLaughlin, N.; He, X.Y. Efficacy of conservation strategies for endangered oriental white storks (Ciconia boyciana) under climate change in Northeast China. *Biol. Conserv.* **2016**, *204*, 367–377. [CrossRef]
45. Yang, Y.; Tilman, D.; Furey, G.; Lehman, C. Soil carbon sequestration accelerated by restoration of grassland biodiversity. *Nat. Commun.* **2019**, *10*, 718. [CrossRef]
46. Conroy, M.; Allen, C.; Peterson, J.; Pritchard, L.; Moore, C. Landscape Change in the Southern Piedmont: Challenges, Solutions, and Uncertainty Across Scales. *Conserv. Ecol.* **2003**, *8*, 17. [CrossRef]
47. Lant, C.L.; Kraft, S.E.; Beaulieu, J.; Bennett, D.; Loftus, T.; Nicklow, J. Using GIS-based ecological–economic modeling to evaluate policies affecting agricultural watersheds. *Ecol. Econ.* **2005**, *55*, 467–484. [CrossRef]
48. Tscharntke, T.; Klein, A.M.; Kruess, A.; Steffan-Dewenter, I.; Thies, C. Landscape perspectives on agricultural intensification and biodiversity–ecosystem service management. *Ecol. Lett.* **2005**, *8*, 857–874. [CrossRef]
49. Jaarsveld, A.S.V.; Biggs, R.; Scholes, R.J.; Bohensky, E.; Reyers, B.; Lynam, T.; Musvoto, C.; Fabricius, C. Measuring conditions and trends in ecosystem services at multiple scales: The Southern African Millennium Ecosystem Assessment (SAfMA) experience. *Philos. Trans. R. Soc. B* **2005**, *360*, 425–441. [CrossRef] [PubMed]
50. Brauman, K.A.; Daily, G.C.; Duarte, T.K.; Mooney, H.A. The nature and value of ecosystem services: An overview highlighting hydrologic services. *Annu. Rev. Env. Resour.* **2007**, *32*, 67–98. [CrossRef]
51. Myers, N.; Mittermeier, R.A.; Mittermeier, C.G.; Fonseca, G.A.B.D.; Kent, J. Biodiversity hotspots for conservation priorities. *Nature* **2000**, *403*, 853–858. [CrossRef] [PubMed]
52. Schröter, M.; Rusch, G.M.; Barton, D.N.; Blumentrath, S.; Nordén, B. Ecosystem services and opportunity costs shift spatial priorities for conserving forest biodiversity. *PLoS ONE* **2014**, *9*, e112557.
53. Ceauşu, S.; Gomes, I.; Pereira, H.M. Conservation planning for biodiversity and wilderness: A real-world example. *Environ. Manage.* **2015**, *55*, 1168–1180. [CrossRef]
54. Martín-López, B.; Gómez-Baggethun, E.; Lomas, P.L.; Montes, C. Effects of spatial and temporal scales on cultural services valuation. *J. Environ. Manag.* **2009**, *90*, 1050–1059. [CrossRef]
55. Lindenmayer, D.B.; Barton, P.S.; Lane, P.W.; Westgate, M.J.; McBurney, L. An Empirical Assessment and Comparison of Species-Based and Habitat-Based Surrogates: A Case Study of Forest Vertebrates and Large Old Trees. *PLoS ONE* **2014**, *9*, e89807.
56. Schulp, C.J.E.; Burkhard, B.; Maes, J.; Vliet, J.V.; Verburg, P.H. Uncertainties in Ecosystem Service Maps: A Comparison on the European Scale. *PLoS ONE* **2014**, *9*, e109643. [CrossRef]

sustainability MDPI

Article

Proposal of a Conceptual Model to Represent Urban-Industrial Systems from the Analysis of Existing Worldwide Experiences

Carmen Ruiz-Puente

INGEPRO Research Group, Department of Transport and Projects and Processes Technology, University of Cantabria, 39005 Santander, Spain; ruizpm@unican.es

Abstract: The adoption of Industrial Symbiosis (IS) practices within urban areas is gaining interest due to the environmental impacts entailed by the development of cities. However, there is still a lack of knowledge about how the relationships between industrial and urban areas can be modelled. In this context, this research aimed at posing a conceptual model to understand and represent Urban-Industrial Systems (UIS). To this end, a set of worldwide previous UIS experiences were overviewed to identify the agents, dynamics, and collaboration opportunities that characterize them. The multi-perspective analysis of these cases indicated that UIS are complex systems, which means that they are autonomous, self-organized, responsive, nonlinear, and willing to consolidate their resilience. As such, Agent-Based Models (ABM) were suggested to be the most suitable approach for their representation.

Keywords: complex systems; industrial ecology; urban-industrial symbiosis; urban-industrial systems; urban metabolism

check for updates

Citation: Ruiz-Puente, C. Proposal of a Conceptual Model to Represent Urban-Industrial Systems from the Analysis of Existing Worldwide Experiences. *Sustainability* **2021**, *13*, 9292. https://doi.org/10.3390/su13169292

Academic Editors: Margarita Martinez-Nuñez and Mª Pilar Latorre-Martínez

Received: 18 July 2021
Accepted: 17 August 2021
Published: 18 August 2021

1. Introduction

Today, around 50% of the world's population lives in cities. This figure is expected to rise up to 70% by 2050 [1]. To face this rapid growth, cities must be planned to harmonize the three pillars of sustainable development, namely economy, environment, and society [2]. The explosion of industrial activity and population over the last two centuries have caused serious environmental degradation [3], putting current societies in a situation in which production processes must be reformulated to be efficient in the use of energy and natural resources. An industrial transformation towards sustainability is of great opportunity for a step change. Digital technologies in flexible manufacturing [4] and in-depth comprehension of the transformation patterns for decision making in resource dependent cities [5] can support this evolution.

Industrial Ecology (IE) and Urban Metabolism (UM) are two concepts closely related to this situation. IE is an interdisciplinary research area aimed at mimicking natural ecosystems in the production of goods and services, thereby trying to close energy and resource cycles [6]. UM concerns the exchange of resource flows and information between urban settlements and their surroundings [7]. As a bridge between both terms, the concept of Urban-Industrial Symbiosis (UISym) emerged to try to close the cycles between industrial and urban areas by promoting resource and energy exchange to each other [8]. The main obstacle behind this interaction is the need for strong investment to implement these symbiotic networks.

Some investigations have been developed throughout the years to address different aspects related to Urban-Industrial Systems (UIS). Both Sun et al. [9] and Dong et al. [10] focused on determining the ecological benefits derived from the implementation of an UIS in Liuzhou (China), obtaining important reductions in resource mining and waste disposal. Sun et al. [11], Bian et al. [12], and Shah et al. [13] extended the scope of these types of studies by introducing the concept of eco-efficiency, thus exploring the economic impacts and geographic feasibility of implementing an UIS in the cities of Shen-yang and Guiyang

(China) and Ulsan (Korea). Other investigations have recently applied more specific tools or methods to design UIS. Fan et al. [14] developed a Pinch Analysis approach for waste integration, and Yong et al. [15] developed an approach for energy integration in UISym sites. The results obtained from each demonstration case study revealed the potential of the extension of the Pinch method for the engineering design of UIS.

The trend of UIS-related research reveals that most studies provide methodological contributions to account for the environmental benefits derived from the analysis of individual case studies while discussing the factors that hinder the implementation of symbiotic practices [16]. In this context, this research analyses the behavioural patterns observed in the main international UIS experiences, whereby the main agents, dynamics, and collaboration types involved in the interrelation between urban and industrial areas are brought together. A methodology based on research on design has been conducted to assess the main international UIS experiences that are in operation nowadays [17]. To this end, Section 2 of this article includes an overview of the symbiotic networks of seven selected real UISym case studies, and Section 3 contains the identification and description of the main features resulting from a multi-perspective analysis of these systems. Then, in Section 4, the fundamentals of a new conceptual model based on the insights obtained from the analysis are raised. Finally, Section 5 contains the main findings of the study, highlighting both their implications for this research field and future lines of action to address the limitations of this study.

2. Overview of Real Urban-Industrial Symbiosis Case Studies

The methodological approach used in this work is based on research on design [17]. With the aim of developing a UIS conceptual model, several worldwide UIS case studies were included in the research for this analysis. These case studies needed to have been successfully created, implemented, and operated. There were seven UIS case studies that met this condition and that were selected to analyse the development and implementation of synergies between industrial and urban areas, focusing on the challenges found for the materialization of this interaction. A total of three of these cases were in Europe (Forth Valley, Kalundborg, and Norrköping), one was in North America (Londonderry), and three were in Asia (Suzhou, Kawasaki, and Liuzhou). Other experiences such as Devens (U.S.), Ginebra (Switzerland), or Salaise-Sablons (France) were not considered due to their early stage of operation and/or limited data availability. First, a characterization of the case studies was conducted in terms of the symbiotic exchanges that occurred in each experience. Table 1 summarizes the main characteristics of the case studies that were overviewed, including their location, starting year, and operation scheme (do-nor(s), flow(s), and recipient(s)). The donor is the agent involved in the symbiotic exchange that supplies any waste flow, secondary outputs, or financial capital. The recipient is the agent involved in the symbiotic exchange that receives any waste flow, secondary outputs, or financial capital. Every exchange connection is denoted by the flow(s) supplied from the donor to the recipient(s). As it can be observed for each case, any agent (e.g., the Suzhou case, WWTP, incineration plant, cogeneration plant) can perform a dual role as a donor or a recipient indistinctly.

The industrial park of Suhzou was selected by the Chinese government to promote clean and renewable energy. There are two main groups of synergies in this park: the symbiosis among wastewater treatment, sludge, and cogeneration plants and the exchanges involving heating, cooling, and electric energy. Recovered water is mainly used for cooling in cogeneration plants. However, treating large volumes of water involves great amounts of sludge, which are generally disposed in landfills due to the absence of standards in this sense. To palliate this, a drying sludge plant was built next to a Wastewater Treatment Plant (WWTP) and a cogeneration plant in 2011. The processing of wet sludge and its subsequent use as fuel for generating electricity through cogeneration saves 12,000 tCO_2eq/year. The ash stemming from the incineration of sludge is used to produce construction materials. During the drying process, 90,000 t/year of condensate are also sent to cogeneration, which serves to save 1 M RMB/year in terms of water and heating costs. The steam generated during cogeneration goes through cooling towers to produce water for the Moon Bay district. Hence, 3390 t CO_2eq, 8000 t CO_2, 70 t SO_2, and 70 t NO_x per year can be saved as well as 50,000,000 RMB/year. associated with maintenance works.

Table 1. Characteristics of the main worldwide experiences on the development of Urban-Industrial Systems (UIS).

Case (Year)	Donor	Flow (s)	Recipient (s)	References
Suzhou, China (2000)	Government	Funding	All others [1]	[18,19]
	Municipality	Wastewater	WWTP [2]	
	WWTP	Wet sludge/Cooling water	Incineration Plant/Cogeneration Plant	
	Incineration Plant	Ash/Dry sludge	Cement Plant/Cogeneration Plant	
	Cogeneration Plant	Electricity/Cooling water	Municipality/SME	
Forth Valley, Edinburgh, Scotland (2001)	Government	Funding	All others	[20–23]
	Municipality	Wastewater/Scrap & Glass	WWTP/Cement Plant	
	WWTP	Wet sludge	Sludge Drying Plant	
	Sludge Drying Plant	Biomass pellets	Electric Power Plant	
	Electric Power Plant	Ash	Ash Treatment Plant	
	Ash Treatment Plant	Treated ash	Cement Plant	
	Biomass Plant	Fertilizers/Electricity	Rural areas/Municipality	
	Rural areas	Bird faeces	Biomass Plant	
Kawasaki, Japan (1997)	Government	Funding	All others	[24,25]
	Plastic Recycling Plant	Fuel	Steelworks	
	Appliance Recycling	Scrap/Plastics	Ironworks & Chemicals Plant/Plastic Recycling Plant	
	Municipality	Appliances/Wastewater/Waste/Wood, tires, plastic, oil/Scrap	Appliance Recycling/WWTP/Incineration Plant/Cement Plant/Steelworks	
	WWTP	Wet sludge/Treated water	Incineration Plant/Paper Plant	
	Incineration Plant	Pellets/Ash	Paper Plant/Cement Plant	
	Steelworks	Slags	Cement Plant	
	Ironworks	Waste Plastics	Plastic Recycling Plant	
	Paper Plant	Wet sludge/Scrap	Incineration Plant/Steelworks	
Londonderry, New Hampshire, U.S. (1996)	Private investor	Funding	All others	[26–28]
	Combined-Cycle Power Plant	Electricity	Municipality & Creamery	
	Municipality	Wastewater	WWTP	
	WWTP	Treated water	Combined-Cycle Power Plant	
	Creamery	Plastic waste	Plastic Recycling Plant	
Kalundborg, Denmark (1960)	Government	Funding	All others	[29,30]
	Municipality	Lake water	Refinery	
	Electric Power Plant	Steam/Calcium sulfate/Residual heat	Refinery & Pharmacy/Cast Plant/Municipality	
	Refinery	Fuel gas/Treated water & Fuel gas	Cast Plant/Electric Power Plant	
	Pharmacy	Biological sludge	Rural areas	
Liuzhou, China (2011)	Government	Tax reduction	All others	[9,31]
	Municipality	Waste	Recycling Plant	
	Electric Power Plant	Residual heat	Municipality & Chemicals Plant	
	Recycling Plant	Scrap/Fuel	Ironworks/Cement Plant	
	Ironworks	Gas & Residual heat/Sulfide	Chemicals Plant/Desulfurizer	
	Desulfurizer	Fertilizers	Rural areas	
Norrköping, Sweden (2011)	Government	Funding	All others	[32,33]
	Municipality	Municipal Solid Waste	Refinery, Incineration & Cogeneration Plants	
	Incineration Plant	Ash/Residual heat	Recycling & Cogeneration Plants/Municipality	
	Bio-industry	Substrate/Distillery slops	Rural areas/Refinery	
	Rural areas	Biomass waste/Grain	Cogeneration Plant/Bio-industry	
	Refinery	Fertilizers	Rural areas	
	Cogeneration Plant	Residual heat/Steam	Municipality/Bio-industry	
	Recycling Plant	Filler, gardening materials	Municipality	

[1] All others: agents involved in each UIS case study; [2] WWTP: Wastewater Treatment Plant.

Forth Valley couples Edinburgh and the petrochemical complex of Grangemouth, which is the greatest industrial area in Scotland. Moreover, Forth Valley includes four large electric plants, a cement plant, two oil companies, and two paper factories. Its synergies include the reuse of shells for roads and inert waste for aggregates and soil as well as the recycling of home appliances. Other synergies related to wastewater, heating, sludge treatment, and energy are also being investigated. The drying of sludge results in pellets, which can provide fuel to supply 30,000 households. Power plants recover and reuse about 500,000 t/year of fly ash and solid ash, resulting in GBP 988,000 in savings in 2004. The cement plant in the park uses 3 M of waste tires and 20,000 t of recycled liquid fuel produced by other companies, which enables the saving of more than 40,000 t of fossil fuels and the reduction nitrogen oxide emissions. A power plant was established to burn 110,000 t/year of bird faeces, generating 81 GWh/year of electricity to supply 20,000 households. The remaining ash stemming from the process was used as high-quality fertilizer.

The Kawasaki case study emerged because of the interest of the Japanese government to form eco-cities. To this end, it funded five facilities related to reuse paper and valorise plastics as inputs for use in both blast furnaces and in the manufacturing of ammonia and concrete moulds. A typical example of by-product exchange in Kawasaki is the use of slags from the production of steel as raw materials in the manufacturing of cement. These steelworks are fed with iron and non-ferrous materials from an appliance recycling facility, whilst cement plants are recycling the sludge from urban wastewater to replace clay as well as wood, plastic, tires, and oil wastes to substitute carbon. Kawasaki has the first paper recycling plant to achieve zero emissions. The city government manages and supervises the collection of Municipal Solid Waste (MSW) and Industrial Water (IW). Non-recyclable MSW are transferred to incineration plants, the ash of which is either reused by cement plants or disposed in a landfill. Besides the incinerators, the government also controls five collection centres and one transportation centre for MSW.

Londonderry is using eco-industrial development to deal with the negative effects of rapid growth. The inhabitants of the city have mobilized to preserve its agricultural heritage and to promote adequate environmental and cultural development. A recycling company approached a creamery to acquire its plastic waste and rinse them using grey water. This was the first step in the Londonderry eco-industrial park project, in which every member would be audited in terms of energy efficiency, water conservation, product management, materials usage, etc. A private investor owned the land and financed the development of the park. A 720 MW combined cycle power plant was installed in this park and was built underground. Soil extracted during its construction was used to develop the regional Manchester airport. In addition, the plant is cooled with 15,140 m^3/day of treated wastewater that is pumped from the WWTP in Manchester. However, the inclusion of companies in using the steam and residual heat from the power plant did not come to fruition, resulting in the power plant going to receivership in 2004.

Kalundborg arose from the premise of conserving natural reserves and improving the economy. Since 1960, it has been an important industrial centre for the country due to its large scale eco-industrial park configuration. In 1961, the power plant of the city decided to replace the use of groundwater with surface water from a lake, prompting a shift in the awareness of resource valuation. The system is formed by five main components: a power plant (600 employees and a carbon-based capacity of 1500 MW), a refinery (250 employees and a capacity of 3.2 Mt/year.), a gypsum board company (160 employees and an annual production of 14 M m^2), an international pharmacy company (1400 employees and annual sales of $2000 M), and the municipality of Kalundborg, which provides heating to 20,000 inhabitants and supplies houses and companies with water. The relationships among these agents resembles a food chain, including actions such as supplying houses, greenhouses, and aquaculture farms with heat obtained from generators and by reusing biological sludge for use as fertilizers or calcium sulphate and fuel gas to manufacture gypsum boards.

Liuzhou is a Chinese city whose economy is headed by the steel and automotive industry. The situation in this city is complex, involving different companies with a variety of economic, environmental, and social interests. This fact hindered the spontaneous creation of a symbiotic network. For these reasons, the government boosted the testing of Liuzhou as a laboratory to assess the potential benefits of IS. A steel company performs as a central node in the symbiotic system, such that it is surrounded by other industries with a high potential for material and energy exchange, such as power plants, cement plants, chemicals plants, etc. There are nine types of materials, energy sources, and wastes that can be exchanged, including blast furnace slags, treated slags, metallurgical gas, waste heat, desulfurization by-products, steel, plastics, tires, and ash. As a result, this UIS can save more than 2.4 M t of materials and 0.9 M t CO_2eq of energy through exchanges, while reducing solid waste by 3.4 M t and CO_2 emissions by 2.3 M t.

Norrköping is characterized by strong renewable energy developments and close cooperation among industrial companies that create self-organized clusters. This municipality was a pioneer in installing and operating an urban heating system. Today, it feeds a cogeneration plant with MSW (25,000 t/year.), which provides urban heat and industrial steam water. A refinery produces distillery slops that become either forage to be used in agriculture or substrate for a biogas plant. The development of this network has been aided by the commitment of the municipality with the environment, the creation of an urban heating system, the development of a cogeneration plant, and the biogas demand by the transport sector. There are future plans for promoting new synergies in Norrköping, such as a sawmill, which would be used to produce wood pellets.

3. Multi-Perspective Analysis of Urban-Industrial Systems

After the characterization of the UIS case studies described in Section 2, a further investigation was done to compare the behavioural patterns of the experiences. A systematic multi-perspective analysis was realized to identify and describe the main features of these systems. This section accounts for the main results, which are summarized in Table 2.

3.1. Types of Urban-Industrial Systems

The main aspect defining the existing types of UIS is the nature of their investor. Private investors fund and facilitate contact among companies to obtain economic benefits. Instead, public investors are national or local governments that use funds or taxes to promote cooperation among companies to bring together the economic, social, and environmental benefits. In both cases, the figure playing the role of the investor can take part in the management of the exchange networks. In the case of public investment, the process can take place with or without the support of a facilitator, who is an intermediary assuming the promotion and management of the network.

According to the case studies in Table 1, public investment (governments as donors) was the approach taken in Suzhou, Forth Valley, Kawasaki, Kalundborg, Liuzhou, and Norrköping, whilst Londonderry was the only city financed through private investment (private investor as a donor). Regardless of the source, it is always necessary to have external funding to boost and maintain UIS. Otherwise, the arrangement of cases such as these is very complicated and is unlikely to happen spontaneously. Investors are usually motivated by environmental needs, such as the valorisation of urban waste generated in municipalities (e.g., Norrköping) or the increasing concern of industrial pollution (e.g., Suzhou).

3.2. Agents in Urban-Industrial Systems

Regardless of the type of UIS, there are common agents in these systems, as demonstrated in Table 2. First is the industry, whose main aim is to transform raw materials into products. SMEs have a similar purpose but work with lower business volumes and mainly perform as waste recipients. The municipality is home to a population, generating Municipal Solid Waste (MSW) and wastewater. Rural areas relate to zones devoted to agriculture and livestock, whilst power plants deal with the production of electric energy.

There are three other agents focused on valorising either materials, water, or energy. The former seeks to provide new uses for wastes, whilst water valorisation concerns treatment plants aimed at purifying wastewater from the industry and the municipality, such that it can be reused without damaging the environment. Finally, energy valorisation stands for both the incineration of MSW and the drying of sludge. Incineration causes a volume reduction in waste, which is transformed into ash that can be used in industry as additives. Thermal energy can also be obtained throughout this process. As for sludge drying, it consists of dehydrating the wet sludge stemming from WWTP and converting it into pellets to be used as fuel by other agents.

Table 2. Characteristics of the agents involved in Urban-Industrial Systems (UIS).

Agent	Type	Role	Aims	Input	Output	Dynamics
Industry	• Cement • Paper • Chemicals • Pharmacy • Iron and steel • Bio-industry	Transformation of raw materials into products (large scale)	To obtain benefits by reusing waste as raw materials, resulting in a reduction of costs; to use treated water and residual thermal energy; to dispose waste that can be used by other industries, SMEs, etc.	• MSW [1] (plastics, etc.) • IW [2] (slags) • Residual thermal energy • Treated water	• IW (fertilizer—biological sludge, slags, scrap, fuel—gas) • Residual thermal energy • IWW [3]	• Initial • Intermediate • Advanced
Municipality	• Cities • Residential areas—less populated • Urban areas—more populated	Production of MSW and UWW [4]	To minimize and manage wastes generated by the population; to optimize energy resources and raw materials	• Residual thermal energy • Treated water • Raw materials	• MSW (plastic, glass, scrap) • UWW	• Initial • Intermediate • Advanced
Energy production	• Cogeneration • Biomass • Combined cycle	Generation of electric energy	To generate energy using wastes from the industry; to use WWTP treated water for cooling such that residual heat can perform as an input for the industry and municipality	• IW (fuel—pellets and gas) • Residual thermal energy • Treated water	• Residual thermal energy • IW (ashes, fertilizers) • Cooling water	• Intermediate • Advanced
Agriculture and livestock (Rural areas)	• Croplands • Farms • Forests	Cultivation of crops and breeding of livestock for human consumption	To obtain benefits by selling organic waste and/or buying fertilizers from the industry	• IW (fertilizers—biological sludge) • Treated water	• MSW (organic waste)	• Intermediate • Advanced
SME	• Offices • Workshops • Shops	Transformation of raw materials into products (small scale)	To obtain benefits by reusing waste as raw materials, resulting in a reduction of costs; to dispose waste that can be used by other industries, SMEs, etc.	• MSW • IW • Cooling water	• IW • MSW	• Intermediate
Material valorization	• Plastics • Home appliances • Ashes • Desulfurizers	Recycling of wastes to enable their reuse	To close the consumption cycle, recovering materials from residual flows to integrate them in the production cycle	• MSW • IW	• Materials	• Intermediate • Advanced
Water valorization	• WWTP [5]	Filtering and purification of waste for reusing purposes	To eliminate waste from water; to convert these wastes into safe sludge that can be dried to be used by the industry or incinerated	• UWW • IWW	• IW (wet sludge) • Treated water	• Intermediate • Advanced
Energy valorization	• Sludge drying • MSW and IW incinerator	Incineration of MSW and drying of sludge from wastewater	To reduce the volume of waste for producing energy; to dry sludge and obtain fuel in the form of pellets	• IW (wet sludge) • MSW	• IW (fuel—pellets, slags) • Residual thermal energy	• Intermediate • Advanced
Public investor	• National • Local	Promotion of companies meeting environmental standards and control of service management	To foster, invest in, develop, and even manage part of the symbiosis network, trying to meet environmental standards	• MSW management revenue	• Tax reduction • Investments	• Initial • Intermediate • Advanced
Private investor	• Private company	Coordination of the creation and maintenance of symbiotic companies and search for funding sources	To obtain benefits from investing in companies, incinerators, and recycling and water treatment plants; to perform as a private manager and mediate in the creation of partners	• Earning of part of the profits	• Investments	• Initial • Intermediate • Advanced

[1] MSW: Municipal Solid Waste; [2] IW: Industrial Water; [3] IWW: Industrial Wastewater; [4] UWW: Urban wastewater; [5] WWTP: Wastewater Treatment Plant.

3.3. Dynamics in Urban-Industrial Systems

The mechanisms to create and successfully operate the UIS analysed in the worldwide experiences took a long-term approach. Overall, the operation of UIS can be divided into initial, intermediate, and advanced stages of development. The initial stage, led by either a public or private investor, serves to encourage the required participants to form a basic exchange network in contact with each other. During the intermediate stage, the symbiotic network grows through the addition of more partakers. In the advanced stage, the system is managed by either public or private investors to maximize its benefits.

The agents become involved in the system at different stages. The components participating in the system from the beginning of the process are only municipality and industry. They are both at the core of UIS not only for conceptual reasons, but also because of the amount of waste they generate, which can be reused by other agents that enter the system in the intermediate stage, such as SMEs, power plants, and rural areas. The latter join the system to strengthen the symbiotic network and obtain benefits from it. Although they are not as essential as those parties who are involved in the initial stage, their role is crucial in terms of association and cooperation.

The agents related to resource valorisation are created complementarily to the existing symbiosis network in the advanced stage to improve it and minimize the amount of waste. They work as intermediaries among agents that could not cooperate otherwise, resulting in new options for the reuse of waste. Despite the differences between the intermediate and advanced stages, this breakdown is not always put into practice. For instance, these steps emerged at the same time in the third case study that was overviewed (Kawasaki) since there was an interest in rapidly densifying the network and closing material loops.

3.4. Collaboration in Urban-Industrial Systems

The most common means of collaboration observed in the main worldwide experiences (Table 1) concern the exchange of materials, water, and energy. This does not have to always be this way; other cases lacking heavy industry might be more active in sharing infrastructure and/or services [34].

The collaboration identified in the case studies can be grouped as to the type of flows exchanged between the donors and recipients summarized in Table 1. These flows were organized and allocated to each type of agent as input and output flows, as presented in Table 2. There are four distinct main categories: materials (MSW—plastic waste, organic waste, scrap, glass; industrial waste—plastics, scrap, ash, slags, fuel, fertilizers, wet sludge), water (urban wastewater, industrial wastewater, treated water, water for cooling purposes), energy (thermal energy), and economy (money, investment, funding, tax reduction, control of service management, benefits from exchanges).

Contacts among the parties involved usually stem from meetings among representatives of the industries promoted by the investor, in which the potential benefits of collaboration are shown. In this sense, computer tools are used to facilitate cooperation, such as synergy databases, where companies can search for potential exchanges. It is also common to form councils involving representatives of both the industry and the municipality to take charge of the development and management of the symbiotic network. As a summary of all the aspects covered in this section, Table 2 shows the agents involved in UIS as well as their role, aims, flows (inputs and outputs), and type.

4. Proposal of a New Conceptual Model

The genesis of the case studies summarized in Figure 1 demonstrates similar characteristics with Planned Eco-Industrial Parks (PEIPs), which seek compatible activities with potential to cause symbiotic networks with the financial support of an external agent [35]. This coincides with the role played by the investors, who promote and coordinate the creation of these experiences. However, this situation would appear to change with the increasing autonomy of park agents, and they start searching for their own benefits with time, to the extent that they end up resembling the dynamics of Self-Organizing Symbiosis (SOS) [36] rather than those of PEIPs.

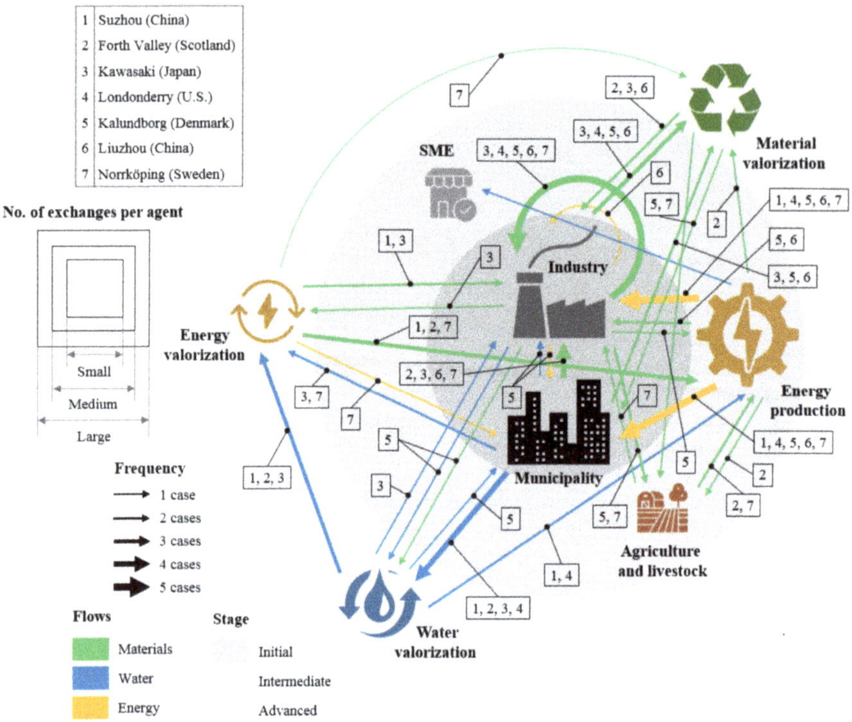

Figure 1. Graphical summary of the type, size, and time of the relationships among the agents involved in the main worldwide Urban-Industrial Systems (UIS).

This unsteady condition highlights the need for elaborating a model for their systemic conceptualization. To this end, the analysis of the behavioural patterns in the main UIS experiences reported worldwide was first, as listed in Figure 1. The dynamics observed in these case studies enabled their subsequent examination as complex systems. Then, a list of indicators was proposed to measure the eco-efficiency of the agents involved in the UIS. The last step consisted of the selection of a modelling method according to the characteristics of UIS as complex systems.

This course of action is in line with the conclusions drawn from recent studies in the field of UIS to emphasize the development of this kind of approach as a key line of research to develop in the future. For instance, Lu et al. [37] pointed out that the lifecycle of a perspective UIS, which is closely related to the evolutionary behaviour of these complex systems with time, must be considered. Moreover, Fan et al. [14] underlined to the importance of implicating all of the agents and flows involved in UIS. In the same vein, Bian et al. [12] referred to the creation of comprehensive systemic models to account for the interactions among UIS sectors while proposing an eco-efficiency indicator approach to account for the effects and contributions in UIS.

4.1. Behavioral Patterns in Urban-Industrial Systems

Figure 1 provides a graphical scheme of the agents identified in Table 2 and their relationships to each other, based on the case studies summarized in Table 1. These agents are represented using different sizes (small, medium, large) depending on the number of exchanges in which they participate, while highlighting the stage of development when they entered the system. Industry, energy production, and municipality agents are depicted as the largest in terms of exchanges, whereas SME, energy valorisation, and agriculture agents are the smallest. Material and water valorisation agents participated in a number of exchanges in between and therefore are represented by as being of medium-sized. The scheme also indicates the frequency with which each flow (materials, water, and energy) is exchanged, specifying its correspondence to the case studies under analysis (e.g., the energy production agent exchanges energy with industry and municipality in five case studies). The inspection of Figure 1 reveals that most of the interactions take place in the last stage. This is due to the ad hoc purpose of the material, energy, and water valorisation agents entering the system in this step since they are aimed at complementing the symbiotic network and increasing the number of relationships in order to reduce waste and close loops.

Most of the cases that were overviewed (2, 5, 6, and 7), where both industry and municipality participate from the initial stage through flow exchanges, are governed by a similar pattern in their dynamics. Their initial stage is deeply rooted due to both the exchange of different waste flows between industry and municipality and the support of public investors. Companies devoted to the production of electric energy are essential in the intermediate stage since they can contribute to increasing the efficiency of the system, either directly (interacting with industry and/or municipality) or indirectly (through other agents, such as in the case of agriculture and livestock in Case 2). Finally, the agents in charge of the valorisation of materials are indispensable in the advanced stage to absorb waste and, therefore, to densify the network. The other agents involved in this stage depend on the priority flows to be reused, resulting in energy and water valorisation plants such as incinerators (e.g., case 7, Norrköping) and WWTP (e.g., case 5, Kalundborg).

As for Case 3, the rush of the public investors to transform Kawasaki into an eco-city provoked a leap in the Urban-Industrial System, whereby the intermediate stage was omitted. Instead, valorisation agents, which usually belong to the advanced stage, were straightforwardly incorporated to intensify the exchange network. This experience further highlights the great importance of these components in consolidating UIS; however, the haste in strengthening the network might lead to some agents, such as power plants, whose role may eventually yield better results in terms of efficiency, being disregarded.

Cases 1 and 4 were not initially conceived as UIS experiences. In the case of Suzhou, public investment was targeted at improving energy efficiency. In the end, this original goal resulted in an ecological urban-industrial development. Instead, Londonderry started with the aim of creating an eco-industrial park through material exchanges in the industry. The municipality adhered to the network in the subsequent steps and did not conduct direct exchanges with the industry. These cases could be defined as late UIS since the incorporation of the municipality takes place after the initial stage and because the exchanges with the industry are indirect.

4.2. Urban-Industrial Systems as Complex Systems

A system can be defined as a set of interactive elements [38]. Complex systems have a structure formed by the following characteristics: (1) autonomous, (2) self-organized, (3) responsive, (4) not governed by linear patterns, and (5) willing to consolidate their resilience [39]. Autonomy is especially relevant for agents in low hierarchical levels because they are responsible for boosting the symbiotic networks. Table 1 highlights how industry plays a different role from the remaining players, setting relationships with a variety of agents, including the link between the refinery industry with the municipality (Kalundborg), the bio-industry with rural areas based on agriculture and livestock (Norrköping), or even between itself as the steelworks and cement plant pair (Kawasaki).

Self-organization stems from the agents within the system that do not require any external support. Thus, every agent belonging to the UIS increases the number and intensity of its relationships with the remaining components in the system [40,41]. In this sense, investors are important in promoting the existence of these cooperation actions. As for their reaction capacity, the basic components of UIS (people, companies, etc.) respond differently to changes in their environment. In the first case study that was overviewed (Shuzou), the network emerged in response to the taxes and fees created by the government due to the high pollution produced in the municipality, which resulted because the bad practices of heavy industries were favoured in the area. Instead, the situation in Norrköping was very different since the goal was to improve the energy efficiency of the system. The agents involved in this case used the support of the government to intensify and promote new cooperative relationships to each other.

Cooperation among companies is boosted by the existence of confidence and previous agreements; however, these aspects do not follow linear patterns, which results in the generation of unpredictable conduct in the system. In turn, this involves additional complexity for the development of interactions [42,43]. The investigation of the case studies compiled in this work suggests that some of the experiences were initiated in a similar manner. However, different networks emerged later on, and the benefits achieved by these networks differed from each other. In addition, many complex systems are also adaptive. The behaviour of the basic components in adaptive systems can evolve with time, providing a certain reaction capacity against changes in the environment through learning mechanisms.

This factor results in the last characteristic of complex systems, which concerns their resilience. UIS are capable of both dealing with technological and social challenges as well as admitting new agents and eliminating exchange flows [44]. An example of resilience in UIS is the collaboration among the agents, which enhances the robustness, adaptability, and flexibility of the system. In the case of Kawasaki, the Japanese government considered this feature and promoted the creation of ad hoc agents to ensure the resilience of the system with time. Based on the examination of all of these characteristics, it can be concluded that UIS are complex and adaptive systems.

4.3. List of Eco-Efficiency Indicators

The concept of eco-efficiency, which arises from the consideration of environmental impacts throughout the lifecycle of products and the willingness to reduce them them, can be useful to value changes in UIS. It seeks to produce more and pollute less [45]. Given the differences in the data quality among economic, environmental, and social parameters established by international organizations to monitor sustainable development as well as the complexity in adapting them to the specifics of UIS, it is difficult to produce consistent aggregations to evaluate the performance of the whole system.

Instead, the World Business Council for Sustainable Development (WBCSD) proposes a series of indicators that can be used to measure the eco-efficiency of a system [46]. Table 3 includes some of these indicators, which were grouped as follows: applicable to the industry and municipality, applicable to the whole UIS, and exclusive for the industry and municipality. The first group is valid for the two main agents in the system (industry and municipality) and can be used to obtain valuable information about the changes produced in any of them thanks to a symbiosis process. They account for energy consumption, the conversion of urban and industrial wastes into by-products, the amount of greenhouse gas emissions produced, and the reduction of waste taxes.

The second group, which concerns all of the agents in the UIS, has a predominantly environmental nature. Energy consumption includes electricity, fossil fuels, biomass, wood, solar and wind power, etc. Water represents all of the freshwater provided by the public network or that is obtained from surface or groundwater sources as well as water used for cooling purposes. The third indicator measures the amount of acid gas or steam emitted into the air as a result of fossil combustion and reactive processes.

Table 3. List of indicators to measure the eco-efficiency of Urban-Industrial Systems (UIS) [46].

Agent	Indicator	Units	Measurement Method	Source
Industry and municipality (applicable)	Energy saving	Currency	Energy consumption	Electricity bill
	MSW and IW reduction	tons	Approach taken by the municipality or industry to measure the amount of waste sent to landfill or incineration	Registry of landfills and incineration plants
	Carbon footprint	CO_2eq	Greenhouse Gas (GHG) Protocol	Emission inventory
	Reduction in waste management taxes/fees	Currency	Checkup of the amount and type of waste disposed	Tax and administrative institutions
All	Energy consumption	GJ	Transformation factors: Higher Heating Value (HHV) for fossil fuels, based on combustion products, water (liquids), CO_2 and nitrogen (gas); electricity and town gas as amount of purchased energy	Purchase and management reports. Energy and fuel use inventories. Bibliography
	Material consumption	tons	Specific approach taken by each agent to account for the amount used	Purchase, management, and cost reports
	Water consumption	m^3	Specific approach taken by each agent to account for the amount used	Purchase, management, and cost reports
	Acid air emissions	tons of SO_2eq	Lists of acids and potential acidification processes from bibliographic sources	Plant control. Acid rain reports. Estimates and/or calculations
Industry (exclusive)	Industrial Production Index (IPI)	%	Laspeyres Price Index	Periodic surveys of the industry
	Amount of products and services	Units or mass	Specific approach taken by the industry	Cost, production, and sale reports. Annual financial reports
	Net sales	Currency	International Accounting Standards Committee (IASC) Generally Accepted Accounting Principles (GAAP)	Annual financial reports
	Net income	Currency	Net sales minus all of the expenses for the study period	Cost, production, and sales reports. Annual financial reports
	Waste generation	tons	Waste definition and disposal methods according to the 1992 Basel Convention	Plant control. Acid rain reports. Estimates and/or calculation
Municipality (exclusive)	Per capita Gross Domestic Product (GDP)	Currency	GDP divided by the population	Statistical institutes
	Disease reduction	No. of patients	Count of treated patients	Archives of health centers and hospitals
	Population increase	No. of inhabitants	Count of the people living in the municipality	Municipal census

The last group considers exclusive indicators for the industry and municipality. The first subgroup accounts for quantitative indicators related to industrial goods, products and services, and substances destined for disposal. In addition, it addresses economic aspects concerning sales, discounts, and income. The subgroup associated with the municipality includes the wealth generated per resident inhabitant, the monitoring of diseases derived from pollution, and the increase in population. In the end, the joint consideration of the indicators in the three groups represents resource flows (materials, water, and energy) and social, environmental, and economic aspects whose management is expected to improve due to the proposed urban-industrial model.

4.4. Selection of Modelling Method

The selection of a modelling method needs to be consistent with the characteristics of the system to be represented [47]. In particular, the creation of a mathematical model to describe UIS is very difficult due to, among other things, the uncertainty, emergence, or adaptability of these complex systems [48]. In this sense, there are some analytical methods used to represent complex systems. A brief description of the most relevant ones and a valuation of their suitability to be applied for modelling UIS can be summarized as follows [49]:

- Statistical models and Bayesian networks serve to model different phenomena through a set of variables and dependence relationships among them. However, they cannot easily represent feedback as it is conducted in UIS;
- Systems dynamics are widely used for modelling complex systems in engineering, economics and business, project and environmental management, etc. Their adaptive capacity is limited, which contrasts with the evolutionary nature of UIS agents;
- Evolutionary models can represent learning and adaptation characteristics; however, they are not adequate to model the heterogeneity and autonomy that define UIS;
- Cellular automata can be defined as a dynamical system formed by a set of simple and homogeneous elements whose aggregation enables modelling complex behaviors. As such, it is unsuitable to model the variety of components in UIS;
- Agent-based models (ABM) enable the simulation of the actions and interactions of autonomous individuals as well as their effects on the system from the micro to macro levels. ABM agents act according to their own interests and can learn and adapt to changes.

Figure 2 summarizes the workflow for selecting the most suitable tool for modelling UIS. As introduced above, the specifics of UIS suggest that ABM is the best option. The first reason supporting this is based on the existence of complex nonlinear discrete interactions among the agents. This means that the actions of an agent can be altered by others, such that their description through traditional methods (e.g., statistical approaches) might be difficult. There is a variety of exchange flows in UIS (Figure 1); however, this does not mean that an agent can only exchange one flow type. For example, the incinerator in the third case study (Kawasaki) only generates ash, whilst this same facility exchanges both material and thermal energy flows in Norrköping.

There are not only different types of agents in UIS, but some of the same kind exhibit different attributes. Traditional modelling approaches represent agents with average characteristics, which is far from the real situation in UIS. Table 2 compiles the ten types of agents involved in UIS. The case of industry is especially enlightening in this sense because it encompasses a variety of different types (cement, chemical, pharmacy, etc.) and has very different relationships to the remaining agents, regardless of their specific field of activity.

The topology of the interactions among the agents in the system is heterogeneous and complex. This is especially relevant for social processes, which involve learning and adaptation. ABM enables realistic topological modelling, which is required to explain the aggregated behaviour of the system. The high number of potential interactions among agents and the variety of flow types that can be exchanged in UIS hinder their modelling.

Finally, UIS have a complex and stochastic behaviour that changes with time, which precludes its forecast in advance. Hence, it cannot be approached using equations or transition rates. This circumstance is clearly represented in the adaptive capacity of the Norrköping case study. In this situation, the government invested in improving a power plant to enable feeding it with biofuel; however, the situation evolved to a greater exploitation of the system with the passage of time, including MSW and densifying the network by promoting different flow exchanges.

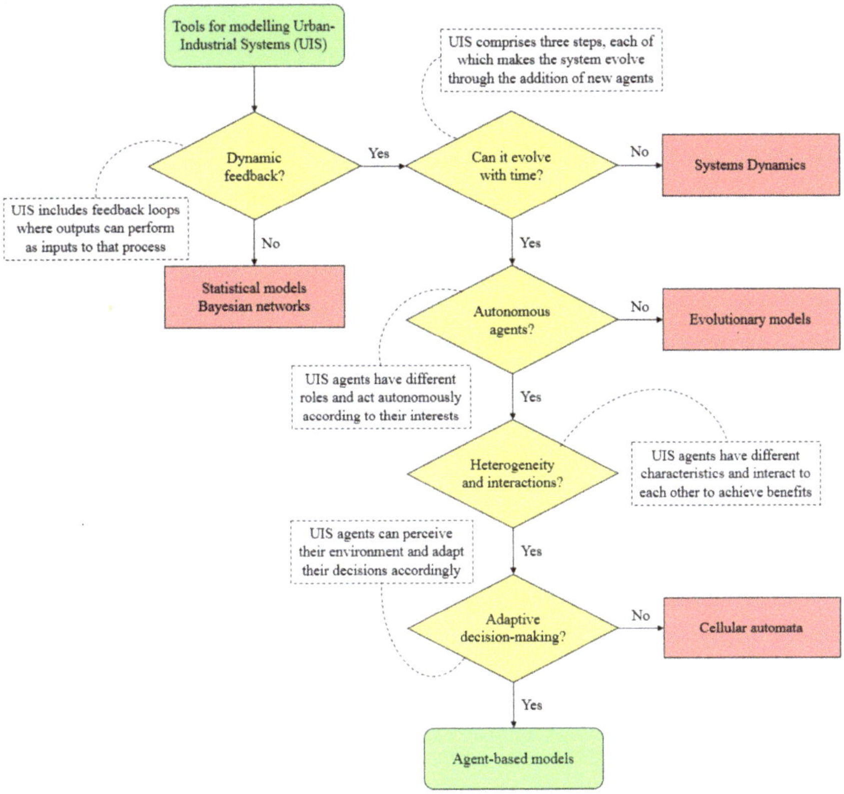

Figure 2. Flowchart for the selection of the most suitable tool for modelling Urban-Industrial Systems (UIS). Adapted from Heckbert et al. [49].

5. Conclusions

In accordance with the aims initially set in this research, a conceptual model to understand and represent Urban-Industrial Systems (UIS) has been presented. To this end, the patterns of different worldwide urban-industrial symbiotic networks were observed. Such patterns served to outline a model capable of accounting for all of the characteristics of UIS as complex systems, including their stages, dynamics, participating agents, and exchange flows. The analysis also emphasized the importance of public and private investment to boost the creation and development of UIS.

The examination of existing UIS enabled the identification of the main agents and flows involved in symbiotic networks as well as their importance. The comprehension of these dynamics led the posing of an analytical model to represent UIS reliably and systematically, giving insight into the relationships among the agents within the system as they join the network throughout its different stages. The proposed model is intended to improve the eco-efficiency of UIS through the consideration of a series of indicators that represent the technical, social, environmental, and economic characteristics of the agents involved.

The trends and characteristics observed in the analysis as well as the conceptualization of UIS modelling are important contributions for encouraging the development of new experiences focused on the creation of UIS. The inferences extracted from the investigation of existing cases provide useful information for potential investors, emphasizing both the profile of the agents required to form a symbiotic network and the time in which each of them should enter the system. Although a preliminary conceptual model has been inferred, its robustness would benefit from widening this systematic research on more successful cases. The representation of UIS through the application of the proposed model is the main line of research that needs to be developed in the future in order to continue building a road map to maximize the benefits of promoting symbiotic interactions between urban and industrial areas in terms of sustainable development. Companies, policy makers, and interested stakeholders could explore the patterns and the effects that their strategies and policies could have on the hopefully successful reconfiguration and implementation of a new urban-industrial model that is able to use waste and secondary outputs as resources. A smart digitalization of industry and cities can become key in supporting and accelerating this transformation towards sustainability.

Funding: This research was funded by the Spanish Ministry of Science, Innovation and Universities, grant number DPI2017-88127-R (AEI/FEDER, UE).

Data Availability Statement: Not applicable.

Conflicts of Interest: The author declares no conflict of interest. The funders had no role in the design of the study; in the collection, analyses, or interpretation of data; in the writing of the manuscript, or in the decision to publish the results.

References

1. Uras, M.; Cossu, R.; Ferrara, E.; Liotta, A.; Atzori, L. PmA: A real-world system for people mobility monitoring and analysis based on Wi-Fi probes. *J. Clean. Prod.* **2020**, *270*, 122084. [CrossRef]
2. Rodrigues, M.; Franco, M. Measuring the urban sustainable development in cities through a composite index: The case of Portugal. *Sustain. Dev.* **2020**, *28*, 507–520. [CrossRef]
3. Saudi, M.H.M.; Sinaga, O.; Roespinoedji, D.; Jabarullah, N.H. Industrial, commercial, and agricultural energy consumption and economic growth leading to environmental degradation. *Ekoloji* **2019**, *28*, 299–310.
4. Margherita, E.G.; Braccini, A.M. Industry 4.0 technologies in flexible manufacturing for sustainable organizational value: Reflections from a multiple case study of Italian manufacturers. *Inf. Syst. Front.* **2020**, 1–22. [CrossRef]
5. Mao, W.; Wang, W.; Sun, H.; Yao, P.; Wang, X.; Luo, D. Urban industrial transformation patterns under natural resource dependence: A rule mining technique. *Energy Policy* **2021**, *156*, 112383. [CrossRef]
6. Gibbs, D.; Deutz, P.; Proctor, A. Industrial ecology and eco-industrial development: A potential paradigm for local and regional development? *Reg. Stud.* **2005**, *39*, 171–183. [CrossRef]
7. Dijst, M.; Worrell, E.; Böcker, L.; Brunner, P.; Davoudi, S.; Geertman, S.; Harmsen, R.; Helbich, M.; Holtslag, A.A.M.; Kwan, M.P.; et al. Exploring urban metabolism—Towards an interdisciplinary perspective. *Resour. Conserv. Recycl.* **2018**, *132*, 190–203. [CrossRef]
8. Van Berkel, R.; Fujita, T.; Hashimoto, S.; Geng, Y. Industrial and urban symbiosis in Japan: Analysis of the eco-town program 1997–2006. *J. Environ. Manag.* **2009**, *90*, 1544–1556. [CrossRef] [PubMed]
9. Sun, L.; Li, H.; Dong, L.; Fang, K.; Ren, J.; Geng, Y.; Fujii, M.; Zhang, W.; Zhang, N.; Liu, Z. Eco-benefits assessment on urban industrial symbiosis based on material flows analysis and energy evaluation approach: A case of Liuzhou city, China. *Resour. Conserv. Recycl.* **2017**, *119*, 78–88. [CrossRef]
10. Dong, L.; Liang, H.; Zhang, L.; Liu, Z.; Gao, Z.; Hu, M. Highlighting regional eco-industrial development: Life cycle benefits of an urban industrial symbiosis and implications in China. *Ecol. Modell.* **2017**, *361*, 164–176. [CrossRef]
11. Sun, L.; Fujii, M.; Li, Z.; Dong, H.; Geng, Y.; Liu, Z.; Fujita, T.; Yu, X.; Zhang, Y. Energy-saving and carbon emission reduction effect of urban-industrial symbiosis implementation with feasibility analysis in the city. *Technol. Forecast. Soc. Chang.* **2020**, *151*, 119853. [CrossRef]
12. Bian, Y.; Dong, L.; Liu, Z.; Zhang, L. A sectoral eco-efficiency analysis on urban-industrial symbiosis. *Sustainability* **2020**, *12*, 3650. [CrossRef]
13. Shah, I.H.; Dong, L.; Park, H.S. Tracking urban sustainability transition: An eco-efficiency analysis on eco-industrial development in Ulsan, Korea. *J. Clean. Prod.* **2020**, *262*, 121286. [CrossRef]
14. Fan, Y.V.; Varbanov, P.S.; Klemeš, J.J.; Romanenko, S.V. Urban and industrial symbiosis for circular economy: Total EcoSite Integration. *J. Environ. Manag.* **2021**, *279*, 111829. [CrossRef]
15. Yong, W.N.; Liew, P.Y.; Woon, K.S.; Wan Alwi, S.R.; Klemeš, J.J. A pinch-based multi-energy targeting framework for combined chilling heating power microgrid of urban-industrial symbiosis. *Renew. Sust. Energ. Rev.* **2021**, *150*, 111482. [CrossRef]
16. Fraccascia, L. Industrial symbiosis and urban areas: A systematic literature review and future research directions. *Procedia Environ. Sci. Eng. Manag.* **2018**, *5*, 73–83.
17. Van den Brink, A.; Bruns, D.; Tobi, H.; Bell, S. *The Relationship between Research and Design. Research in Landscape Architecture: Methods and Methodology*, 1st ed.; Routledge: London, UK, 2016. [CrossRef]

18. Wen, Z.; Meng, X. Quantitative assessment of industrial symbiosis for the promotion of circular economy: A case study of the printed circuit boards industry in China's Suzhou New District. *J. Clean. Prod.* **2015**, *90*, 211–219. [CrossRef]
19. Yu, C.; Dijkema, G.P.J.; De Jong, M.; Shi, H. From an eco-industrial park towards an eco-city: A case study in Suzhou, China. *J. Clean. Prod.* **2015**, *102*, 264–274. [CrossRef]
20. Berkel, R.; Bossilkov, A.; Corder, G. *Regional Synergies for Sustainable Resource Processing: A Status Report*; Technical Report; Centre for Sustainable Resource Processing: London, UK, 2005.
21. Golev, A.; Corder, G.D. Developing a classification system for regional resource synergies. *Miner. Eng.* **2012**, *29*, 58–64. [CrossRef]
22. Harris, S. Drivers and Barriers to Industrial Ecology in the UK. Ph.D. Thesis, The University of Edinburgh, Edinburgh, UK, 2004.
23. Harris, S.; Berkel, R.; Kurup, B. Fostering Industrial Symbiosis for Regional Sustainable Development Outcomes. In Proceedings of the Corporate Responsibility Research Conference (CRRC), Belfast, UK, 7–9 September 2008; pp. 1–21.
24. Geng, Y.; Tsuyoshi, F.; Chen, X. Evaluation of innovative municipal solid waste management through urban symbiosis: A case study of Kawasaki. *J. Clean. Prod.* **2010**, *18*, 993–1000. [CrossRef]
25. Ohnishi, S.; Dong, H.; Geng, Y.; Fujii, M.; Fujita, T. A comprehensive evaluation on industrial & urban symbiosis by combining MFA, carbon footprint and energy methods—Case of Kawasaki, Japan. *Ecol. Indic.* **2017**, *73*, 315–324. [CrossRef]
26. Chertow, M.R. Industrial symbiosis: Literature and taxonomy. *Annu. Rev. Energy Environ.* **2000**, *25*, 313–337. [CrossRef]
27. Deutz, P.; Gibbs, D. Eco-industrial development and economic development: Industrial ecology or place promotion? *Bus. Strategy Environ.* **2004**, *13*, 347–362. [CrossRef]
28. Van Berkel, R. Regional resource synergies for sustainable development in heavy industrial areas: An overview of opportunities and experiences. *Perth WA Aust.* **2006**, *135*, 139.
29. Cervantes Torre-Marín, G.; Granados, R.S.; Herrera, G.R.; Martínez, F.R. Ecología industrial y desarrollo sustentable. *Ingeniería* **2009**, *13*, 63–70.
30. Ehrenfeld, J.; Gertler, N. Industrial ecology in practice: The evolution of interdependence at Kalundborg. *J. Ind. Ecol.* **1997**, *1*, 67–79. [CrossRef]
31. Dong, L.; Zhang, H.; Fujita, T.; Ohnishi, S.; Li, H.; Fujii, M.; Dong, H. Environmental and economic gains of industrial symbiosis for Chinese iron/steel industry: Kawasaki's experience and practice in Liuzhou and Jinan. *J. Clean. Prod.* **2013**, *59*, 226–238. [CrossRef]
32. Hatefipour, S.; Baas, L.; Eklund, M. The Händelö Area in Norrköping, Sweden Does It Fit for Industrial Symbiosis Development? In *World Renewable Energy Congress*; Linköping University Electronic Press: Linköping, Sweden, 2011; p. 8.
33. Rehn, S. Influencing Industrial Symbiosis Development: A Case Study of Händelö and Northern Harbour Industrial Areas. Master's Thesis, Linköping University, Linköping, Sweden, 2013.
34. Neves, A.; Godina, R.; GAzevedo, S.; Pimentel, C.; COMatias, J. The potential of industrial symbiosis: Case analysis and main drivers and barriers to its implementation. *Sustainability* **2019**, *11*, 7095. [CrossRef]
35. Albino, V.; Fraccascia, L.; Giannoccaro, I. Exploring the role of contracts to support the emergence of self-organized industrial symbiosis networks: An agent-based simulation study. *J. Clean. Prod.* **2016**, *112*, 4353–4366. [CrossRef]
36. Tao, Y.; Evans, S.; Wen, Z.; Ma, M. The influence of policy on industrial symbiosis from the Firm's perspective: A framework. *J. Clean. Prod.* **2019**, *213*, 1172–1187. [CrossRef]
37. Lu, C.; Wang, S.; Wang, K.; Gao, Y.; Zhang, R. Uncovering the benefits of integrating industrial symbiosis and urban symbiosis targeting a resource-dependent city: A case study of Yongcheng, China. *J. Clean. Prod.* **2020**, *255*, 120210. [CrossRef]
38. Von Bertalanffy, L. *General System Theory: Foundations, Development, Applications*, 1st ed.; George Braziller: New York, NY, USA, 1969.
39. Holland, J.H. Studying complex adaptive systems. *J. Syst. Sci. Complex.* **2006**, *19*, 1–8. [CrossRef]
40. Allen, P.M. *Cities and Regions as Self-Organizing Systems Models of Complexity*; Taylor & Francis: London, UK, 1997.
41. Boons, F. Self-organization and sustainability: The emergence of a regional industrial ecology. *Emerg. Complex. Organ.* **2008**, *10*, 41.
42. Boccara, N. *Modeling Complex Systems*; Springer: Berlin/Heidelberg, Germany, 2004.
43. Jato-Espino, D.; Ruiz-Puente, C. Bringing facilitated industrial symbiosis and game theory together to strengthen waste exchange in industrial parks. *Sci. Total Environ.* **2021**, *771*, 145400. [CrossRef] [PubMed]
44. Walker, B.; Holling, C.; Carpenter, S.; Kinzig, A. Resilience, Adaptability and Transformability in Social–Ecological Systems. Retrieved 2 June 2021. *Ecol. Soc.* **2004**, *9*. Available online: http://www.jstor.org/stable/26267673 (accessed on 18 August 2021). [CrossRef]
45. Koskela, M.; Vehmas, J. Defining eco-efficiency: A case study on the Finnish forest industry. *Bus. Strateg. Environ.* **2012**, *21*, 546–566. [CrossRef]
46. Verfaillie, H.A.; Bidwell, R. Measuring eco-efficiency: A guide to reporting company performance. *Ginebra* **2000**, *3*, 16–21.
47. Romero, E.; Ruiz, M.C. Proposal of an agent-based analytical model to convert industrial areas in industrial eco-systems. *Sci. Total Environ.* **2014**, *468–469*, 394–405. [CrossRef]
48. Romero, E.; Ruiz, M.C. Framework for applying a complex adaptive system approach to model the operation of eco-industrial parks. *J. Ind. Ecol.* **2013**, *17*, 731–741. [CrossRef]
49. Heckbert, S.; Baynes, T.; Reeson, A. Agent-based modeling in ecological economics. *Ann. N. Y. Acad. Sci.* **2010**, *1185*, 39–53. [CrossRef] [PubMed]

sustainability

MDPI

Article

Understanding Hazardous Waste Exports for Disposal in Europe: A Contribution to Sustainable Development

Carmen Callao [1,*], M. Pilar Latorre [2] and Margarita Martinez-Núñez [3]

1 Legal Department, Universidad San Jorge, 50830 Zaragoza, Spain
2 Department of Business & Administration, Facultad de Ciencias Sociales y del Trabajo, University of Zaragoza, 50009 Zaragoza, Spain; latorrep@unizar.es
3 Department of Management Engineering, Business Administration and Statistics, ETSI Sistemas de Telecomunicación, Technical University of Madrid, 28040 Madrid, Spain; margarita.martinez@upm.es
* Correspondence: ccallao@usj.es; Tel.: +34-976-060-100

Abstract: The concept of sustainable development was introduced in Europe by the Treaty of Amsterdam (1997) and was extended to waste management in the Waste Framework Directive. In order to achieve sustainable development, hazardous waste (HW) must be managed safely and in accordance with regulations. This also applies to worldwide HW transport, especially when HW is shipped for disposal. The United Nations, through the Basel Convention, aims to prevent the export of HW from developed countries to developing countries for disposal. In Europe, HW shipments are regulated by Regulation (EC) No. 1013/2006 of the European Parliament and by the Council of 14 June 2006 on shipments of waste. Additionally, all HW shipments must be in accordance with two principles contained in the Waste Framework Directive: proximity and self-sufficiency. Using data from 2014 and network analysis methodology, this paper fills the gaps in the scientific literature by looking at how shipments of HW travel for disposal in Europe, how the regulations affect these shipments and how GDP per capita influences the shipment of waste. The results show that countries with a high GDP per capita play an important role in the network (having the highest in-degree) and that the absence of landfill taxes for HW does not influence HW shipments for disposal. Therefore, countries in the EU act in accordance with the proximity and self-sufficiency principles.

Keywords: hazardous waste shipment; network analysis; gross domestic product per capita; disposal; proximity principle; self-sufficiency principle

check for updates

Citation: Callao, C.; Latorre, M.P.; Martinez-Núñez, M. Understanding Hazardous Waste Exports for Disposal in Europe: A Contribution to Sustainable Development. *Sustainability* **2021**, *13*, 8905. https://doi.org/10.3390/su13168905

Academic Editor: Silvia Fiore

Received: 7 June 2021
Accepted: 31 July 2021
Published: 9 August 2021

1. Introduction

Sustainable development is a fundamental objective of the EU and was included in the 1997 Treaty of Amsterdam [1]. Since then, the Sustainable Development Strategy has gone through a great revolution.

Sustainable development includes waste management as the Waste Framework Directive (WFD) urges Member States to "promote and support sustainable production and consumption models" and introduces the United Nations Sustainable Development Goals in its objectives, showing the relation between waste management and sustainability.

Although sustainability is an aim of the European Union, the European waste management industry shows a weak model of sustainability [2].

The study of efficient hazardous waste (HW) management in relation to legislation can be a driving force towards the achievement of sustainable development [3]. It should be noted that, in order to achieve sustainability, certain regulations and directives must be met and fulfilled to ensure safe and environmentally sound practices are implemented. The implementation of the two environmental principles included in the WFD, proximity and self-sufficiency, affect HW exports in Europe. However, HW is not only a European concern but also a worldwide concern, especially regarding its disposal. HW disposal must be carried out safely and controlling HW shipments is essential to determine where and how such disposal takes place.

In the analysis of HW shipments transported for disposal, one must take into account how they affect sustainability. On the one hand, shipments of HW affect sustainability through carbon emissions, as road transport has a great impact on greenhouse gas (GHG) emissions [4–6]. However, an analysis of the regulations applying to the shipment of waste is also important, as transboundary waste shipments contribute to efficient waste management [7], and trade policies also affect environmental quality [8] and shipments.

Understanding these regulations and policy implications is key to achieving the Sustainable Development Goals [9].

Finally, disposal in landfill sites or in incineration facilities has a great impact on the environment and therefore on sustainability [10–14].

The main research problem is the lack of information on how HW is transported and disposed of in Europe, and the relationship between HW disposal shipments and compliance with European environmental laws. This article analyses different aspects of sustainability related to HW management and HW shipments and contributes by: (1) deepening the analysis of waste shipments in Europe and the way in which landfill taxes affect waste shipments; (2) analysing HW regulations and the adherence to the principles of proximity and self-sufficiency, and; (3) presenting a qualitative analysis of HW shipments for disposal using different variables – GDP, HW generation and the amount of HW shipped for disposal. To fulfil these aims, network analysis is used to display the importance of networks, giving us a picture of HW shipments, and showing the communities which arise in relation to HW exports for disposal.

Before the methodology is set out, there is a review of key aspects of HW shipments for disposal: a legislative review, including logistics, and a brief analysis of costs and capacity as they relate to the disposal of HW.

2. Objectives and Scope to Present Legislative and Literature Review

2.1. Objectives

Research literature shows how hazardous waste travels worldwide for disposal from rich countries to poor countries [15], making GDP an important element in HW shipments. European countries cannot export HW for disposal to countries outside the Organization for Economic Co-operation and Development. In order to achieve their objectives and fulfill regulations, countries can use different policies and landfill taxes.

In this research we answer the following questions, not yet analyzed by researchers:

- Does HW travel within Europe for disposal to countries with a low or high GDP?
- How do countries interact to fulfill self-sufficiency and proximity principles?

With these questions we try to fill the gap in the research literature about how HW travels and the relationship between HW shipments and legal compliance with European environmental laws.

2.2. Scope

The geographical scope of the research is Europe.

Regarding waste management the scope is Hazardous Waste exports for disposal. The year analyzed is 2014.

It must be pointed out that the scope of this research has several limitations. Firstly, due the research time, only 2014 data of hazardous waste shipments for disposal have been used in the network analysis. Secondly, not all European countries are studied, as Eurostat only has data for the countries included in the table. Thirdly, no data on number of landfills in each country have been used. Finally, the network is made from legal and official data from Eurostat, even if data from illegal shipments are important to understand the impact on sustainability.

2.3. Literature Review: The Origin of the Restriction of HW Exports

The Basel Convention controls HW transport for disposal worldwide. The EU, as part of the Basel Convention, has incorporated its provisions through the European Waste Shipment Regulation (EWSR) [16].

The EWSR was modified in 2014 by Regulation (EU) No 660/2014 of the European Parliament and by the Council of 15 May 2014, amending Regulation (EC) No 1013/2006 on shipments of waste and aiming to strengthen inspections of waste shipments, in order to discourage illegal shipping. Even after this amendment, loopholes in the legislation have been found [17].

Recently, and after China's plastic waste ban, the EWSR has been modified by the Commission Delegated Regulation (EU) 2020/2174 of 19 October 2020 amending Annexes IC, III, IIIA, IV, V, VII and VIII of Regulation (EC) No 1013/2006 of the European Parliament and of the Council on shipments of waste. China's ban was caused by plastic pollution [18] and will affect the plastic waste trade networks which have been hereto established [19], as well as the global circular economy [20].

Besides the Basel Convention and the EWSR, Directive 2008/98/EC on waste management includes two principles connected with waste shipments, as described in Article 16 of the WFD: self-sufficiency and proximity. The self-sufficiency principle indicates that "Member States shall take appropriate measures, in cooperation with other Member States where this is necessary or advisable, to establish an integrated and adequate network of waste disposal installation. The network shall be designed to enable the Community as a whole to become self-sufficient in waste disposal." The proximity principle states that "the network shall enable waste to be disposed of or recovered in one of the nearest appropriate installations by means of the most appropriate methods and technologies to ensure a high level of protection for the environment and public health."

The proximity and self-sufficiency principles can increase the market power of local disposers [21], as Reggiani and Silvestri state, but these principles are also analyzed because of their legal importance [22,23].

Compliance with both sets of regulations, the EWSR and the WFD and its principles, should lead to fewer exports of HW for disposal, and better control the illegal traffic in waste to poor countries [17,18,24,25].

The application of the proximity principle decreases the dangers in the transport of HW [26–28] and the GHG emissions caused by the transport of waste by road, and the self-sufficiency principle can lead countries and companies to innovate in order to comply with the regulations [29–32].

2.4. HW Management: Costs and GDP

Waste management costs have been indicated as one of the reasons for illegal shipments [33,34] and a barrier to a circular economy in which recovery is prioritized over disposal [35,36].

GDP is an important variable in this analysis for two reasons: on the one hand, there is a link between GDP and waste generation [37,38], and, on the other, as HW travel worldwide from rich countries to poor countries, it is important to know how GDP affects the export of waste in Europe, and if HW is disposed in countries with a high or a low GDP.

2.5. HW Shipment for Disposal and Disposal Taxes

Disposal operations are classified by the WFD into 15 categories, identified with the codes D1 to D15. In research on disposal taxes and their effects, not all papers distinguish between different disposal operations [39–41]. Instead, they discuss disposal in general. However, Sigman's analysis of HW taxes [42] establishes a difference between landfill disposal and incineration. Dinan [39] proposed the taxation of disposal and the establishment of a reuse subsidy. Levinson [40] studied the effect of disposal taxes on HW shipments for disposal, finding that HW disposal taxes increase the decentralisation of HW disposal. The literature on this topic has developed widely, studying not only the impact of landfill and disposal taxes [43–45] but also the impact of environmental taxes [46–48].

It is important to determine what kind of disposal operations should be taxed if the right effect is to be achieved and there is to be sustainable development. Incineration (D10) increases in countries with landfill taxes, which causes landfills (D1) to decrease [49–51]. Taxes and regulations that ban the landfill disposal of some types of waste have allowed the Netherlands to reduce its number of landfills [52]. Interestingly, Scharff (2014) points out that "underground storage" in Germany is in a grey area between disposal and recovery, while others recognise underground storage as a common disposal practice [53].

According to a study on landfill taxes in Europe [54], landfill prices vary among and within European countries according to waste classification (e.g. HW, non-HW and municipal solid waste). Bulgaria, Finland and Norway have no landfill taxes for HW, while in the Belgian region of Wallonia, HW is partially banned. In other countries HW is taxed, with rates ranging from less than 10 euros/tonne (Portugal) to more than 60 euros/tonne (UK, Ireland, Denmark, Czech Republic and Estonia), which may be one of the causes of the shipment of waste for disposal. Some of these countries have also established taxes on incineration to promote waste recycling (Austria, Denmark, France, The Netherlands and Norway), whereas in countries where only landfill disposal is taxed, incineration has increased.

The variation of landfill taxes is also verified by the information provided by the Confederation of European Waste-to-Energy Plants (CEWEP) [55], updated to December 2017.

3. Materials and Methods

3.1. *Network Analysis for HW Shipment for Disposal in Europe*

This research uses network analysis to determine the relationships among nodes (countries) and uses Gephi to show the relevance of these nodes in the network or the communities formed by the countries, in the framework of HW exports for disposal among EU Member States. Gephi is used not only to create a trade network to discover whether the self-sufficiency and proximity principles are being adhered to, but also to relate trade/shipments to GDP per capita and HW production. This paper analyzes how these variables affect HW shipments for disposal in Europe.

The stages of this research are presented in the following diagram (Figure 1):

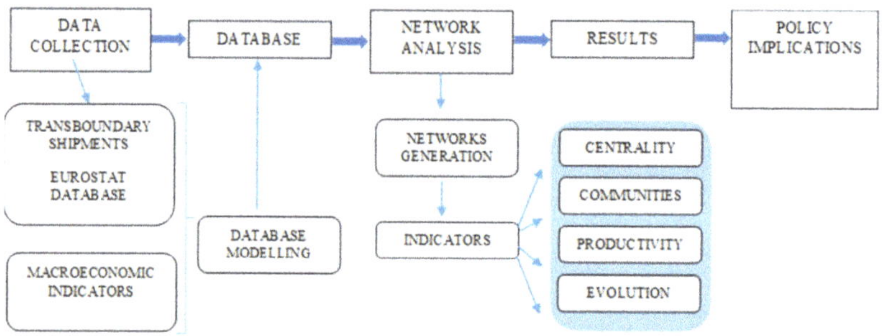

Figure 1. Block diagram.

Gephi is a piece of software designed to explore networks and has previously been used in scientific studies related to waste management. Lepawsky [56,57] used Gephi to evaluate e-waste trade and to determine its evolution and patterns. In his research he used e-waste data import transactions reported by territories and available from United Nations Comtrade. Chen et al. [58] and Wang et al. [59] used Gephi in an analysis of the literature related to waste. While the former used data from the WoS Collection Database on the most cited publications on construction and demolition waste, the latter used data on the literature on incinerating waste to produce energy.

3.2. *Network Model*

Different metrics, like degree centrality, eigenvector centrality and modularity, are used to analyze how European countries are linked to HW shipments for disposal. In the networks with origins in European countries in 2014, V represents the set of countries from Europe and E represents the shipments for disposal. Let (v_i,v_j)∈E, with v_i,v_j ∈V as an edge (i.e., export) in G, representing HW shipments among countries v_i and v_j. This analysis assumes that countries' relationships are unidirectional—that is, from exporter to importer—and, therefore, the graph is directed.

3.2.1. Centrality Network Metrics

Centrality metrics measure how important a country is in the European network. In this analysis, centrality shows the importance and the role of a given country in HW exports for disposal. Centrality includes 'micro' measures that show how a given node relates to the overall network [60,61]. Knowing the importance of countries (i.e., nodes) in the generated network indicates the relationships between these countries in the shipment of HW for disposal.

Degree Centrality

Degree centrality [62] represents the number of links each country/node has in the network, using the following formula:

$$DC^{v_i} = \frac{d(v_i)}{|V| - 1} \tag{1}$$

where $d(v_i)$ denotes the degree centrality (DC) of node v_i in the network. This metric counts the direct links of each country in the network.

Eigenvector Centrality

Eigenvector centrality was proposed by Bonacich [63], as follows:

$$\lambda \cdot EC^{v_i}(G) = \sum_{v_j} g_{ij} EC^{v_j}(G) \tag{2}$$

in which g_{ij} takes the value 1 if $(v_i, v_j) \in E$ and 0 otherwise (retrievable if G is represented using an adjacency matrix) and λ is a proportional factor (i.e., the eigenvalue).

Eigenvector centrality measures the influence of a node on a network. In other words, nodes that have high values of this measurement are well connected. Also, in this sense, they are good connectors as waste exporters and importers from a large number of countries and in large amounts. When the degree of centrality of the eigenvectors is greater, the cohesion of the group is greater.

3.2.2. Structural Analysis of the Network through Modularity

Modularity is another technique used to observe the relationships of HW shipments among European countries. This notion of community partition using modularity was first proposed by Newman and Girvan in [64]. The vertices in networks create groups or communities, which means that some countries in the analyzed network have many edges (exports) while other countries have few edges. Countries in the same community are better connected, while those in different communities are less likely to be connected.

$$M(\Pi, G) = \sum_{\pi \in \Pi} e_{\pi\pi}(G) - \sum_{\pi \in \Pi,\ \pi' \in \Pi, \pi'' \in \Pi} e_{\pi\pi'}(G) e_{\pi'\pi''}(G)$$

where Π represents any community structure and $e_{\pi\pi}$ (G) represents the fraction of all edges in the network that connect nodes in π to nodes in π'.

4. Results

The network analysis was performed with the disposal data obtained through Eurostat for the year 2014. As established in Regulation (EC) No. 2150/2002 of the European Parliament and the council of 25 November 2002 on waste statistics, Member States are obliged to provide data to Eurostat. The main reasons for analyzing the year 2014 are that in 2014 (1) the Circular Economy Package was presented and (2) the EWSR was modified. The Circular Economy Package was the starting point for legislative modifications in the Directives to regulate different waste streams and try to increase recovery and recycling. It is a key year to give a picture of HW shipments before the implementation of new recycling targets and new regulatory changes.

Table 1 shows tonnes exported for disposal from 2011 to 2015.

Table 1. Tonnes exported for disposal 2010–2015.

Year	Tonnes Exported for Disposal
2011	1,712,608
2012	1,509,190
2013	1,480,184
2014	1,528,391
2015	1,025,445

In 2015, a decrease in the quantities exported can be observed. The reason for this may be the change in landfill tax policies in some Member States, as CEWEP shows [55]. The Netherlands reintroduced its landfill tax, Norway repealed its landfill tax and Sweden established a fee in 2015. These changes in landfill taxes may have affected the exports of HW for disposal.

This study analyzes the shipments made in 2014 on the basis of the following scientific assumptions: countries are adhering to the proximity and self-sufficiency principles; as has been shown in research, there are difficulties in finding sites for HW facilities for disposal in the case of landfills and incinerators [65–67] because these must meet environmental, economic and social criteria; and countries must find the best routing model for their exports to minimize transportation costs and risks [26].

4.1. From Data to Network Generation

Taking the current network model (Section 4.1), let G = (V,E) be the graph representing the network for European waste disposal analysis, in which V is the set of operating countries and E is the set of existing shipments among them.

The figures show two different networks. Figure 2 shows the network based on the effective shipments of waste and the GDP per capita and Figure 3 shows the communities formed in the network.

In the export analysis, Italy (573,614 tonnes), Germany (237,777 tonnes) and the Netherlands (195,969 tonnes) were the countries with the greatest amounts of HW exported for disposal. The countries that generated the most HW were Germany (21,812,660 tonnes), Bulgaria (12,206,169 tonnes) and France (10,783,405 tonnes).

Table 2 shows that countries with the highest GDP per capita or with a GDP per capita above 40,000 euros in 2014, according to Eurostat, exported the most HW for disposal to other countries with a high GDP per capita.

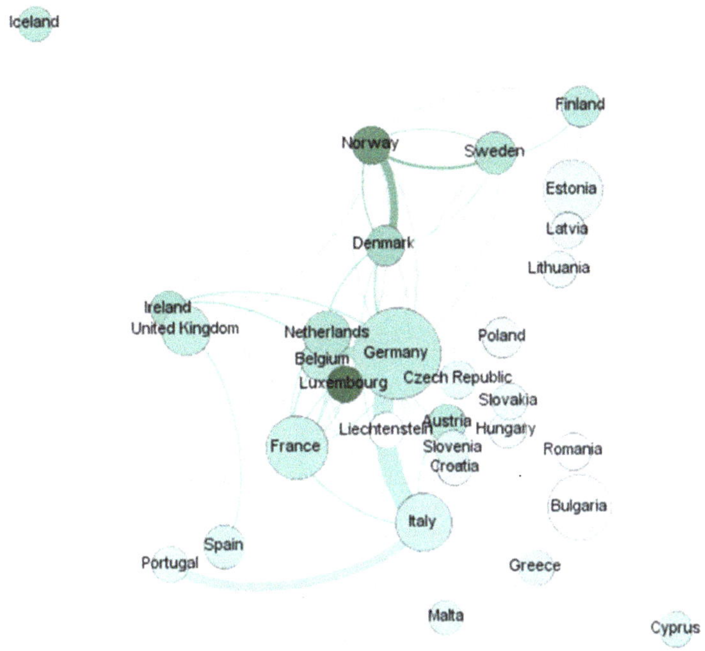

Figure 2. Network of HW shipments for disposal in Europe.

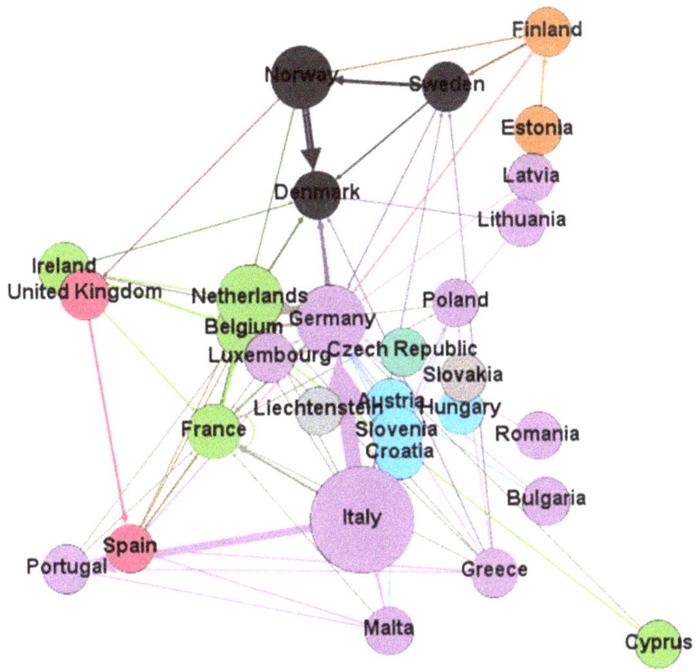

Figure 3. Network for HW shipments for disposal in Europe (2014), displayed by modularity.

Table 2. HW exports from high-income countries.

Exporting Countries	Importing Countries
Denmark	Germany Norway
Ireland	Belgium Germany France Denmark The Netherlands UK
Luxembourg	Belgium Germany France The Netherlands
Sweden	Denmark Germany Finland
Norway	Denmark Germany France Finland Sweden UK

The countries with the lowest GDP per capita or with GDP per capita lower than 15,000 euros did not receive HW for disposal, except Lithuania, which received HW for disposal from Latvia. Only two countries exported as much as 19% of the HW they produced (Malta and Slovenia), while the countries with the next highest exports exported under 10% of the waste they produced.

No data for HW landfill taxes were found from 2014. It is therefore not possible to assess whether these influenced HW shipments. However, in 2012, only three countries (Bulgaria, Finland and Norway) did not have landfill taxes for HW, and this did not appear to affect waste shipments—that is, European countries did not look to export to countries with no landfill taxes.

For degree centrality, three main nodes were considered (Germany with 26 relationships, France with 18 relationships and Belgium with 15 relationships). These countries are in central Europe, and, following the proximity principle, the logistics connectivity for these countries may have been greater. Furthermore, these three countries correspond to the highest in-degree values.

The results are shown in Tables 3 and 4 and Figures 2 and 3. Table 3 shows the amount of HW produced, the GDP per capita, the amount of HW exported, the in-degree (from how many countries waste is received or imported), the out-degree (to how many countries HW is exported), the degree (in-degree + out-degree) and the ratio of exports to generation.

Figure 2 highlights the waste tonnage generated by each node (i.e., the node size corresponds to the tonnage generated). The nodes are green, with their shades varying according to GDP per capita (a darker color corresponds to a higher GDP per capita). Finally, the thickness of the line corresponds to the amount of export flow between the countries.

Figure 3 shows the network displayed by modularity; each color represents a different community.

Table 3. Results from the Network Analysis for HW Shipments for Disposal in Europe in 2014.

Label	Export	GDP per Capita	Generated	Ratio Exp/Gen (%)	Indegree	Outdegree	Degree
Belgium	98,391	33,800	2,946,195	3.34	11	4	15
Bulgaria	1157	5500	12,206,169	0.01	0	2	2
Czech Republic	100	15,400	1,162,342	0.01	0	0	0
Denmark	2637	44,900	1,718,394	0.15	8	2	10
Germany	237,777	34,000	21,812,660	1.09	20	6	26
Estonia	11,504	13,200	10,410,321	0.11	1	1	2
Ireland	50,738	41,300	482,907	10.51	0	6	6
Greece	10,759	17,000	221,041	4.87	0	9	9
Spain	2984	22,300	2,984,518	0.10	5	4	9
France	69,386	31,300	10,783,405	0.64	12	6	18
Croatia	12,393	10,300	130,239	9.52	0	3	3
Italy	573,614	25,400	8,923,548	6.43	1	9	10
Cyprus	67	20,400	173,377	0.04	0	2	2
Latvia	107	10,300	104,142	0.10	0	2	2
Lithuania	765	11,300	165,477	0.46	1	3	4
Luxembourg	14,934	80,600	237,180	6.30	1	4	5
Hungary	174	10,700	596,554	0.03	0	1	1
Malta	6997	17,900	36,654	19.09	0	6	6
The Netherlands	195,969	38,600	4,830,495	4.06	8	4	12
Austria	7854	36,200	1,272,288	0.62	6	1	7
Poland	21	10,500	1,679,051	0.00	5	1	6
Portugal	1596	16,300	701,228	0.23	3	3	6
Romania	69	7000	590,300	0.01	0	1	1
Slovenia	29,628	17,500	155,229	19.09	0	2	2
Slovakia	100	13,600	371,214	0.03	0	0	0
Finland	15,036	34,200	1,998,693	0.75	4	2	6
Sweden	11,503	40,500	2,568,154	0.45	4	4	8
United Kingdom	29,597	31,000	5,755,258	0.51	2	1	3
Iceland	100	33,800	1000	10.00	0	0	0
Liechtenstein	100	1000	1204	8.31	0	0	0
Norway	142,734	67,400	1,368,049	10.43	3	6	9

Table 4. Communities in the European Disposal Network.

Modularity Class	Countries (Eigencentrality)
0	Belgium (0.764), Ireland (0), France (0.968), Cyprus (0), The Netherlands (0.715)
1	Czech Republic (0)
2	Denmark (0.524), Norway (0.388), Sweden (0.173)
3	Bulgaria (0), Germany (1), Greece (0), Italy (0.002), Latvia (0), Lithuania (0.002), Luxembourg (0.170), Malta (0), Poland (0.224), Portugal (0.008), Romania (0.008)
4	Estonia (0.004), Finland (0.360)
5	Spain (0.246), United Kingdom (0.087)
6	Croatia (0), Hungary (0), Austria (0.015), Slovenia (0)
7	Slovakia (0)
8	Iceland (0)
9	Liechtenstein (0)

The node size is proportional to the waste tonnage that the country exports. The thickness of the line corresponds to the size of the export flow between the countries.

The modularity shows groups/communities in the network. These groups account for GDP per capita and show how the European countries apply the proximity and self-sufficiency principles.

Table 4 shows the communities formed in the network. The reasons for these communities are discussed in Section 6.

The largest community is the third (purple), which is composed of 10 countries: Bulgaria, Germany, Greece, Italy, Latvia, Lithuania, Luxembourg, Malta, Poland, Portugal and Romania. Germany has the highest eigencentrality score but the other members of this community have a score of nearly 0. Germany is also the country with the highest in-degree, which indicates that it receives the highest volume of HW. The second most important community is the first (green), which is composed of five countries: Belgium, Ireland, France, Cyprus and the Netherlands. Community six (blue) consists of Croatia, Hungary, Austria and Slovenia. The other communities are small.

5. Discussion

This paper analyses HW shipments for disposal, the effect of the regulations on the shipment of waste, the application of the two principles contained in the Waste Framework Directive—proximity and self-sufficiency—and the way in which GDP affects these shipments.

The adherence to these principles shows a low density network, while HW for recovery in the same year shows a high density network [68]. The density of the networks represents the links between the nodes, showing there are many fewer shipments for disposal than shipments for recovery.

The communities formed by some of the countries show that there is one country with a higher eigencentrality value [63], that is, a country that has a bigger relevance to the network.

The centrality shown by Germany can also be seen in the literature, as there has been a thorough analysis of waste treatment facilities in this country [69].

The countries with the highest in-degree (Germany 20, France 12, Belgium 15 and the Netherlands 12) are, except Belgium, countries with a high incineration capacity. These countries also have a GDP per capita above 30,000 euros.

In contrast with the "Pollution Havens" described in the research literature, in which waste travels from rich to poor countries [70,71], in Europe, HW is sent to be disposed of in countries with a high GDP.

This shows that high GDP makes these countries more able to use the best available techniques for wastes incineration [13].

The countries with high incineration capacity (France, Germany, Sweden, Denmark, the Netherlands, Austria and Finland) have a GDP per capita above 30,000 euros. These countries also have an important value for in-degree. It must be taken into account that, Sora [72] states that the opening of the incineration market threatens the application of the proximity principle.

6. Conclusions

One of the novelties of this study was the use of network analysis to fill in the gaps in the research literature about how HW travels for disposal in Europe and the relationship between HW shipments and legal compliance with environmental laws in relation to sustainability and sustainable development in Europe.

Network analysis is a useful tool to answer these research questions and to find out if HW travels for disposal to European countries with a low or high GDP and how countries interact to fulfil the principles of self-sufficiency and proximity.

HW is shipped for disposal to countries with a high GDP and high incineration capacities, which means that when countries must apply proximity and self-sufficiency principles, waste is shipped to countries with a high GDP, because these countries have better treatment facilities. This demonstrates how GDP is a determining factor in the export of waste.

Countries with a high GDP per capita have more incineration facilities; they are better prepared for the disposal of HW.

Good practices for the environment and for sustainable development are demonstrated by networks, showing coherence in the fulfilment of the principles of self-sufficiency and proximity, and the adherence to legal regulations.

The absence of a landfill tax does not affect the export of waste; countries with no landfill tax did not have higher in-degrees than countries that applied a landfill tax.

The network analysis demonstrated the relationships between countries when HW is shipped for disposal, and the association between countries generated from the adherence to the proximity and self-sufficiency principles.

Degree centrality demonstrated that countries in central Europe (Germany, France and Belgium) were the main nodes. Following the proximity principle, this may be because of better logistics connectivity. The application of these principles helps to improve efficiency in HW management systems, since it minimizes emissions from HW transport and indicates that countries have sufficient capacity for the disposal of the HW they generate.

Further research should be undertaken to establish the quantities of HW exported and imported for landfill and incineration in each country. Additionally, two circumstances may affect HW shipments: (1) the exit of the UK from the EU may affect waste shipments to and from this country and (2) the plastic waste ban imposed by China. New data may show how these circumstances affect waste trade and the stability of the communities.

The control of compliance with the analyzed regulations will be fundamental to avoid illegal waste trafficking and to protect the environment and citizens' health.

The capacities for waste management in Europe (i.e., landfill and incineration capacities) should also be determined. In future, Europe should establish appropriate regulations that take into account all these circumstances, in order to make a better contribution to sustainable development.

Author Contributions: Conceptualization, C.C., M.P.L.; methodology, M.P.L.; software, M.P.L.; validation, M.M.-N. and M.P.L.; formal analysis, M.P.L., M.M.-N., C.C.; investigation, C.C.; resources, C.C.; data curation, C.C.; writing—original draft preparation, C.C.; writing—review and editing, M.P.L., M.M.-N. All authors have read and agreed to the published version of the manuscript.

Funding: This research received no external funding.

Institutional Review Board Statement: Not applicable.

Informed Consent Statement: Not applicable.

Data Availability Statement: The data presented in this study are openly available in https://ec.europa.eu/eurostat/cache/metadata/en/env_wasship_esms.htm (accessed on 25 June 2019).

Conflicts of Interest: The authors declare no conflict of interest.

References

1. EUR-Lex-11997D/TXT-EN-EUR-Lex. Available online: https://eur-lex.europa.eu/legal-content/EN/TXT/?uri=CELEX:11997D/TXT (accessed on 10 July 2021).
2. Peter, J.; Daphne, C. Sustainability and the European Waste Management Industry. *Adv. Environ. Stud.* **2019**, *3*, 198–208. [CrossRef]
3. Callao, C.; Martinez-Nuñez, M.; Latorre, M.P. European Countries: Does common legislation guarantee better hazardous waste performance for European Union member states? *Waste Manag.* **2019**, *84*, 147–157. [CrossRef]
4. Enzmann, J.; Ringel, M. Reducing road transport emissions in Europe: Investigating a demand side driven approach †. *Sustainability* **2020**, *12*, 7594. [CrossRef]
5. Kazancoglu, Y.; Ozbiltekin-Pala, M.; Ozkan-Ozen, Y.D. Prediction and evaluation of greenhouse gas emissions for sustainable road transport within Europe. *Sustain. Cities Soc.* **2021**, *70*, 102924. [CrossRef]
6. Santos, G. Road transport and CO_2 emissions: What are the challenges? *Transp. Policy* **2017**, *59*, 71–74. [CrossRef]
7. Parajuly, K.; Fitzpatrick, C. Understanding the impacts of transboundary waste shipment policies: The case of plastic and electronic waste. *Sustainability* **2020**, *12*, 2412. [CrossRef]
8. Alola, A.A.; Bekun, F.V.; Sarkodie, S.A. Dynamic impact of trade policy, economic growth, fertility rate, renewable and non-renewable energy consumption on ecological footprint in Europe. *Sci. Total Environ.* **2019**, *685*, 702–709. [CrossRef] [PubMed]
9. Adedoyin, F.F.; Alola, A.A.; Bekun, F.V. An assessment of environmental sustainability corridor: The role of economic expansion and research and development in EU countries. *Sci. Total Environ.* **2020**, *713*, 136726. [CrossRef]

10. Sauve, G.; van Acker, K. The environmental impacts of municipal solid waste landfills in Europe: A life cycle assessment of proper reference cases to support decision making. *J. Environ. Manag.* **2020**, *261*, 110216. [CrossRef]
11. Duan, Z.; Scheutz, C.; Kjeldsen, P. Trace gas emissions from municipal solid waste landfills: A review. *Waste Manag.* **2021**, *119*, 39–62. [CrossRef]
12. Wang, D.; Tang, Y.T.; Long, G.; Higgitt, D.; He, J.; Robinson, D. Future improvements on performance of an EU landfill directive driven municipal solid waste management for a city in England. *Waste Manag.* **2020**, *102*, 452–463. [CrossRef] [PubMed]
13. Best Available Techniques (BAT) Reference Document for Waste Incineration-Publications Office of the EU. Available online: https://op.europa.eu/en/publication-detail/-/publication/075477b7-329a-11ea-ba6e-01aa75ed71a1/language-en (accessed on 10 July 2021).
14. Eriksson, O.; Finnveden, G. Energy recovery from waste incineration—The importance of technology data and system boundaries on CO_2 emissions. *Energies* **2017**, *10*, 539. [CrossRef]
15. Bernard, S. North–south trade in reusable goods: Green design meets illegal shipments of waste. *J. Environ. Econ. Manag.* **2015**, *69*, 22–35. [CrossRef]
16. European Parliament, REGULATION (EC) No 1013/2006 OF THE EUROPEAN PARLIAMENT AND OF THE COUNCIL, EURLEX. 2007. Available online: https://eur-lex.europa.eu/legal-content/EN/TXT/PDF/?uri=CELEX:02006R1013-20160101&qid=1454069470717&from=EN (accessed on 3 March 2019).
17. Morganti, M.; Favarin, S.; Andreatta, D. Illicit waste trafficking and loopholes in the European and Italian legislation. *Eur. J. Crim. Policy Res.* **2020**, *26*, 105–133. [CrossRef]
18. Khan, S.A. Symposium on global plastic pollution. Clearly hazardous, obscurely regulated: Lessons from the basel convention on waste trade. *AJIL Unbound* **2020**, *114*, 200–205. [CrossRef]
19. Wang, C.; Zhao, L.; Lim, M.K.; Chen, W.Q.; Sutherland, J.W. Structure of the global plastic waste trade network and the impact of China's import Ban. *Resour. Conserv. Recycl.* **2020**, *153*, 104591. [CrossRef]
20. Qu, S.; Guo, Y.; Ma, Z.; Chen, W.-Q.; Liu, J.; Liu, G.; Wang, Y.; Xu, M. Implications of China's foreign waste ban on the global circular economy. *Resour. Conserv. Recycl.* **2019**, *144*, 252–255. [CrossRef]
21. Reggiani, C.; Silvestri, F. Municipal solid waste, market competition and the EU Policy. *Environ. Resour. Econ.* **2017**, *71*, 457–474. [CrossRef]
22. Jans, J.H. The Status of the Self-sufficiency and Proximity Principles with regard to the Disposal and Recovery of Waste in the European Community. *J. Environ. Law* **1999**, *11*, 121–156. [CrossRef]
23. Reese, M. *The Proximity Principle*; Edward Elgar Publishing Limited: Cheltenham, UK, 2018; Chapter VI.17; pp. 219–233.
24. Bisschop, L. *Crime, Law, and Social Change*; Kluwer Academic Publishers: New York, NY, USA, 2012.
25. Reganati, F.; Pittiglio, R.; Toschi, L. How to detect illegal waste shipments? The case of the international trade in polyethylene waste. *Econ. Bull.* **2017**, *37*, 2625–2640.
26. Yilmaz, O.; Kara, Y.; Yetis, U. Hazardous waste management system design under population and environmental impact considerations. *J. Environ. Manag.* **2017**, *203*, 720–731. [CrossRef] [PubMed]
27. ReVelle, C.; Cohon, J.; Shobrys, D. Simultaneous siting and routing in the disposal of hazardous wastes. *Transp. Sci.* **1991**, *25*, 138–145. [CrossRef]
28. Alumur, S.; Kara, B.Y. A new model for the hazardous waste location-routing problem. *Comput. Oper. Res.* **2007**, *34*, 1406–1423. [CrossRef]
29. Stoever, J.; Weche, J.P. Environmental regulation and sustainable competitiveness: Evaluating the role of firm-level green investments in the context of the porter hypothesis. *Environ. Resour. Econ.* **2017**, *702*, 429–455. [CrossRef]
30. Cecere, G.; Corrocher, N. Stringency of regulation and innovation in waste management: An empirical analysis on EU countries. *Ind. Innov.* **2016**, *23*, 625–646. [CrossRef]
31. Porter, M.E.; van der Linde, C. Toward a new conception of the environment-competitiveness relationship. In *Economic Costs and Consequences of Environmental Regulation*; Taylor and Francis, 2018; pp. 413–434. Available online: https://www.aeaweb.org/articles?id=10.1257/jep.9.4.97 (accessed on 18 June 2020).
32. Porter, M.E.; Advantage, C. Creating and sustaining superior performance. *Compet. Advant.* **1985**, *167*, 167–206.
33. Bernard, S. *Transboundary Movement of Waste: Second-Hand Markets and Illegal Shipments*; SSRN Electron. J; 2012; Available online: https://ssrn.com/abstract=1999005 or http://dx.doi.org/10.2139/ssrn.1999005; (accessed on 25 June 2020).
34. Boudier, F.; Bensebaa, F. Hazardous waste management and corporate social responsibility: Illegal trade of electrical and electronic waste. *Bus. Soc. Rev.* **2011**, *116*, 29–53. [CrossRef]
35. Chen, R.; Chen, R.J.C. An integrated sustainable business and development system: Thoughts and opinions. *Sustainability* **2014**, *6*, 6862–6871. [CrossRef]
36. Golev, A.; Corder, G.D.; Giurco, D.P. Barriers to industrial symbiosis: Insights from the use of a maturity grid. *J. Ind. Ecol.* **2015**, *19*, 141–153. [CrossRef]
37. Kusch, S.; Hills, C.D. The link between e-waste and GDP—New insights from data from the pan-European region. *Resources* **2017**, *6*, 15. [CrossRef]
38. Awasthi, A.K.; Cucchiella, F.; D'Adamo, I.; Li, J.; Rosa, P.; Terzi, S.; Wei, G.; Zeng, X. Modelling the correlations of e-waste quantity with economic increase. *Sci. Total Environ.* **2018**, *613–614*, 46–53. [CrossRef]

39. Dinan, T.M. Economic efficiency effects of alternative policies for reducing waste disposal. *J. Environ. Econ. Manag.* **1993**, *25*, 242–256. [CrossRef]
40. Levinson, A. NIMBY taxes matter: The case of state hazardous waste disposal taxes. *J. Public Econ.* **1999**, *74*, 31–51. [CrossRef]
41. Palmer, K.; Walls, M. Optimal policies for solid waste disposal Taxes, subsidies, and standards. *J. Public Econ.* **1997**, *65*, 193–205. [CrossRef]
42. Sigman, H. The effects of hazardous waste taxes on waste generation and disposal. *J. Environ. Econ. Manag.* **1996**, *30*, 199–217. [CrossRef]
43. Heijnen, P.; Elhorst, J.P. The diffusion of local differentiated waste disposal taxes in the Netherlands. *De Econ.* **2018**, *166*, 239–258. [CrossRef]
44. Nicolli, F.; Mazzanti, M. Landfill diversion in a decentralized setting: A dynamic assessment of landfill taxes. *Resour. Conserv. Recycl.* **2013**, *81*, 17–23. [CrossRef]
45. Hoogmartens, R.; Eyckmans, J.; van Passel, S. Landfill taxes and Enhanced Waste Management: Combining valuable practices with respect to future waste streams. *Waste Manag.* **2016**, *55*, 345–354. [CrossRef]
46. Hoerner, J.A.; Bosquet, B.; Andrew Hoerner-Director, J. Environmental tax reform: The European experience about the authors. *The European Experience*. 2001. Available online: https://citeseerx.ist.psu.edu/viewdoc/download?doi=10.1.1.474.2253&rep=rep1&type=pdf (accessed on 20 June 2020).
47. Chiroleu-Assouline, M.; Fodha, M. From regressive pollution taxes to progressive environmental tax reforms. *Eur. Econ. Rev.* **2014**, *69*, 126–142. [CrossRef]
48. Bosquet, B. Environmental tax reform: Does it work? A survey of the empirical evidence. *Ecol. Econ.* **2000**, *34*, 19–32. [CrossRef]
49. Bartelings, H.; Linderhof, V. *Effective Landfill Taxation: A Case Study for the Netherlands*; 2006; pp. 1–25. Available online: https://citeseerx.ist.psu.edu/viewdoc/download?doi=10.1.1.400.3598&rep=rep1&type=pdf. (accessed on 20 June 2020).
50. Bartelings, H.; van Beukering, P.; Kuik, O.; Linderhof, V.; Oosterhuis, F. Effectiveness of Landfill Taxation. 2005. Available online: https://www.osti.gov/etdeweb/biblio/20755664 (accessed on 20 June 2020).
51. Olofsson, M.; Sahlin, J.; Ekvall, T.; Sundberg, J. Driving forces for import of waste for energy recovery in Sweden. *Waste Manag. Res.* **2005**, *23*, 3–12. [CrossRef] [PubMed]
52. Scharff, H. Landfill reduction experience in The Netherlands. *Waste Manag.* **2014**, *34*, 2218–2224. [CrossRef] [PubMed]
53. Yaramanci, U. Geoelectric exploration and monitoring in rock salt for the safety assessment of underground waste disposal sites. *J. Appl. Geophys.* **2000**, *44*, 181–196. [CrossRef]
54. Fischer, C.; Lehner, M.; Mckinnon, D.L. Overview of the use of landfill taxes in Europe. *ETC/SCP* **2012**, 96. Available online: http://www.embopar.pt/folder/documento/99_Landfill%20taxes%20in%20Europe.pdf (accessed on 15 December 2019).
55. CEWEP, Landfill Taxes and Bans Overview. Available online: https://www.cewep.eu/wp-content/uploads/2017/12/Landfill-taxes-and-bans-overview.pdf (accessed on 4 July 2020).
56. Lepawsky, J. Are we living in a post-Basel world? *Area* **2015**, *47*, 7–15. [CrossRef]
57. Lepawsky, J. The changing geography of global trade in electronic discards: Time to rethink the e-waste problem. *Geogr. J.* **2015**, *181*, 147–159. [CrossRef]
58. Chen, J.; Su, Y.; Si, H.; Chen, J. Managerial Areas of Construction and Demolition Waste: A Scientometric Review. *Int. J. Environ. Res. Public Health* **2018**, *15*, 2350. [CrossRef] [PubMed]
59. Wang, Y.; Lai, N.; Zuo, J.; Chen, G.; Du, H. Characteristics and trends of research on waste-to-energy incineration: A bibliometric analysis, 1999–2015. *Renew. Sustain. Energy Rev.* **2016**, *66*, 95–104. [CrossRef]
60. Freeman, L.C. A set of measures of centrality based on betweenness. *Sociometry* **1977**, *40*, 35–41. [CrossRef]
61. Jackson, M.O. *Social and Economic Networks*; Princeton University Press: Princeton, NJ, USA, 2008.
62. Clark, R.A.; McGuire, C. Sociographic Analysis of Sociometric Valuations. *Child. Dev.* **1952**, *23*, 129. [CrossRef]
63. Bonacich, P. Power and centrality: A family of measures. *Am. J. Sociol.* **1987**, *92*, 1170–1182. [CrossRef]
64. Newman, M.E.J.; Girvan, M. Finding and evaluating community structure in networks. *Phys. Rev.* **2004**, *69*, 026113. [CrossRef] [PubMed]
65. De Feo, G.; de Gisi, S. Using MCDA and GIS for hazardous waste landfill siting considering land scarcity for waste disposal. *Waste Manag.* **2014**, *34*, 2225–2238. [CrossRef]
66. Hariz, H.A.; Dönmez, C.Ç.; Sennaroglu, B. Siting of a central healthcare waste incinerator using GIS-based Multi-Criteria Decision Analysis. *J. Clean. Prod.* **2017**, *166*, 1031–1042. [CrossRef]
67. Zakaria, B.; Abdullah, R.; Ramli, M.F.; Latif, P.A. Selection criteria using the Delphi method for siting an integrated hazardous waste disposal facility in Malaysia. *J. Environ. Plan. Manag.* **2013**, *56*, 512–530. [CrossRef]
68. Latorre, M.P.; Martinez-Nuñez, M.; Callao, C. Modelling and analysing the relationship between innovation and the European Regulations on hazardous waste shipments. *Int. Environ. Agreem. Polit. Law Econ.* **2021**, 1–20. Available online: https://link.springer.com/article/10.1007/s10784-021-09536-5 (accessed on 5 June 2020).
69. Weber, K.; Quicker, P.; Hanewinkel, J.; Flamme, S. Status of waste-to-energy in Germany, Part I–Waste treatment facilities. *Waste Manag. Res.* **2020**, *38* (Suppl. 1), 23–44. [CrossRef]
70. Cotta, B. What goes around, comes around? Access and allocation problems in Global North–South waste trade. *Int. Environ. Agreem. Polit. Law Econ.* **2020**, *20*, 255–269. [CrossRef]

71. Walters, R.; Loureiro, M.A.F. Waste crime and the global transference of hazardous substances: A Southern Green perspective. *Crit. Criminol.* **2020**, *28*, 463–480. [CrossRef]
72. Sora, M.J.; Ventosa, I.P. *Fiscalidad Ambiental E Instrumentos De Financiación De La Economía Verde*; Fundación Forum Ambiental: Barcelona, Spain, 2014.

sustainability

Article

Sustainable Urban Drainage Systems in Spain: Analysis of the Research on SUDS Based on Climatology

Ana Isabel Abellán García [1,*], Noelia Cruz Pérez [2] and Juan C. Santamarta [2]

[1] Escuela Técnica Superior de Ingeniería de Montes, Forestal y del Medio Natural, Universidad Politécnica de Madrid (UPM), 28040 Madrid, Spain
[2] Departamento de Ingeniería Agraria, Náutica, Civil y Marítima, Universidad de La Laguna (ULL), 38206 Tenerife, Spain; ncruzper@ull.edu.es (N.C.P.); jcsanta@ull.edu.es (J.C.S.)
* Correspondence: ana.abellan.garcia@alumnos.upm.es

Abstract: Sustainable urban drainage systems (SUDS), or urban green infrastructure for stormwater control, emerged for more sustainable management of runoff in cities and provide other benefits such as urban mitigation and adaptation to climate change. Research in Spain began a little over twenty years ago, which was later than in other European countries, and it began in a heterogeneous way, both in the SUDS typology and spatially within the peninsular geography. The main objective of this work has been to know through bibliographic review the state of the art of scientific research of these systems and their relationship with the different types of climates in the country. These structures have a complex and sensitive dependence on the climate, which in the Iberian Peninsula is mostly type B and C (according to the Köppen classification). This means little water availability for the vegetation of some SUDS, which can affect the performance of the technique. To date, for this work, research has focused mainly on green roofs, their capabilities as a sustainable construction tool, and the performance of different plant species used in these systems in arid climates. The next technique with the most real cases analyzed is permeable pavements in temperate climates, proving to be effective in reducing flows and runoff volumes. Other specific investigations have focused on the economic feasibility of installing rainwater harvesting systems for the laundry and the hydraulic performance of retention systems located specifically in the northeast of the Iberian Peninsula. On the contrary, few scientific articles have appeared that describe other SUDS with vegetation such as bioretention systems or green ditches, which are characteristic of sustainable cities, on which the weather can be a very limiting factor for their development.

Keywords: sustainable urban drainage systems; green infrastructures; stormwater green infrastructure; Mediterranean climate; arid climate; template climate; Spain

Citation: Abellán García, A.I.; Cruz Pérez, N.; Santamarta, J.C. Sustainable Urban Drainage Systems in Spain: Analysis of the Research on SUDS Based on Climatology. *Sustainability* **2021**, *13*, 7258. https://doi.org/10.3390/su13137258

Academic Editors: Margarita Martinez-Nuñez and Mª Pilar Latorre-Martínez

Received: 29 April 2021
Accepted: 25 June 2021
Published: 29 June 2021

1. Introduction

A new approach to urban stormwater management emerged in the 1980s and 1990s, introducing a holistic and environmentalist approach to urban hydrology, and which is increasingly spreading around all cities of the world [1]. This methodology reproduces, as faithfully as possible, the natural hydrological cycle to minimize the impact of urban development. It aims to reduce the negative impacts in terms of quantity and quality of runoff, as well as maximize the landscape integration and the social and environmental value of the elements involved in urban stormwater management [2]. This new way of treating urban stormwater took different names around the world. A very widespread one is Sustainable Urban Drainage Systems (SUDS), those elements of the infrastructure (urban–hydraulic–landscaping) whose mission is capture, filter, retain, transport, store, and infiltrate the urban runoff, trying to reproduce as close as possible the natural water cycle [3]. This definition is similar to green infrastructures in the United States: an approach to hydrological cycle that uses soils and vegetation to enhance and/or mimic the natural hydrologic cycle processes of infiltration, evapotranspiration, and reuse [4]. On the other hand, in Europe, the concept of green infrastructure is broader, encompassing all those elements that provide connectivity to ecosystems, provide ecosystem services, and contribute to the mitigation and adaptation to climate change; these are classified in different scales: the local level, municipal level, and regional or state level [5]. Therefore, the SUDS would be urban green infrastructures to be implemented at the local–municipal level. In addition, since some SUDS are characterized by the use of vegetation (for example: green roofs, bioswale, artificial wetlands, bioretention areas ...), they can also be included within the so-called Nature-based Solutions (NbS), which include those elements in Nature that inspire facing new social challenges efficiently and responsibly with the environment [6].

SUDS cover a wide variety of elements or techniques such as [7] rainwater harvesting systems, green roofs, permeable surfaces, bioretention systems, vegetated swales, filter strips, infiltration systems, and detention–retention systems.

In Spain, SUDS appeared later than in other countries such as the UK or USA and were not as widely distributed or studied [8]. Thus, the objective of this article is to determine the state of the art in Spain and identify possible deficiencies in the research and/or experiences (if there are any of them); more specifically, it is to identify which techniques are the most analyzed and if they depend on the climate of the area or not.

Study Area

According to the Spanish State Agency of Meteorology (AEMET), in the Iberian Peninsula, there are mainly three types of climates in agreement with the Köppen classification [9]: (i) Dry climates (type B): BWh (warm desert) and BWk (cold desert), corresponding to the provinces of Almería, Murcia, and Alicante, where minimal rainfall occurs, and BSh (warm steppe) and BSk (cold steppe) for Extremadura and the Balearic Islands; (ii) Temperate climates (type C): Csa (temperate with dry and hot summer, known as Mediterranean climate) is found in approximately 40% of the surface of the Iberian Peninsula and the Balearic Islands, being the most common climate, it extends over almost all the southern half and much of the Mediterranean shoreline, Csb (temperate with hot and dry summer) in most of the northwest of the Peninsula and inland mountainous areas, Cfa (temperate without dry season with hot summer) in the northeast of the peninsula and in a strip of medium altitude in the Pyrenees, Cfb (temperate without dry season with mild summer) in the Cantabrian region; (iii) Cold climates (type D): Dsb (cold with dry and temperate summer) and Dsc (cold with dry and cool summer), Dfb (cold without dry season and mild summer) and Dfc (cold with dry summer and cool summer) in high mountain areas of the Pyrenees, the Cantabrian Mountains, and the Iberian System. Figure 1 shows the spatial distribution of the different climatic classes in Spain.

Figure 1. Köppen Climate Classification. The province's written names are the places where studies of real cases have been carried out. Source: Adaptation of Mapas climáticos de España (1981–2010) Y ETo (1996–2016) [10].

The weather in most of the country is characterized by temperate temperatures and rainfall regimes divided into two periods: one maximum in autumn and the other secondary in spring (except for the west and south of the peninsula, where the rainiest periods are autumn. and winter) [11]. This irregularity in the distribution can affect the development of plants [12] (many SUDS, such as bioretention areas or green roofs, have vegetation) and the performance of drainage elements as well [13]. So, the climate is a key element to consider in SUDS operation. Another characteristic of the rainfall in the Mediterranean and dry areas is the high intensity of rainfall events that are expected to increase in the future because of climate change [14].

SUDS are solutions for climate change adaptation and mitigation [15], and for this, they appeared as recommendations in publications for urban sustainability [16]. However, as we know the importance of the weather, doubts arise about the performance of these solutions in regions with different weather conditions, and therefore, there are concerns as to whether they are translatable. For this reason, this analysis intends not only to elucidate the state of the art of research on these techniques in Spain (if it is homogeneous throughout the national territory or not, if all the techniques arouse equal interest, what are the parameters, characteristics, or functionalities most analyzed) but also to find out the operation of the different sustainable drainage technologies under different climatic conditions.

So, although the main question to answer in this article was about the state of the art in scientific research on sustainable drainage systems in Spain, there have also been attempts to answer other questions in this regard, such as: Is there any relationship between climatology and the techniques studied? Since they are multifaceted structures, does the research focus on hydrological–hydraulic performance, or are other potential benefits evaluated?

2. Materials and Methods

The scientific publications compiled in SCOPUS on the sustainable treatment of urban runoff in Spain were the starting point for this analysis. SCOPUS was selected because it was a search engine that includes a greater number of journals compared to others such as Web of Science, and its citation analysis was faster [17]. Previously, we verified that the journals where researchers of specialized university centers (such as GITECO (https://www.giteco.unican.es/SUDSlab/inicio.shtml (accessed on 22 February 2021)) or IIAMA (https://www.iiama.upv.es/iiama/es/ (accessed on 22 February 2021)) published could be found indexed to SCOPUS.

In this bibliographic review, those publications made in scientific journals with DOI (and indexed in JCR or SJR) have been considered, looking for scientific evidence that shows the performance of these techniques in their different facets under the climatological characteristics of Spain [9].

The development of the methodology followed has had the corresponding steps: (i) Bibliographic search in SCOPUS, to find any paper about SUDS in Spain; (ii) Selection of the bibliography found, which was focused on obtaining the necessary data to answer the research question in this article (What is the state of scientific research on SUDS in Spain?); and (iii) Obtaining information from the selected documents focused mainly on knowing the temporal evolution of research in this field in Spain, if there were more theoretical studies than empirical cases, what techniques were most analyzed according to the different climates of the country or if there was some type of stormwater green infrastructure that has not been studied or monitored. Each of these points is detailed below.

2.1. Bibliographic Search in Scopus

The search in the SCOOPUS database included publications of any nature, without a time limit and geographically affiliated with Spain that contain the following key concepts (Table 1), since one of the main objectives of this article was to know the status of the scientific research on sustainable urban drainage, the years of experience in this area, and the amount of research carried out.

Table 1. Keywords used in the search for articles related to SUDS in Spain. Source: Prepared by the authors.

Keywords	Search String
Sustainable Urban Drainage	(TITLE-ABS-KEY (sustainable AND urban AND drainage) AND AFFILCOUNTRY (Spain))
Stormwater Green Infrastructure	(TITLE-ABS-KEY (stormwater AND green AND infrastructure) AND AFFILCOUNTRY (Spain))
Nature-Based Solutions Rainwater/Stormwater	(TITLE-ABS-KEY (nature AND based AND solutions) AND AFFILCOUNTRY (Spain) AND TITLE-ABS-KEY (stormwater OR rainwater))
Permeable Pavement	(TITLE-ABS-KEY (permeable AND pavement) AND AFFILCOUNTRY (Spain))
Green Roof	(TITLE-ABS-KEY (green AND roof) AND AFFILCOUNTRY (Spain))
Soakaway	(TITLE-ABS-KEY (soakaway) AND AFFILCOUNTRY (Spain))
Bioretention	(TITLE-ABS-KEY (bioretention) AND AFFILCOUNTRY (Spain))
Infiltration Drainage/Sustainable/Urban Stormwater	(TITLE-ABS-KEY (infiltration) AND TITLE-ABS-KEY (drainage OR sustainable OR urban AND stormwater) AND AFFILCOUNTRY (Spain))
Detention Drainage/Sustainable/Urban Stormwater	(TITLE-ABS-KEY (detention) AND TITLE-ABS-KEY (drainage OR sustainable OR urban AND stormwater) AND AFFILCOUNTRY (Spain))
Retention Drainage/Sustainable/Urban Stormwater	(TITLE-ABS-KEY (retention) AND TITLE-ABS-KEY (urban AND drainage) AND AFFILCOUNTRY (Spain))

Table 1. *Cont.*

Keywords	Search String
Artificial Wetland/Urban Drainage	(TITLE-ABS-KEY (artificial AND wetland) AND AFFILCOUNTRY (Spain) AND TITLE-ABS-KEY (stormwater OR rainwater OR drainage))
Bioswale	(TITLE-ABS-KEY (bioswale) AND AFFILCOUNTRY (Spain))
Vegetated Swale	(TITLE-ABS-KEY (vegetated AND swale) AND AFFILCOUNTRY (Spain))
Filter Strips Drainage/Sustainable/Urban Stormwater	(TITLE-ABS-KEY (filter AND strips) AND TITLE-ABS-KEY (drainage OR sustainable OR (urban AND stormwater)) AND AFFILCOUNTRY (Spain))
Rainwater Harvesting	(TITLE-ABS-KEY (rainwater AND harvesting) AND AFFILCOUNTRY (Spain))
Urban Green Infrastructure Drainage/Rainwater/Stormwater	(TITLE-ABS-KEY (urban AND green AND infrastructure) AND TITLE-ABS-KEY (stormwater OR drainage OR rainwater) AND AFFILCOUNTRY (Spain))
Blue Green Infrastructure	(TITLE-ABS-KEY (blue AND green AND infrastructure) AND AFFILCOUNTRY (Spain))

2.2. Selection of Bibliography

The initial search provided a total of 424 records, including articles, book chapters, lectures, and reviews. However, several articles with different keywords appeared with different search criteria, so there were some duplicate items. The screening process consisted in the elimination of duplicates, exclusion of publications without DOI, and we reviewed and read the papers to ensure that the practical cases were located within the Spanish geography (without include the work of Spanish researchers or Spanish entities in foreign locations). Thus, the number of documents to be analyzed became 137, of which 116 were articles, 5 were reviews, 9 were books or book chapters, and 7 were conference papers. This analysis considered the information contained in articles reviews and conferences. Most of the papers consulted belong to journals indexed in prestigious scientific databases, such as Journal Report Citation (JCR), SJR, and SCOPUS.

2.3. Extraction of Information

To answer the questions related to the state of SUDS research in Spain, which is one of the objectives of this study (Which techniques are the most analyzed? What are the most studied parameters and the main characteristics or functionalities? Is the study distributed evenly throughout the country? Does the study of these techniques arouse interest over time?), the minimum information collected included the following:

- Exposed drainage technique or techniques, according to a typical classification [3,7]: green roofs, rainwater harvesting, permeable pavement, detention systems, green channels, infiltration systems, retention systems, artificial wetlands, permeable pavements. The generalities have been included in a group called Sustainable Urban Drainage (SUDS in Figures).
- Year of publication.
- Type of study carried out: analysis of real cases (study of both structural and non-structural SUDS projects), laboratory tests (testing of a technique or any of its components in the laboratory), bibliographic review (studies of previous publications on the subject or comparisons of existing cases from other studies data) and application of models (use of hydrological, hydraulic, economic, or environmental models to simulate the operation of SUDS projected in a location but that do not exist in reality).
- Parameters analyzed in the articles: hydrological (in relation to flows and runoff volumes), structural (to evaluate the structure or typical structural components of each technique), ecological (to consider the biota involved in the performance of the techniques), energy (refer to the ability of SUDS to serve as an insulator or improve

urban thermal comfort), economic (cost–benefit studies and life cycle analysis of systems), social (citizens' perceptions about urban drainage and related urban policies), and planning (proposals for the inclusion of SUDS at the urban level, urban drainage design and management methodologies).

The information to delve into the analysis of the SUDS based on the weather included the following:

- The location of the study cases (included only those studies of SUDS or its components operating outdoors under the normal climatic conditions) to determine the climatology.
- Within the real cases of study were differentiated: testing new technologies (TNT), checking the performance of an alternative component or a novel structure; comparative (C), comparison of results between the performance of SUDS between variants or with gray infrastructures; data collection (DC), SUDS results monitored over time; model application (MA), creation of a model with previously obtained results; and survey (S), interviews on various aspects of SUDS.
- For the analysis of the case studies, the subject matter specified were: C, Component, or system layer (the article deals with one or more specific components of a SUDS); EP, Energy performance (the object of the research is to evaluate the potential energy benefits of a technique); EE, Economic study; ES, Ecosystem services (research to evaluate the potential ecosystem services of one or more techniques); HP, Hydrological performance (hydrological performance of SUDS or any of its components); HHP, Hydrological–hydraulic performance (hydrological performance and hydraulic operation of SUDS or any of its components); LCA, Life Cycle Analysis (economic–environmental analysis tool to analyze the suitability of a long-term technique); UP, Urban policies (article showing different policies and ways of managing urban water); RQ, Runoff quality (research focuses on runoff water quality and the ability of SUDS to manage it); SP, Social perception (how citizens perceive some of the sustainable drainage techniques); V, Vegetation (study focused on the plants that make up some of the SUDS).

3. Results

A total of 128 publications met the selection requirements. Figure 2 shows the total number of articles according to the technique analyzed and the evolution in the number of publications since the first explicit reference to the SUDS in the 1990s [18]. Under the title of several techniques are the articles that deal with projects that contemplated the operation (hydrological and quality of runoff) of several techniques simultaneously. With SUDS, we refer to those articles that deal with the sustainable management of urban runoff in a general way and not always using that term, also, green infrastructures.

Table 2 shows in a more concrete way the classification of the articles according to type of technique, type of study, and main subject studied in each case. The numbers indicate the number of publications found in this regard and in the last column of the table, the references of the classified articles.

Figure 3 schematically shows the articles grouped and counted according to the analyzed study parameters and also to the type of study (see Section 2.3).

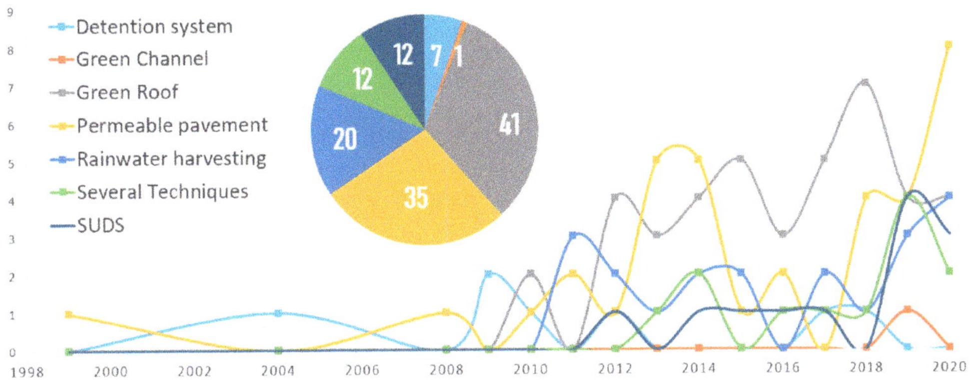

Figure 2. Number of articles according to the analyzed technique and temporal evolution of the number of publications. Source: Prepared by the authors.

Table 2. Types of studies and subjects analyzed according to the exposed drainage technique or techniques in the papers. Source: Prepared by the authors.

Type of Sustainable Drainage Technique	Type of Study	Subject Studied		References
Detention System (7)	Model application (5)	Hydrology	5	[19–23]
	Real case (2)	Hydrology	1	[24]
		Runoff quality	1	[25]
Green Channel (1)	Laboratory test (1)	Energy	1	[26]
Green Roof (41)	Bibliographical review (4)	Ecology	2	[27,28]
		Energy	2	[29,30]
	Laboratory test (2)	Ecology	1	[31]
		Energy	1	[32]
	Model application (9)	Economy	5	[33–37]
		Energy	3	[38–40]
		Social	1	[41]
	Real Case (26)	Ecology	7	[42–48]
		Economy	1	[49]
		Energy	13	[50–62]
		Hydrology	2	[63,64]
		Runoff quality	1	[65]
		Social	2	[66,67]
Permeable Pavement (35)	Bibliographical review (3)	Hydrology	1	[68]
		Structural	2	[8,69]
	Laboratory test (21)	Energy	1	[70]
		Hydrology	5	[71–75]
		Hydrology/ Runoff quality	1	[76]
		Runoff quality	2	[77,78]

Table 2. *Cont.*

Type of Sustainable Drainage Technique	Type of Study	Subject Studied		References
		Structural	12	[18,79–89]
	Model application (2)	Structural	2	[90,91]
	Real case (9)	Energy	2	[92,93]
		Hydrology	5	[94–98]
		Hydrology/ Runoff quality	1	[99]
		Runoff quality	1	[100]
Rainwater Harvesting (20)	Model application (12)	Economy	9	[101–109]
		Hydrology	2	[110,111]
		Structural	1	[112]
	Real case (8)	Economy	2	[113,114]
		Runoff quality	2	[115,116]
		Social	4	[117–120]
Several Techniques (12)	Bibliographical review (1)	Energy	1	[121]
	Model application (7)	Economy	2	[122,123]
		Hydrology	5	[124–128]
	Real case (4)	Hydrology	2	[129,130]
		Hydrology/Runoff quality	1	[131]
		Runoff quality	1	[132]
Sustainable Urban Drainage (12)	Bibliographical review (4)	Planning	3	[133–135]
		Structural	1	[136]
	Model application (4)	Economy	1	[137]
		Planning	2	[138,139]
		Social	1	[140]
	Real case (4)	Planning	2	[141,142]
		Social	2	[143,144]

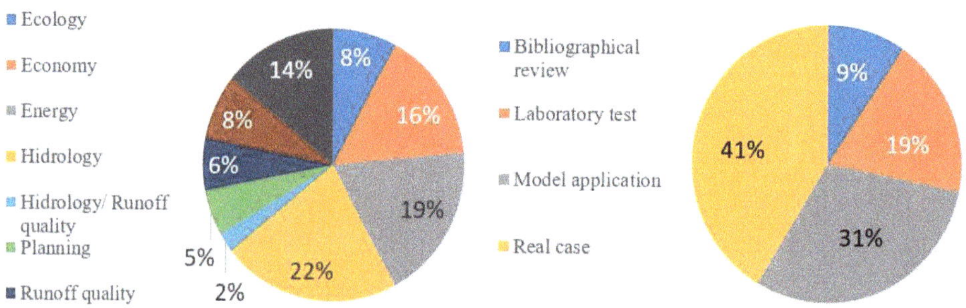

Figure 3. The graph on the left shows the percentage of study of the main subjects covered in the articles, and the graph on the right shows the proportion of papers according to type of study. Source: Prepared by the authors.

3.1. Sustainable Urban Drainage

This classification includes those articles that talk about sustainable drainage and green infrastructures in a global way, how they should be implemented, projects executed, and the benefits that these systems provide.

It also includes a recent bibliographic review of the SUDS in Spain from 2019: "The potential of sustainable urban drainage systems (SUDS) as an adaptive strategy to climate change in the Spanish Mediterranean" [136]. This paper is a compilation of some of the techniques implemented in Spain, particularly in Alicante.

Since these were not real case studies that could be affected by the weather, we did not delve further into the content of this classification.

3.2. Projects With Several SUDS

In addition to real cases of independent SUDS, publications that contemplated the simultaneous operation of different SUDS also appeared in the search: a pair referring to the AQUAVAL project [129–131] in Valencia province (Csa climate) and out in the north of the country (Cfb climate), whose conclusions appear in Table 3.

Table 3. Main conclusions of the studies on several SUDS together. Abbreviations used in Kind of Study: C, Comparative; DC, Data collection. Abbreviations used in Subject Studied: HHP, Hydrological–hydraulic performance; RQ, Runoff quality. Source: Prepared by the authors.

Year	Title	Climate	SUDS Studied	Kind of Study	Subject Studied	Main Conclusions
2013	The sustainable management of surface water at the building scale: Preliminary results of case studies in the UK and Spain [129]	Csa	Green roof, permeable pavement, rainwater harvesting	C	HHP	Comparison of the hydrological performance of SUDS in the United Kingdom and Xátiva (AQUAVAL project). The monitored elements revealed good hydrological–hydraulic performance of these systems.
2014	Comparative analysis of the outflow water quality of two sustainable linear drainage systems [132]	Cfb	Green channel, French drain	C	RQ	The results of measurements of water quality parameters (dissolved oxygen, TSS, pH, electrical conductivity, turbidity, and total hydrocarbons) showed fewer pollutants at the outlet of SUDS than the outlet of conventional drainage systems.
2014	SuDS Efficiency during the Start-Up Period under Mediterranean Climatic Conditions [130]	Csa	Infiltration pond, green channel; green roof	C	HHP-RQ	AQUAVAL project in Benaguasil: The hydrological and water quality results for the infiltration pond and green channel showed a significant attenuation of flows, volumes, and pollutants. However, the water quality of the green roof was worse than the conventional one.

Table 3. *Cont.*

Year	Title	Climate	SUDS Studied	Kind of Study	Subject Studied	Main Conclusions
2017	The role of monitoring sustainable drainage systems for promoting transition towards regenerative urban built environments: a case study in the Valencian region, Spain [131]	Csa	Green channel; green roof, rainwater harvesting, detention and infiltration systems	DC	HHP-RQ	AQUAVAL project in Benaguasil: The results of the monitored SDUS proved good hydraulic performance in a typical Mediterranean climate and an improvement in water quality in green channels and infiltration systems.

In addition to the real cases, there were also investigations that propose models to evaluate the suitability of the use of these techniques in flood control (giving positive results) [124–128], improving adaptation to change climate [122,125], and providing another environmental benefits [123].

3.3. Green Roofing

Green roofs, with 41 publications, were the most analyzed techniques, and 24 of these articles showed the results of monitored roofs. Table 4 summarizes the main conclusions obtained in investigations of real cases in Spain in a semi-arid climate (type B) and in a Mediterranean climate (type Csa).

Table 4. Main conclusions of the studies on green roofs. Abbreviations used in Kind of Study, TNT, Test new technologies and/or materials; S: Survey; C: Comparative; DC: Data collection; MA, Model application. Abbreviations used in Subject Studied: HP, Hydrological performance; EP, Energy performance; V, Vegetation; C, Component, or system layer; LCA, Life Cycle Analysis; ES, Ecosystem services; UP, Urban policies; SP, Social perception. Source: Prepared by the authors.

Year	Title—Reference	Climate	Kind of Study	Subject Studied	Main Conclusions
2012	Use of rubber crumbs as drainage layer in experimental green roofs [63]	BSk	TNT	HP	The use of rubber crumbs as a drainage layer material in extensive green roofs was feasible
2012	Use of rubber crumbs as drainage layer in green roofs as potential energy improvement material [51]	BSk	TNT	EP	The use of rubber crumbs for the energy improvement of a building did not give better results than other typical green roof components
2012	Green roofs as passive system for energy savings when using rubber crumbs as drainage layer [50]	BSk	TNT	HP—EP	The use of these techniques using rubber crumbs showed an improvement in these yields compared to a normal ceiling during the monitoring period
2013	Green roof systems: A study of public attitudes and preferences in southern Spain [66]	Csa	S	SP	Sociodemographic characteristics and environmental background of childhood influenced the green roof type preferences of citizens.
2014	Environmental performance of recycled rubber as drainage layer in extensive green roofs. A comparative Life Cycle Assessment [49]	BSk	TNT	LCA	The rubber crumbs produced less environmental impact than pozzolan, which is an element that can be substituted in green roofs.

Table 4. *Cont.*

Year	Title—Reference	Climate	Kind of Study	Subject Studied	Main Conclusions
2014	Photovoltaic-green roofs: An experimental evaluation of system performance [53]	BSk	C	EP	The green roof systems (Gazania rigens and Sedum clavatum) with photovoltaic panels showed a considerably lower roof surface temperature compared to the photovoltaic panel–gravel configuration
2015	A critical analysis of factors affecting photovoltaic-green roof performance [54]	BSk	C	EP	The results revealed that the increase in photovoltaic production provided by photovoltaic green roofs depended on several factors, and among the plant species studied, Sedum clavatum showed the best interaction with photovoltaics and the building.
2015	Evaluating the growth of several Mediterranean endemic species in artificial substrates: Are these species suitable for their future use in green roofs? [42]	BSh	C	V	Study of the growth of Silene vulgaris, Silene secundiflora, Crithmum maritimum, Lagurus ovatus, Asteriscus maritimus, and Lotus creticus on three artificial substrates. S. vulgaris and L. ovatus showed greater germination and growth than the other species.
2015	Plant cover and floristic composition effect on thermal behaviour of extensive green roofs [55]	BSk	TNT	EP	Study of the thermal performance of an extensive green roof according to coverage and floristic composition (Sedum species) that compares the behavior of the system with low (10%) and high (80%) vegetation coverage. There were not significant changes between both.
2015	The inorganic component of green roof substrates impacts the growth of Mediterranean plant species as well as the C and N sequestration potential [43]	BSh	TNT	C: Substrate	Lotus creticus and Asteriscus maritimus were planted and evaluated for 10 months in four substrates with the same compost but several inorganic materials in different proportions. The study demonstrated that the composition of the substrate impacts on native plant growth and C and N sequestration.
2015	The thermal behaviour of extensive green roofs under low plant coverage conditions [56]	BSk	DC	EP substrate	The results of a monitored green roof study focused on analyzing the thermal behavior of the substrate layer with the growing plants (10% vegetation cover). Where plants provide little shade, the thermal performance of the roof depended on the characteristics of the lower layers, especially the substrate.

Table 4. *Cont.*

Year	Title—Reference	Climate	Kind of Study	Subject Studied	Main Conclusions
2016	The composition and depth of green roof substrates affect the growth of Silene vulgaris and Lagurus ovatus species and the C and N sequestration under two irrigation conditions [44]	BSh	C	C: Substrate	The authors analyzed the sequestration of C and N and their state with irrigation at 40% of the ETP values in two different substrates with Silene vulgaris and Lagurus ovatus and concluded that this irrigation allowed an adequate vegetal cover.
2016	Thermal assessment of extensive green roofs as passive tool for energy savings in buildings [58]	BSk	C	EP	Extensive green roof samples provided lower energy consumption than conventional roofs during hot periods, while they consumed higher energy during heating periods.
2017	Sustainable earth-based vs. conventional construction systems in the Mediterranean climate: Experimental analysis of thermal performance [59]	BSk	C	E	Seven cubicles with the same inner dimensions and orientation but different construction systems are thermally tested at a real experimental scale. Similar thermal behavior can be achieved by using sustainable and environmentally friendly construction systems instead of the current high embodied energy conventional ones.
2018	Performance evaluation of five Mediterranean species to optimize ecosystem services of green roofs under water-limited conditions [46]	BSk	C	V	An experiment evaluated the growth and coverage of Brachypodium phoenicoides, Crithmum maritimum, Limonium virgatum, Sedum sediforme, and Sporobolus pungens, with irrigation at 50% and 25% of the ET0 values. All species survived and showed an adequate aesthetic performance and plant cover, although not equally between them.
2018	Thermal regulation capacity of a green roof system in the Mediterranean region: The effects of vegetation and irrigation level [60]	BSk	DC	EP	Quantification of the contribution of the vegetation cover and the effect of the volume of irrigation water on the thermal efficiency of a green roof. The presence of vegetation reduced the thermal variations. Sedum sediforme behaved as a better insulator than Brachypodium phoenicoides during the experimental period (spring and summer).

Table 4. *Cont.*

Year	Title—Reference	Climate	Kind of Study	Subject Studied	Main Conclusions
2018	Hydrological performance of green roofs at building and city scales under Mediterranean conditions [64]	Csa	C	HP	The authors monitored a green and a conventional roof for comparison and created hydrological models that were calibrated and validated. Green roofs provided a good hydrological performance.
2018	Mediterranean green roof simulation in Caldes de Montbui (Barcelona): Thermal and hydrological performance test of Frankenia laevis L., Dymondia margaretae Compton, and Iris lutescens Lam [45]	Csa	DC	HP-EP plants	The authors evaluated the thermal and hydrological behavior of Frankenia laevis, Dymondia margaretae, and Iris lutescens with a minimum irrigation contribution and a dry land treatment. The results showed that the most influential factors recorded were the relationship between air and water in the substrate and the interaction between the green layer and the substrate. In particular, D. margaretae conserved more water than the other species in both summer and winter.
2018	Risk assessment by percolation leaching tests of extensive green roofs with fine fraction of mixed recycled aggregates from construction and demolition waste [65]	Csa	S	C: Substrate	The aim of this study was the environmental risk of contaminant release in leachates from different substrate mixtures based on recycled construction waste aggregates. Records were lower compared to laboratory test data, showing how laboratory conditions may overestimate the potential contaminating effect of these materials.
2019	Evaluating the establishment performance of six native perennial Mediterranean species for use in extensive green roofs under water-limiting conditions [47]	BSk	C	V	The authors cultivated Asteriscus maritimus, Brachypodium phoenicoides, Crithmum maritimum, Limonium virgatum, Sedum sediforme, and Sporobolus pungens under good irrigation and water deficit conditions to evaluate the effects of water deficit on their growth. Sedum sediforme appeared to be the species best adapted to water deficit and Brachypodium phoenicoides and Limonium virgatum showed a satisfactory aesthetic performance in water deficit conditions.

Table 4. *Cont.*

Year	Title—Reference	Climate	Kind of Study	Subject Studied	Main Conclusions
2019	Long-term experimental analysis of thermal performance of extensive green roofs with different substrates in Mediterranean climate [61]	Csa	C	EP substrate	The thermal performance over two years of three green roofs with different types of substrates (commercial and recycled materials) and a traditional ballasted gravel roof. The results of a comparison between the thermal performance over two years of three green roofs with different types of substrates (commercial and recycled materials) and a traditional ballasted gravel roof indicated that for hot and dry weather conditions, the greater capacity to retain water in the substrate provided a greater cooling capacity.
2020	Evaluation of the development of five Sedum species on extensive green roofs in a continental Mediterranean climate [48]	BSk	C	V	The paper reflected the growth and development patterns of Sedum album, S. sediforme, S. sexangulare, Sedum spurium, and Sedum spurium and concluded that Sedum album, S. sediforme, and S. sexangulare were species recommended for use in extensive green roofs, while S. spurium presented some limitations for their use.
2020	Creating urban green infrastructure where it is needed—A spatial ecosystem service-based decision analysis of green roofs in Barcelona [67]	Csa	S	ES	The authors created a decision tool for the implementation of green roofs based on their potential ecosystem services from models and the opinions of the participants in workshops held within the Naturvation (https://naturvation.eu/ (accessed on 25 February 2021)).
2020	Study of the improvement on energy efficiency for a building in the Mediterranean area by the installation of a green roof system [62]	Csa	MA	EP	A model created with TRNSYS and calibrated with experimental data from a monitored green roof resulted in a substantial improvement in the building's cooling energy demand, a 30% reduction in energy demand for a standard summer day, and 15% for a winter day.

Given the importance of climate in the development and maintenance of vegetation and that the climate of much of the national territory is characterized by long periods of drought, an important part of the research carried out has focused on the functioning of species such as *Brachypodium phoenicoides*, *Crithmum maritimum*, *Limonium virgatum*, *Sedum sediforme*, *Sporobolus pungens*, [46,47] and *Asteriscus maritimus* [47], which were studied in the Balearic Islands; Sedums such as *lbum*, *sexangulare*, and *spurium* [48], which were observed in Lleida; and *Silene vulgaris, Silene secundiflora, Crithmum maritimum, Lagurus ovatus, Asteriscus maritimus,* and *Lotus creticus* in Murcia [42].

3.4. Permeable Pavements

Permeable pavements were the second technique with the most publications, with a total of 35 papers. Nine of them reflected the results of experimental installations, and 20 were laboratory tests. Table 5 shows the most representative conclusions of the study cases in locations with a temperate mesothermal climate (type Cfb) and in places with a Mediterranean climate (Csa).

Table 5. Main conclusions of the studies on permeable pavements. Abbreviations used in Kind of Study: C, Comparative; DC, Data collection. Abbreviations in Subject Studied: HP, Hydrological performance; EP, Energy performance; LCA, Life Cycle Analysis; S, Survey; RQ, Runoff quality; C, Component. Source: Prepared by the authors.

Year	Title—Reference	Climate	Kind of Study	Subject Studied	Main Conclusions
2010	Performance of pervious pavement parking bays storing rainwater in the north of Spain [94]	Cfb	C	HP	The comparison of the performance of three types of permeable pavements, two with geotextiles (Inbitex and One Way) and another without it, did not show differences in the storage capacity of the SUDS.
2011	Analysis and contrast of different pervious pavements for management of stormwater in a parking area in Northern Spain [95]	Cfb	C	HP	The materials of the surface layer of the permeable pavements tested had a greater effect than the geotextile layer in their storage capacity; although the behavior was different in the three types of permeable pavements identified, the differences in their ability to retain water were not significant.
2011	Design and construction of an experimental pervious paved parking area to harvest reusable rainwater [99]	Cfb	C	HP-RQ	The materials of the surface layer of the permeable pavements tested had a greater effect than the geotextile layer in their storage capacity; although the behavior was different in the three types of permeable pavements identified, the differences in their ability to retain water were not significant. The quality of the stored water was suitable, although in the conditions of the first flush, it did not give good results and neither did it comply with some parameters of the Spanish legislation.
2013	Monitoring and evaluation of the thermal behavior of permeable pavements for energy recovery purposes in an experimental parking lot: Preliminary results [92]	Cfb	C	EP (Ground Source Heat Pumps)	The temperature of the subbase was different from the air temperature during the study period, which showed that the subbase is less affected by air temperature than the pavement bedding because of the insulating capacity of the permeable pavements.
2013	Temperature performance of different pervious pavements: Rainwater harvesting for energy recovery purposes [93]	Cfb	C	EP (Ground Source Heat Pumps)	The temperature of the subbase was different from the air temperature during the study period, which shows that the subbase was less affected by air temperature than the pavement bedding because of the insulating capacity of the permeable pavements. The rainwater tank did not represent a health risk associated with the appearance of Legionellae (in case the permeable pavement worked in geothermal air conditioning).
2014	Water quality and quantity assessment of pervious pavements performance in experimental car park areas [100]	Cfb	C	RQ	The results of three field studies demonstrated important correlations between sub-base materials and outlet water quality parameters. The polymer-modified porous concrete surface course in combination with limestone aggregate performed better than basic oxygen furnace slag.

Table 5. *Cont.*

Year	Title—Reference	Climate	Kind of Study	Subject Studied	Main Conclusions
2018	The long-term hydrological performance of permeable pavement systems in Northern Spain: An approach to the "end-of-life" concept [97]	Cfb	DC	C	Despite suffering a significant reduction in permeability after 10 years of operation, the permeable pavements analyzed showed a high rate of infiltration, although there were spatial variations in the reduction of infiltration capacity due to static loads from vehicles.
2018	A study of the application of permeable pavements as a sustainable technique for the mitigation of soil sealing in cities: A case study in the south of Spain [96]	Csa	C	HP	The efficiencies of the maximum flow reduction of the monitored pavements exceed 95% and, in relation to the volumetric efficiencies, they were higher than 80%.
2020	Middle-term evolution of efficiency in permeable pavements: A real case study in a Mediterranean climate [98]	Csa	DC	C	The pavements tested did not suffer from obstructions in the medium term, and the variability in efficiency could be due to the Mediterranean climate, the variations in the behavior of the pavement seemed to be more influenced by the initial saturation of the soil than by possible obstructions in the first years of operation.

Regarding the use of this technique for adaptation to climate change, the Life CERSUDS project [91] has investigated the capacity of these forms of permeable surfaces made from ceramic elements systems to mitigate the expected effects.

3.5. Rainwater Harvesting

There were also several studies of rainwater harvesting and potential uses (Table 6), which were all located in places with Mediterranean climatology (Barcelona and Girona).

Table 6. Main conclusions of the studies on rainwater harvesting. Abbreviations used in Type of Study: S, Survey; C, Comparative; DC, Data collection; MA, Model application. Abbreviations used in Subject Studied: UP, Urban policies; RQ, Runoff quality; EE, Economic study; SP, Social perception.

Year	Title—Reference	Climate	Kind of Study	Subject Studied	Main Conclusions
2011	A comparative appraisal of the use of rainwater harvesting in single and multi-family buildings of the Metropolitan Area of Barcelona (Spain): Social experience, drinking water savings, and economic costs [113]	Csa	C	SP	Rainwater was rarely used for flushing toilets and cleaning clothes despite giving favorable results in the Metropolitan Area of Barcelona. The perception about these systems was that the environmental benefit exceeded the pecuniary. The main drawback identified was the long payback period of these systems.
2011	Cost-efficiency of rainwater harvesting strategies in dense Mediterranean neighbourhoods [114]	Csa	C	EE	The strategies for collecting and reusing rainwater in dense urban areas of the Mediterranean were economically advantageous only if they were carried out at an appropriate scale allowing economies of scale and considering the expected evolution of water prices.
2011	Roof selection for rainwater harvesting: Quantity and quality assessments in Spain [115]	Csa	C	RQ	The quality of rainwater runoff in the study area appeared to be better than the average quality found in the literature review. Smooth sloping roofs have performed better in terms of runoff quality and therefore may be preferable for stormwater catchment.

Table 6. *Cont.*

Year	Title—Reference	Climate	Kind of Study	Subject Studied	Main Conclusions
2015	Watering the garden: Preferences for alternative sources in suburban areas of the Mediterranean coast [117]	Csa	MA	SP	The analysis of the sizes of rain collection tanks for irrigation, in suburban areas of Girona, concluded that many had been oversized.
2017	Urban rainwater runoff quantity and quality—A potential endogenous resource in cities? [116]	Csa	DC	RQ	The study of the quantity and quality of runoff collected from different urban surfaces according to use (pedestrian or motorized mobility) and materials (concrete, asphalt and slabs) showed that precast concrete slabs provided a better-quality runoff.
2020	Diverse pathways-common phenomena: Comparing transitions of urban rainwater harvesting systems in Stockholm, Berlin and Barcelona [119]	Csa	C	UP	Urban planning could be decisive in the development of urban rainwater harvesting systems. Its socio-environmental benefits could bring sustainability and resilience to cities.
2020	A breakthrough in urban rain-harvesting schemes through planning for urban greening: Case studies from Stockholm and Barcelona [118]	Csa	DC	UP	The lack of inclusive and democratic procedures, of a long-term commitment in the implementation of these systems (which require proper design and monitoring) could cause significant challenges in a fairer and more sustainable stormwater management.
2020	Non-conventional resources for the coming drought: the development of rainwater harvesting systems in a Mediterranean suburban area [120]	Csa	S	UP	The rainwater harvesting systems in Catalonia and Spain turned out to be very marginal. The article concluded that important changes in water policies were needed for the implementation of rainwater harvesting systems, such as determining their obligation or offering subsidies.

A large portion of the articles are economic analyses on the use of rainwater among this type of analysis, including the creation of a software, Plugrisost [105], to evaluate the profitability and environmental impact of rainwater tanks, which has been used to estimate water prices (for different uses) from which it is economically profitable to install a rainwater harvesting system [106] and to carry out an environmental and economic analysis of rainwater storage systems that supply water for laundry [109].

3.6. *Green Channel*

There was hardly exhaustive research on green channels, although it is necessary to mention a laboratory investigation [26] focused on the temperature variations in the different layers of a green channel.

The hydrological behavior of green channels was effective from the hydrological point of view [129–131] and improving the runoff quality [131,132].

3.7. *Detention Systems*

The studies of detention systems observed were mainly hydrological–hydraulic models applied in Barcelona [24], Granada [19,20], and Valencia [21,22,25], cities with a Csa climate, and Cantabria [21] with a Cfb climate.

3.8. *Research by Climate*

The studies found from the end of the 1990s to the end of the year 2020, although they were not homogeneously distributed throughout the Spanish geography, they cover a large part of the national territory based on the climate. In more arid climates (type B), it seems that research (there are 16 articles of empirical studies carried out in arid climate) has focused more on energy performance (eight papers of 16) and optimizing vegetation selection (six papers of 16), while in temperate climates (type C), it has focused more on hydrology (eight papers of 29 based on C climate). Figure 4 graphically shows the amounts of articles dedicated, on the one hand, to each subject of study and on the other to each type of specific technique.

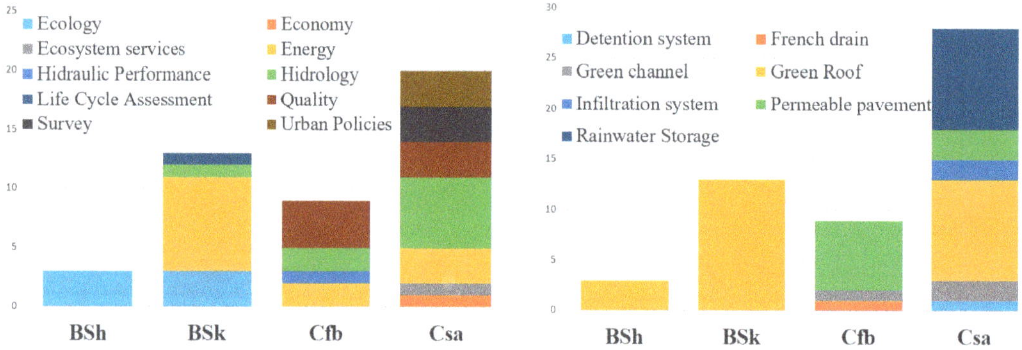

Figure 4. The graph on the left shows the number of studies of the main subjects covered in the articles about real cases according to climate, and the graph on the right shows the techniques studied in the papers. The authors have separately accounted for each of the techniques reflected in the articles that contemplated several simultaneously. Source: Prepared by the authors.

4. Discussion

A difference between the recent bibliographic review [136] (that compares the implementation of SUDS in other countries with respect to Spain) and the recent publication Sustainable Urban Drainage Systems in Spain: A Diagnosis [145] (an exhaustive compilation of implemented techniques) is that this paper only considers those cases in which a scientific investigation process had been carried out.

The usefulness of SUDS as effective and sustainable management of urban runoff in different climatic regions of Spain is widely demonstrated in several papers [64,96,129–131] analyzed in this review, as well as their potential in other fields such as mitigation of climate change in cities [136,139], but the success rates of local–regional SUDS in Spanish different climatology are still not validated.

By far, the most studied techniques are green roofs and permeable surfaces (Figure 2), followed by rainwater harvesting and detention systems. In contrast, typical green street techniques [146] such as bioswales, bioretention areas, or filtering strips providers of several ecosystem services [147] have hardly been analyzed.

The study of SUDS is unequally distributed throughout the Spanish geography; Catalonia and Cantabria are the regions with the greatest number of studies of these techniques, their components, and their operation. In Cantabria, the GITECO research group has carried out a large number of investigations [8], but these have almost entirely focused on permeable surfaces and their hydraulic–hydrological performance. In Catalonia, research centers with different objects of study, such as ICTA (https://www.uab.cat/web/icta-1345819904158.html (accessed on 8 February 2021)) or CREAF (http://www.creaf.cat/es (accessed on 9 February 2021)) have investigated mainly green roofs and rainwater harvesting systems from different points of view (not only dealing with the hydrological and hydraulic performance, but energy, biological, and economic).

The establishment of vegetation is essential for the correct long-term operation of a green infrastructure [148], and it depends directly on the weather. Since some areas of Spain are predominantly dry [9], one of the main concerns could be the selection of species that can withstand water scarcity. Perhaps this is the reason why the regions where vegetative growth and development have been most investigated are the Balearic Islands, Lleida, and Murcia, which are characterized by their low rainfall [9]. The deductions that can be drawn after observing Table 4 is that for the prevailing dry climate in the country, it is advisable to use a mixture of perennial and annual plants with porous and light substrates [42]; the presence of vegetation is decisive for the functions of thermal insulation [56] and water retention with the characteristic rainfall regime of dry areas [45]. However, not all vegetation is equally effective [48]; species such as Sedum sediforme [60] give better results than others [47,48,60]. Regarding the hydrological operation, the effectiveness of green roofs has been demonstrated, but the results of improving the quality of runoff are not satisfactory [130], so further tests in real facilities are recommended, since the results differ from those obtained in the laboratory.

However, more interest seems to focus on the condition of insulation against the heat of these techniques due to the number of publications in this regard (see Table 2).

Permeable pavements work well hydrologically regardless of the climate in which they have been analyzed (see Table 5); although their performance in quality management depends largely on the composition [77–79], there are no records of its operation in arid climates.

Although bioretention systems and green channels or bioswales are some of the green infrastructure solutions recommended at the urban level due to their multifunctionality [149], there are not plentiful investigations, as occurs with other techniques. Its multifunctional performance depends largely on the biota [150], which derives from factors such as location and climate (predominantly dry and with little precipitation in Spain [9]); plant selection and plant conditioning factors can be a limiting factor [151]. Therefore, it would be advisable to investigate further which plant elements and components are the ones that would work best under long-term peninsular climatic conditions, since ecosystem services will depend on plants, such as urban biodiversity or CO_2 reduction, and maintenance costs, among others [152].

5. Conclusions

The SUDS study includes different disciplines, hydrology, edaphology, ecology, economics, etc. [7]. However, in Spain, the study is highly polarized; that is, the papers with various techniques and those about permeable surfaces deal with the hydrology, while green roofs papers are focused on the improvement of the energy efficiency of buildings, and rainwater harvesting investigations show their economic performance. This can be associated with the fact that the studies are carried out by specialists who tend to prioritize their own fields without considering the important impacts of other fields [153].

There are many more types of SUDS than those found in this research, such as filter strings, trenches or infiltration wells, artificial wetlands, etc. However, although there is evidence that they have been implemented in the Spanish geography, there are no studies that evaluate its operation: neither the hydrological–hydraulic performance nor its potential components or possible secondary benefits.

It is interesting to mention that the most analyzed techniques in Spain are "in situ" control. That means there is too much to investigate about local and regional control SUDS—in other words, techniques that manage runoff from streets, municipalities, or large areas. This may be because it is easy to install a green roof or a permeable pavement in a university building or research centers, but it is more complicated to follow and monitor techniques located in the public space. In this case, it is necessary to have a collaboration between the researchers, the public administration, and citizens.

It would be advisable to carry out more interdisciplinary studies or a holistic analysis of these techniques in their operation in urban areas. Especially SUDS such as bioswales or bioretention systems that develop populations of living beings are limited in their growth by the rainfall regimes of the country.

Author Contributions: Conceptualization, J.C.S.; methodology, A.I.A.G.; formal analysis, N.C.P. All authors have read and agreed to the published version of the manuscript.

Funding: This research received no external funding.

Institutional Review Board Statement: Not applicable.

Informed Consent Statement: Not applicable.

Data Availability Statement: Some or all data, models, or code that support the findings of this study are available from the corresponding author upon reasonable request.

Acknowledgments: To all the scientists and researchers whose contribution to research into a new approach to runoff management has made this article possible.

Conflicts of Interest: The authors declare no conflict of interest.

References

1. Abrahams, J.C.; Coupe, S.J.; Sañudo-Fontaneda, L.A.; Schmutz, U. The Brookside Farm Wetland Ecosystem Treatment (WET) System: A Low-Energy Methodology for Sewage Purification, Biomass Production (Yield), Flood Resilience and Biodiversity Enhancement. *Sustainability* **2017**, *9*, 147. [CrossRef]
2. Monberg, R.J.; Howe, A.G.; Ravn, H.P.; Jensen, M.B. Exploring structural habitat heterogeneity in sustainable urban drainage systems (SUDS) for urban biodiversity support. *Urban Ecosyst.* **2018**, *21*, 1159–1170. [CrossRef]
3. Woods-Ballard, B.; Kellagher, R.; Martin, P.; Jefferies, C.; Bray, R.; Shaffer, P. *The SUDS Manual*; CIRIA C697: London, UK, 2007.
4. United States Environmental Protection Agency. Managing Wet Weather with Green Infrastructure. Action Strategy. 2008. Available online: https://nepis.epa.gov/Exe/ZyPDF.cgi/P1008SI8.PDF?Dockey=P1008SI8.PDF (accessed on 17 February 2021).
5. European Environment Agency. *Green Infrastructure and Territorial Cohesion*; Tecnical Report (Number 18); Publications Office of the European Union: Luxembourg, 2011; pp. 1–138. [CrossRef]
6. European Commission. *Towards an EU Research and Innovation Policy Agenda for Nature-Based Solutions & Re-Naturing Cities*; Directorate-General for Research and Innovation Climate Action, Environment, Resource Efficiency and Raw Materials EN; Publications Office of the European Union: Luxembourg, 2015. [CrossRef]
7. Woods-Ballard, B.; Ashley, R.; Illman, S.; Scott, T.; Wilson, S. *The SuDS Manual*; C753; CIRIA: London, UK, 2015.
8. Castro-Fresno, D.; Andrés-Valeri, V.C.; Sañudo-Fontaneda, L.A.; Rodriguez-Hernandez, J. Sustainable Drainage Practices in Spain, Specially Focused on Pervious Pavements. *Water* **2013**, *5*, 67–93. [CrossRef]
9. AEMET. Atlas Climático Ibérico. Temperatura del Aire y Precipitación (1971–2000). Agencia Estatal de Meteorología, Ministerio de Medio Ambiente y Medio Rural y Marino. 2011, p. 79. Available online: http://scholar.google.com/scholar?hl=en&btnG=Search&q=intitle:ATLAS+CLIM?TICO+IB?RICO+IBERIAN+CLIMATE+ATLAS#0 (accessed on 10 March 2021).
10. Chazarra, A.; Flórez, E.; Peraza, B.; Tohá, T.; Lorenzo, B.; Criado, E.; Moreno, J.V.; Romero, R.; Botey, R. *Mapas Climáticos de España (1981–2010) y ETo (1996–2016)*; Agencia Estatal de Meteorología: Madrid, Spain, 2018.
11. Climatología Hispagua. Available online: https://hispagua.cedex.es/datos/climatologia (accessed on 1 June 2021).
12. Kletter, A.; von Hardenberg, J.; Meron, E.; Provenzale, A. Patterned vegetation and rainfall intermittency. *J. Theor. Biol.* **2009**, *256*, 574–583. [CrossRef]
13. Aronica, G.T.; Freni, G.; Oliveri, E. Uncertainty analysis of the influence of rainfall time resolution in the modelling of urban drainage systems. *Hydrol. Process.* **2005**, *19*, 1055–1071. [CrossRef]
14. Ribas, A.; Olcina, J.; Sauri, D. More exposed but also more vulnerable? Climate change, high intensity precipitation events and flooding in Mediterranean Spain. *Disaster Prev. Manag. Int. J.* **2020**, *29*, 229–248. [CrossRef]
15. Charlesworth, S.M. A review of the adaptation and mitigation of global climate change using sustainable drainage in cities. *J. Water Clim. Chang.* **2010**, *1*, 165–180. [CrossRef]
16. TECNALIA. *'Soluciones Naturales' para la Adaptación al Cambio Climático en el Ámbito Local de la Comunidad Autónoma del País Vasco*; Guía Metodológica para su Identificación y Mapeo. Caso de Estudio Donostia-San Sebastián; Ihobe, Sociedad Pública de Gestión Ambiental Departamento de Medio Ambiente, Planificación Territorial y Vivienda, Gobierno Vasco: Bilbao, Spain, 2016.
17. De Granda-Orive, J.I.; Alonso-Arroyo, A.; Roig-Vázquez, F. ¿Qué base de datos debemos emplear para nuestros análisis bibliográficos? Web of Science versus SCOPUS. *Arch. Bronconeumol.* **2011**, *47*, 213. [CrossRef] [PubMed]
18. Pindado, M.Á.; Aguado, A.; Josa, A. Fatigue behavior of polymer-modified porous concretes. *Cem. Concr. Res.* **1999**, *29*, 1077–1083. [CrossRef]
19. Osorio, F.; Muhaisen, O.; García, P.A. Copula-Based Simulation for the Estimation of Optimal Volume for a Detention Basin. *J. Hydrol. Eng.* **2009**, *14*, 1378–1382. [CrossRef]
20. Muhaisen, O.S.; Osorio, F.; García, P.A. Two-copula based simulation for detention basin design. *Civ. Eng. Environ. Syst.* **2009**, *26*, 355–366. [CrossRef]
21. Andrés-Doménech, I.; Montanari, A.; Marco, J.B. Stochastic rainfall analysis for storm tank performance evaluation. *Hydrol. Earth Syst. Sci.* **2010**, *14*, 1221–1232. [CrossRef]
22. Andrés-Doménech, I.; Montanari, A.; Marco, J.B. Efficiency of Storm Detention Tanks for Urban Drainage Systems under Climate Variability. *J. Water Resour. Plan. Manag.* **2012**, *138*, 36–46. [CrossRef]
23. Sánchez-Beltrán, H.; Rodríguez, C.M.; Triviño, J.B.; Iglesias-Rey, P.; Valderrama, J.S.; Martínez-Solano, F. Characterization of Modular Deposits for Urban Drainage Networks Using CFD Techniques. *Procedia Eng.* **2017**, *186*, 84–92. [CrossRef]
24. Cembrano, G. Optimal control of urban drainage systems. A case study. *Control. Eng. Pract.* **2004**, *12*, 1–9. [CrossRef]

25. Andrés-Doménech, I.; Hernández-Crespo, C.; Martín, M.; Valeri, V.C.A. Characterization of wash-off from urban impervious surfaces and SuDS design criteria for source control under semi-arid conditions. *Sci. Total. Environ.* **2018**, *612*, 1320–1328. [CrossRef] [PubMed]
26. Rey-Mahía, C.; Sañudo-Fontaneda, L.A.; Andrés-Valeri, V.C.; Álvarez-Rabanal, F.P.; Coupe, S.J.; Roces-García, J. Evaluating the Thermal Performance of Wet Swales Housing Ground Source Heat Pump Elements through Laboratory Modelling. *Sustainability* **2019**, *11*, 3118. [CrossRef]
27. Fernandez, R.; Gonzalez, P. Green Roofs as a Habitat for Birds: A Review. *J. Anim. Veter-Adv.* **2010**, *9*, 2041–2052. [CrossRef]
28. Bureš, S. A view beyond traditional growing media USES. *Acta Hortic.* **2013**, *1013*, 109–116. [CrossRef]
29. Chanampa, M.; Rivas, P.V.; Ojembarrena, J.A.; Olivieri, F. Systems of Vegetal Façade and Green Roofs used as a Sustainable Option in Architecture. *Des. Princ. Pract. Int. J. Annu. Rev.* **2010**, *4*, 1–10. [CrossRef]
30. Cascone, S.; Coma, J.; Gagliano, A.; Pérez, G. The evapotranspiration process in green roofs: A review. *Build. Environ.* **2019**, *147*, 337–355. [CrossRef]
31. Ondoño, S.; Bastida, F.; Moreno-Ortego, J.L. Microbiological and biochemical properties of artificial substrates: A preliminary study of its application as Technosols or as a basis in Green Roof Systems. *Ecol. Eng.* **2014**, *70*, 189–199. [CrossRef]
32. Coma, J.; De Gracia, A.; Chàfer, M.; Pérez, G.; Cabeza, L.F. Thermal characterization of different substrates under dried conditions for extensive green roofs. *Energy Build.* **2017**, *144*, 175–180. [CrossRef]
33. Rivela, B.; Cuerda, I.; Olivieri, F.; Bedoya, C.; Neila, J. Análisis de Ciclo de Vida para el ecodiseño del sistema Intemper TF de cubierta ecológica aljibe. *Mater. Construcción* **2012**, *63*, 131–145. [CrossRef]
34. Lamnatou, C.; Chemisana, D. Photovoltaic-green roofs: A life cycle assessment approach with emphasis on warm months of Mediterranean climate. *J. Clean. Prod.* **2014**, *72*, 57–75. [CrossRef]
35. Foudi, S.; Spadaro, J.V.; Chiabai, A.; Polanco-Martínez, J.M.; Neumann, M.B. The climatic dependencies of urban ecosystem services from green roofs: Threshold effects and non-linearity. *Ecosyst. Serv.* **2017**, *24*, 223–233. [CrossRef]
36. Guzmán-Sánchez, S.; Jato-Espino, D.; Lombillo, I.; Diaz-Sarachaga, J.M. Assessment of the contributions of different flat roof types to achieving sustainable development. *Build. Environ.* **2018**, *141*, 182–192. [CrossRef]
37. Carretero-Ayuso, M.J.; De Extremadura, U.; Sanz-Calcedo, J.G. Comparison between building roof construction systems based on the LCA. *Rev. Construcción* **2018**, *17*, 123–136. [CrossRef]
38. Herrera-Gomez, S.S.; Quevedo-Nolasco, A.; Pérez-Urrestarazu, L. The role of green roofs in climate change mitigation. A case study in Seville (Spain). *Build. Environ.* **2017**, *123*, 575–584. [CrossRef]
39. Alvarez-Vázquez, L.J.; Fernández, F.J.; Martínez, A.; Vázquez-Méndez, M.E. Urban Heat Island Effect in Metropolitan Areas: An Optimal Control Perspective. *Lect. Notes Comput. Sci. Eng.* **2019**, *126*, 829–837. [CrossRef]
40. Tello, J.I.; Tello, L.; Vilar, M.L. On the Existence of Solutions of a Two-Layer Green Roof Mathematical Model. *Mathematics* **2020**, *8*, 1608. [CrossRef]
41. Briz-De-Felipe, T.; De Felipe-Boente, I. A methodological approach for urban green areas: A case study in Madrid. *Rev. Chapingo Ser. Cienc. For. Ambient.* **2017**, *23*, 315–328. [CrossRef]
42. Ondoño, S.; Martínez-Sánchez, J.; Moreno, J. Evaluating the growth of several Mediterranean endemic species in artificial substrates: Are these species suitable for their future use in green roofs? *Ecol. Eng.* **2015**, *81*, 405–417. [CrossRef]
43. Ondoño, S.; Martínez-Sánchez, J.; Moreno, J. The inorganic component of green roof substrates impacts the growth of Mediterranean plant species as well as the C and N sequestration potential. *Ecol. Indic.* **2016**, *61*, 739–752. [CrossRef]
44. Ondoño, S.; Martínez-Sánchez, J.; Moreno, J. The composition and depth of green roof substrates affect the growth of Silene vulgaris and Lagurus ovatus species and the C and N sequestration under two irrigation conditions. *J. Environ. Manag.* **2016**, *166*, 330–340. [CrossRef] [PubMed]
45. Vestrella, A.; Biel, C.; Savè, R.; Bartoli, F. Mediterranean Green Roof Simulation in Caldes de Montbui (Barcelona): Thermal and Hydrological Performance Test of *Frankenia laevis* L., Dymondia margaretae Compton and Iris lutescens Lam. *Appl. Sci.* **2018**, *8*, 2497. [CrossRef]
46. Azeñas, V.; Janner, I.; Medrano, H.; Gulías, J. Performance evaluation of five Mediterranean species to optimize ecosystem services of green roofs under water-limited conditions. *J. Environ. Manag.* **2018**, *212*, 236–247. [CrossRef]
47. Azeñas, V.; Janner, I.; Medrano, H.; Gulías, J. Evaluating the establishment performance of six native perennial Mediterranean species for use in extensive green roofs under water-limiting conditions. *Urban For. Urban Green.* **2019**, *41*, 158–169. [CrossRef]
48. Pérez, G.; Chocarro, C.; Juárez, A.; Coma, J. Evaluation of the development of five Sedum species on extensive green roofs in a continental Mediterranean climate. *Urban For. Urban Green.* **2020**, *48*, 126566. [CrossRef]
49. Rincón, L.; Coma, J.; Pérez, G.; Castell, A.; Boer, D.; Cabeza, L.F. Environmental performance of recycled rubber as drainage layer in extensive green roofs. A comparative Life Cycle Assessment. *Build. Environ.* **2014**, *74*, 22–30. [CrossRef]
50. Pérez, G.; Coma, J.; Solé, C.; Castell, A.; Cabeza, L.F. Green roofs as passive system for energy savings when using rubber crumbs as drainage layer. *Energy Procedia* **2012**, *30*, 452–460. [CrossRef]
51. Pérez, G.; Vila, A.; Rincón, L.; Solé, C.; Cabeza, L.F. Use of rubber crumbs as drainage layer in green roofs as potential energy improvement material. *Appl. Energy* **2012**, *97*, 347–354. [CrossRef]
52. Coma, J.; Pérez, G.; Castell, A.; Solé, C.; Cabeza, L.F. Green roofs as passive system for energy savings in buildings during the cooling period: Use of rubber crumbs as drainage layer. *Energy Effic.* **2014**, *7*, 841–849. [CrossRef]

53. Chemisana, D.; Lamnatou, C. Photovoltaic-green roofs: An experimental evaluation of system performance. *Appl. Energy* **2014**, *119*, 246–256. [CrossRef]
54. Lamnatou, C.; Chemisana, D. A critical analysis of factors affecting photovoltaic-green roof performance. *Renew. Sustain. Energy Rev.* **2015**, *43*, 264–280. [CrossRef]
55. Bevilacqua, P.; Coma, J.; Pérez, G.; Chocarro, C.; Juárez-Escario, A.; Solé, C.; De Simone, M.; Cabeza, L.F. Plant cover and floristic composition effect on thermal behaviour of extensive green roofs. *Build. Environ.* **2015**, *92*, 305–316. [CrossRef]
56. Pérez, G.; Vila, A.; Solé, C.; Coma, J.; Castell, A.; Cabeza, L.F. The thermal behaviour of extensive green roofs under low plant coverage conditions. *Energy Effic.* **2015**, *8*, 881–894. [CrossRef]
57. Alcazar, S.S.; Olivieri, F.; Neila, J. Green roofs: Experimental and analytical study of its potential for urban microclimate regulation in Mediterranean–continental climates. *Urban Clim.* **2016**, *17*, 304–317. [CrossRef]
58. Coma, J.; Pérez, G.; Solé, C.; Castell, A.; Cabeza, L.F. Thermal assessment of extensive green roofs as passive tool for energy savings in buildings. *Renew. Energy* **2016**, *85*, 1106–1115. [CrossRef]
59. Serrano, S.; De Gracia, A.; Pérez, G.; Cabeza, L.F. Sustainable earth-based vs. conventional construction systems in the Mediterranean climate: Experimental analysis of thermal performance. *IOP Conf. Series Mater. Sci. Eng.* **2017**, *251*, 12007. [CrossRef]
60. Azeñas, V.; Cuxart, J.; Picos, R.; Medrano, H.; Simó, G.; López-Grifol, A.; Gulías, J. Thermal regulation capacity of a green roof system in the mediterranean region: The effects of vegetation and irrigation level. *Energy Build.* **2018**, *164*, 226–238. [CrossRef]
61. Porcaro, M.; de Adana, M.R.; Comino, F.; Peña, A.; Martín-Consuegra, E.; Vanwalleghem, T. Long term experimental analysis of thermal performance of extensive green roofs with different substrates in Mediterranean climate. *Energy Build.* **2019**, *197*, 18–33. [CrossRef]
62. Peñalvo-López, E.; Cárcel-Carrasco, J.; Alfonso-Solar, D.; Valencia-Salazar, I.; Hurtado-Pérez, E. Study of the Improvement on Energy Efficiency for a Building in the Mediterranean Area by the Installation of a Green Roof System. *Energies* **2020**, *13*, 1246. [CrossRef]
63. Vila, A.; Pérez, G.; Sole, C.; Fernández, A.; Cabeza, L.F. Use of rubber crumbs as drainage layer in experimental green roofs. *Build. Environ.* **2012**, *48*, 101–106. [CrossRef]
64. Andrés-Doménech, I.; Perales-Momparler, S.; Morales-Torres, A.; Escuder-Bueno, I. Hydrological Performance of Green Roofs at Building and City Scales under Mediterranean Conditions. *Sustainability* **2018**, *10*, 3105. [CrossRef]
65. López-Uceda, A.; Galvín, A.P.; Ayuso, J.; Jiménez, J.R.; Vanwalleghem, T.; Peña, A. Risk assessment by percolation leaching tests of extensive green roofs with fine fraction of mixed recycled aggregates from construction and demolition waste. *Environ. Sci. Pollut. Res.* **2018**, *25*, 36024–36034. [CrossRef] [PubMed]
66. Cañero, R.F.; Emilsson, T.; Fernandez-Barba, C.; Machuca, M.; Ángel, H. Green roof systems: A study of public attitudes and preferences in southern Spain. *J. Environ. Manag.* **2013**, *128*, 106–115. [CrossRef]
67. Langemeyer, J.; Wedgwood, D.; McPhearson, T.; Baró, F.; Madsen, A.L.; Barton, D.N. Creating urban green infrastructure where it is needed—A spatial ecosystem service-based decision analysis of green roofs in Barcelona. *Sci. Total. Environ.* **2020**, *707*, 135487. [CrossRef]
68. Tziampou, N.; Coupe, S.J.; Sañudo-Fontaneda, L.A.; Newman, A.P.; Castro-Fresno, D. Fluid transport within permeable pavement systems: A review of evaporation processes, moisture loss measurement and the current state of knowledge. *Constr. Build. Mater.* **2020**, *243*, 118179. [CrossRef]
69. Gupta, A.; Rodriguez-Hernandez, J.; Castro-Fresno, D. Incorporation of Additives and Fibers in Porous Asphalt Mixtures: A Review. *Materials* **2019**, *12*, 3156. [CrossRef]
70. Del-Castillo-García, G.; Borinaga-Treviño, R.; Sañudo-Fontaneda, L.A.; Pascual-Muñoz, P. Influence of pervious pavement systems on heat dissipation from a horizontal geothermal system. *Eur. J. Environ. Civ. Eng.* **2013**, *17*, 956–967. [CrossRef]
71. Rodriguez-Hernandez, J.; Valeri, V.C.A.; Ascorbe-Salcedo, A.; Castro-Fresno, D. Laboratory Study on the Stormwater Retention and Runoff Attenuation Capacity of Four Permeable Pavements. *J. Environ. Eng.* **2016**, *142*, 04015068. [CrossRef]
72. Andrés-Valeri, V.C.; Marchioni, M.; Sañudo-Fontaneda, L.A.; Giustozzi, F.; Becciu, G. Laboratory Assessment of the Infiltration Capacity Reduction in Clogged Porous Mixture Surfaces. *Sustainability* **2016**, *8*, 751. [CrossRef]
73. Andres-Valeri, V.C.; Juli-Gandara, L.; Jato-Espino, D.; Rodriguez-Hernandez, J. Characterization of the Infiltration Capacity of Porous Concrete Pavements with Low Constant Head Permeability Tests. *Water* **2018**, *10*, 480. [CrossRef]
74. Madrazo-Uribeetxebarria, E.; Garmendia-Antín, M.; Almandoz-Berrondo, J.; Andrés-Doménech, I. Hydraulic performance of permeable asphalt and picp in swmm, validated by laboratory data. *Sustain. City XIII* **2019**, *238*, 569–579. [CrossRef]
75. Brugin, M.; Marchioni, M.; Becciu, G.; Giustozzi, F.; Toraldo, E.; Andrés-Valeri, V.C. Clogging potential evaluation of porous mixture surfaces used in permeable pavement systems. *Eur. J. Environ. Civ. Eng.* **2017**, *24*, 620–630. [CrossRef]
76. Hernández-Crespo, C.; Fernández-Gonzalvo, M.; Martín, M.; Andrés-Doménech, I. Influence of rainfall intensity and pollution build-up levels on water quality and quantity response of permeable pavements. *Sci. Total. Environ.* **2019**, *684*, 303–313. [CrossRef]
77. López, J.; Echeverría, J.; Martín, I.S.; Delgado, O. Dynamic testing in columns for soil heavy metal removal for a car park SUDS. *Sci. Total. Environ.* **2020**, *738*, 140229. [CrossRef]
78. Fathollahi, A.; Coupe, S.J.; El-Sheikh, A.H.; Sañudo-Fontaneda, L.A. The biosorption of mercury by permeable pavement biofilms in stormwater attenuation. *Sci. Total. Environ.* **2020**, *741*, 140411. [CrossRef]

79. González-Angullo, N.; Castro, D.; Rodriguez-Hernandez, J.; Davies, J. Runoff infiltration to permeable paving in clogged conditions. *Urban Water J.* **2008**, *5*, 117–124. [CrossRef]
80. Rodriguez-Hernandez, J.; Castro-Fresno, D.; Fernández-Barrera, A.H.; Vega-Zamanillo, Á. Characterization of Infiltration Capacity of Permeable Pavements with Porous Asphalt Surface Using Cantabrian Fixed Infiltrometer. *J. Hydrol. Eng.* **2012**, *17*, 597–603. [CrossRef]
81. Sañudo-Fontaneda, L.A.; Rodriguez-Hernandez, J.; Vega-Zamanillo, A.; Castro-Fresno, D. Laboratory analysis of the infiltration capacity of interlocking concrete block pavements in car parks. *Water Sci. Technol.* **2013**, *67*, 675–681. [CrossRef]
82. Nnadi, E.; Coupe, S.; Sañudo-Fontaneda, L.A.; Rodriguez-Hernandez, J. An evaluation of enhanced geotextile layer in permeable pavement to improve stormwater infiltration and attenuation. *Int. J. Pavement Eng.* **2014**, *15*, 925–932. [CrossRef]
83. Sañudo-Fontaneda, L.A.; Rodriguez-Hernandez, J.; Calzada-Perez, M.A.; Castro-Fresno, D. Infiltration Behaviour of Polymer-Modified Porous Concrete and Porous Asphalt Surfaces used in SuDS Techniques. *CLEAN Soil Air Water* **2013**, *42*, 139–145. [CrossRef]
84. Bernat-Maso, E.; Gil, L.; Roca, P.; Sarrablo, V.; Puigvert, F. Mechanical characterisation of Textile Ceramic plates. Testing on elastic foundations. *Eng. Struct.* **2014**, *74*, 193–204. [CrossRef]
85. Rodriguez-Hernandez, J.; Andrés-Valeri, V.C.; Calzada-Pérez, M.A.; Vega-Zamanillo, Á.; Castro-Fresno, D. Study of the Raveling Resistance of Porous Asphalt Pavements Used in Sustainable Drainage Systems Affected by Hydrocarbon Spills. *Sustainability* **2015**, *7*, 16226–16236. [CrossRef]
86. Valeri, V.C.A.; Rodriguez-Torres, J.; Calzada-Perez, M.A.; Rodriguez-Hernandez, J. Exploratory study of porous asphalt mixtures with additions of reclaimed tetra pak material. *Constr. Build. Mater.* **2018**, *160*, 233–239. [CrossRef]
87. Skaf, M.; Pasquini, E.; Revilla-Cuesta, V.; Ortega-López, V. Performance and Durability of Porous Asphalt Mixtures Manufactured Exclusively with Electric Steel Slags. *Materials* **2019**, *12*, 3306. [CrossRef]
88. García-Casuso, C.; Lapeña-Mañero, P.; Blanco-Fernández, E.; Vega-Zamanillo, Á.; Montenegro-Cooper, J.M. Laboratory Assessment of Water Permeability Loss of Geotextiles Due to Their Installation in Pervious Pavements. *Water* **2020**, *12*, 1473. [CrossRef]
89. Elizondo-Martinez, E.-J.; Tataranni, P.; Rodriguez-Hernandez, J.; Castro-Fresno, D. Physical and Mechanical Characterization of Sustainable and Innovative Porous Concrete for Urban Pavements Containing Metakaolin. *Sustainability* **2020**, *12*, 4243. [CrossRef]
90. Jato-Espino, D.; Rodriguez-Hernandez, J.; Valeri, V.C.A.; Ballester-Muñoz, F. A fuzzy stochastic multi-criteria model for the selection of urban pervious pavements. *Expert Syst. Appl.* **2014**, *41*, 6807–6817. [CrossRef]
91. Arbones, E.D.M.; González, E.F.; Peidro, J.M.; García, J.C. LIFE CERSUDS: UNA PROPUESTA PARA ADAPTAR NUESTRAS CIUDADES AL CAMBIO CLIMÁTICO. *Proy. Progreso Arquit.* **2020**, 102–117. [CrossRef]
92. Novo, A.V.; Bayon, J.R.; Castro-Fresno, D.; Rodriguez-Hernandez, J. Monitoring and Evaluation of the Thermal Behavior of Permeable Pavements for Energy Recovery Purposes in an Experimental Parking Lot: Preliminary Results. *J. Energy Eng.* **2013**, *139*, 230–237. [CrossRef]
93. Novo, A.V.; Bayon, J.R.; Castro-Fresno, D.; Rodriguez-Hernandez, J. Temperature Performance of Different Pervious Pavements: Rainwater Harvesting for Energy Recovery Purposes. *Water Resour. Manag.* **2013**, *27*, 5003–5016. [CrossRef]
94. Gomez-Ullate, E.; Bayon, J.R.; Coupe, S.; Castro-Fresno, D. Performance of pervious pavement parking bays storing rainwater in the north of Spain. *Water Sci. Technol.* **2010**, *62*, 615–621. [CrossRef]
95. Gomez-Ullate, E.; Castillo-Lopez, E.; Castro-Fresno, D.; Bayon, J.R. Analysis and Contrast of Different Pervious Pavements for Management of Storm-Water in a Parking Area in Northern Spain. *Water Resour. Manag.* **2010**, *25*, 1525–1535. [CrossRef]
96. Rodríguez-Rojas, M.; Huertas-Fernández, F.; Moreno, B.; Martínez, G.; Grindlay, A. A study of the application of permeable pavements as a sustainable technique for the mitigation of soil sealing in cities: A case study in the south of Spain. *J. Environ. Manag.* **2018**, *205*, 151–162. [CrossRef]
97. Sañudo-Fontaneda, L.A.; Andres-Valeri, V.C.; Costales-Campa, C.; Cabezon-Jimenez, I.; Cadenas-Fernandez, F. The Long-Term Hydrological Performance of Permeable Pavement Systems in Northern Spain: An Approach to the "End-of-Life" Concept. *Water* **2018**, *10*, 497. [CrossRef]
98. Rodríguez-Rojas, M.I.; Huertas-Fernández, F.; Moreno, B.; Martínez, G. Middle-Term Evolution of Efficiency in Permeable Pavements: A Real Case Study in a Mediterranean Climate. *Int. J. Environ. Res. Public Health* **2020**, *17*, 7774. [CrossRef]
99. Gomez-Ullate, E.; Novo, A.V.; Bayon, J.R.; Rodriguez-Hernandez, J.; Castro-Fresno, D. Design and construction of an experimental pervious paved parking area to harvest reusable rainwater. *Water Sci. Technol.* **2011**, *64*, 1942–1950. [CrossRef] [PubMed]
100. Sañudo-Fontaneda, L.A.; Charlesworth, S.; Castro-Fresno, D.; Valeri, V.C.A.; Rodriguez-Hernandez, J. Water quality and quantity assessment of pervious pavements performance in experimental car park areas. *Water Sci. Technol.* **2014**, *69*, 1526–1533. [CrossRef]
101. Angrill, S.; Farreny, R.; Gasol, C.M.; Gabarrell, X.; Viñolas, B.; Josa, A.; Rieradevall, J. Environmental analysis of rainwater harvesting infrastructures in diffuse and compact urban models of Mediterranean climate. *Int. J. Life Cycle Assess.* **2012**, *17*, 25–42. [CrossRef]
102. Morales-Pinzón, T.; Lurueña, R.; Rieradevall, J.; Gasol, C.M.; Gabarrell, X. Financial feasibility and environmental analysis of potential rainwater harvesting systems: A case study in Spain. *Resour. Conserv. Recycl.* **2012**, *69*, 130–140. [CrossRef]
103. Vargas-Parra, M.V.; Villalba, G.; Gabarrell, X.; Durany, X.G. Applying exergy analysis to rainwater harvesting systems to assess resource efficiency. *Resour. Conserv. Recycl.* **2013**, *72*, 50–59. [CrossRef]

104. Vargas-Parra, M.V.; Rovira, M.R.; Gabarrell, X.; Villalba, G. Cost-effective rainwater harvesting system in the Metropolitan Area of Barcelona. *J. Water Supply Res. Technol.* **2014**, *63*, 586–595. [CrossRef]
105. Gabarrell, X.; Morales-Pinzón, T.; Rieradevall, J.; Rovira, M.R.; Villalba, G.; Josa, A.; Martínez-Gasol, Y.C.; Dias, A.C.; Martínez-Aceves, D.X. Plugrisost: A model for design, economic cost and environmental analysis of rainwater harvesting in urban systems. *Water Pract. Technol.* **2014**, *9*, 243–255. [CrossRef]
106. Morales-Pinzón, T.; Rieradevall, J.; Gasol, C.M.; Gabarrell, X. Modelling for economic cost and environmental analysis of rainwater harvesting systems. *J. Clean. Prod.* **2015**, *87*, 613–626. [CrossRef]
107. Angrill, S.; Segura-Castillo, L.; Boix, A.P.; Rieradevall, J.; Gabarrell, X.; Josa, A. Environmental performance of rainwater harvesting strategies in Mediterranean buildings. *Int. J. Life Cycle Assess.* **2016**, *22*, 398–409. [CrossRef]
108. Petit-Boix, A.; Devkota, J.; Phillips, R.; Vargas-Parra, M.V.; Josa, A.; Gabarrell, X.; Rieradevall, J.; Apul, D. Life cycle and hydrologic modeling of rainwater harvesting in urban neighborhoods: Implications of urban form and water demand patterns in the US and Spain. *Sci. Total. Environ.* **2018**, *621*, 434–443. [CrossRef]
109. Vargas-Parra, M.V.; Rovira-Val, M.R.; Gabarrell, X.; Villalba, G. Rainwater harvesting systems reduce detergent use. *Int. J. Life Cycle Assess.* **2019**, *24*, 809–823. [CrossRef]
110. Zubelzu, S.; Rodríguez-Sinobas, L.; Andrés-Domenech, I.; Castillo-Rodríguez, J.; Perales-Momparler, S. Design of water reuse storage facilities in Sustainable Urban Drainage Systems from a volumetric water balance perspective. *Sci. Total. Environ.* **2019**, *663*, 133–143. [CrossRef]
111. Villar-Navascués, R.; Pérez-Morales, A.; Gil-Guirado, S. Assessment of Rainwater Harvesting Potential from Roof Catchments through Clustering Analysis. *Water* **2020**, *12*, 2623. [CrossRef]
112. Cheng, A.L.; Silva, L.M.; Buenano, M.R.; Vega, N.L. Development of an Adaptive Rainwater-Harvesting System for Intelligent Selective Redistribution. In *Proceedings of the 2019 IEEE Fourth Ecuador Technical Chapters Meeting (ETCM)*; Institute of Electrical and Electronics Engineers (IEEE): Guayaquil, Ecuador, 2019; pp. 1–5.
113. Domènech, L.; Saurí, D. A comparative appraisal of the use of rainwater harvesting in single and multi-family buildings of the Metropolitan Area of Barcelona (Spain): Social experience, drinking water savings and economic costs. *J. Clean. Prod.* **2011**, *19*, 598–608. [CrossRef]
114. Farreny, R.; Gabarrell, X.; Rieradevall, J. Cost-efficiency of rainwater harvesting strategies in dense Mediterranean neighbourhoods. *Resour. Conserv. Recycl.* **2011**, *55*, 686–694. [CrossRef]
115. Farreny, R.; Morales-Pinzón, T.; Guisasola, A.; Tayà, C.; Rieradevall, J.; Gabarrell, X. Roof selection for rainwater harvesting: Quantity and quality assessments in Spain. *Water Res.* **2011**, *45*, 3245–3254. [CrossRef]
116. Angrill, S.; Petit-Boix, A.; Morales-Pinzón, T.; Josa, A.; Rieradevall, J.; Gabarrell, X. Urban rainwater runoff quantity and quality—A potential endogenous resource in cities? *J. Environ. Manag.* **2017**, *189*, 14–21. [CrossRef] [PubMed]
117. Garcia, X.; Llausàs, A.; Ribas, A.; Saurí, D. Watering the garden: Preferences for alternative sources in suburban areas of the Mediterranean coast. *Local Environ.* **2014**, *20*, 548–564. [CrossRef]
118. Suleiman, L.; Olofsson, B.; Saurí, D.; Palau-Rof, L. A breakthrough in urban rain-harvesting schemes through planning for urban greening: Case studies from Stockholm and Barcelona. *Urban For. Urban Green.* **2020**, *51*, 126678. [CrossRef]
119. Suleiman, L.; Olofsson, B.; Saurí, D.; Palau-Rof, L.; Soler, N.G.; Papasozomenou, O.; Moss, T. Diverse pathways—common phenomena: Comparing transitions of urban rainwater harvesting systems in Stockholm, Berlin and Barcelona. *J. Environ. Plan. Manag.* **2019**, *63*, 369–388. [CrossRef]
120. Saurí, D.; Garcia, X. Non-conventional resources for the coming drought: The development of rainwater harvesting systems in a Mediterranean suburban area. *Water Int.* **2020**, *45*, 125–141. [CrossRef]
121. Senosiain, J. Urban Regreeneration: Green Urban Infrastructure as a Response to Climate Change Mitigation and Adaptation. *Int. J. Des. Nat. Ecodynamics* **2020**, *15*, 33–38. [CrossRef]
122. Radinja, M.; Comas, J.; Corominas, L.; Atanasova, N. Multi-criteria Evaluation of Sustainable Urban Drainage Systems. *Smart Sustain. Plan. Cities Reg.* **2018**, 269–274. [CrossRef]
123. Locatelli, L.; Guerrero, M.; Russo, B.; Martínez-Gomariz, E.; Sunyer, D.; Martínez, M. Socio-Economic Assessment of Green Infrastructure for Climate Change Adaptation in the Context of Urban Drainage Planning. *Sustainability* **2020**, *12*, 3792. [CrossRef]
124. Jato-Espino, D.; Charlesworth, S.M.; Bayon, J.R.; Warwick, F. Rainfall–Runoff Simulations to Assess the Potential of SuDS for Mitigating Flooding in Highly Urbanized Catchments. *Int. J. Environ. Res. Public Health* **2016**, *13*, 149. [CrossRef]
125. Rodríguez-Sinobas, L.; Zubelzu, S.; Perales-Momparler, S.; Canogar, S. Techniques and criteria for sustainable urban stormwater management. The case study of Valdebebas (Madrid, Spain). *J. Clean. Prod.* **2018**, *172*, 402–416. [CrossRef]
126. Radinja, M.; Comas, J.; Corominas, L.; Atanasova, N. Assessing stormwater control measures using modelling and a multi-criteria approach. *J. Environ. Manag.* **2019**, *243*, 257–268. [CrossRef]
127. Tuomela, C.; Jato-Espino, D.; Sillanpää, N.; Koivusalo, H. Modelling Stormwater Pollutant Reduction with LID Scenarios in SWMM. *Smart Sustain. Plan. Cities Reg.* **2018**, 96–101. [CrossRef]
128. García-Terán, C.; Tejero-Monzón, I.; Gil-Díaz, J.L. Sustainable urban drainage systems: A tool to adapt combined sewer systems to climate change. *Proc. Inst. Civ. Eng. Munic. Eng.* **2019**, *172*, 175–184. [CrossRef]
129. Charlesworth, S.M.; Perales-Momparler, S.; Lashford, C.; Warwick, F. The sustainable management of surface water at the building scale: Preliminary results of case studies in the UK and Spain. *J. Water Supply: Res. Technol.* **2013**, *62*, 534–544. [CrossRef]

130. Perales-Momparler, S.; Hernández-Crespo, C.; Vallés-Morán, F.; Martín, M.; Andrés-Doménech, I.; Álvarez, J.A.; Jefferies, C. SuDS Efficiency during the Start-Up Period under Mediterranean Climatic Conditions. *CLEAN Soil. Air. Water* **2013**, *42*, 178–186. [CrossRef]
131. Perales-Momparler, S.; Andrés-Doménech, I.; Hernández-Crespo, C.; Vallés-Morán, F.; Martín, M.; Escuder-Bueno, I.; Andreu, J. The role of monitoring sustainable drainage systems for promoting transition towards regenerative urban built environments: A case study in the Valencian region, Spain. *J. Clean. Prod.* **2017**, *163*, S113–S124. [CrossRef]
132. Valeri, V.C.A.; Castro-Fresno, D.; Sañudo-Fontaneda, L.A.; Rodriguez-Hernandez, J. Comparative analysis of the outflow water quality of two sustainable linear drainage systems. *Water Sci. Technol.* **2014**, *70*, 1341–1347. [CrossRef]
133. Rodríguez-Rojas, M.I.; Cuevas, M.M.; Martinez, G.; Moreno, B. Planning criteria for Water Sensitive Urban Design. *The Sustainable City IX* **2014**, *1*, 1579–1591. [CrossRef]
134. Rodríguez-Rojas, M.I.; Cuevas-Arrabal, M.M.; Escobar, B.M.; Montes, G.M. El cambio de paradigma de la gestión del drenaje urbano desde la perspectiva del planeamiento. Una propuesta metodológica. *BAGE* **2017**, *2017*, 55–74. [CrossRef]
135. Ramírez-Agudelo, N.; Anento, R.P.; Villares, M.; Roca, E. Nature-Based Solutions for Water Management in Peri-Urban Areas: Barriers and Lessons Learned from Implementation Experiences. *Sustainability* **2020**, *12*, 9799. [CrossRef]
136. Arahuetes, A.; Cantos, J.O. The potential of sustainable urban drainage systems (SuDS) as an adaptive strategy to climate change in the Spanish Mediterranean. *Int. J. Environ. Stud.* **2019**, *76*, 764–779. [CrossRef]
137. Morales-Torres, A.; Escuder-Bueno, I.; Andrés-Doménech, I.; Perales-Momparler, S. Decision Support Tool for energy-efficient, sustainable and integrated urban stormwater management. *Environ. Model. Softw.* **2016**, *84*, 518–528. [CrossRef]
138. Andersson, E.; Langemeyer, J.; Borgström, S.; McPhearson, T.; Haase, D.; Kronenberg, J.; Barton, D.N.; Davis, M.; Naumann, S.; Röschel, L.; et al. Enabling Green and Blue Infrastructure to Improve Contributions to Human Well-Being and Equity in Urban Systems. *BioScience* **2019**, *69*, 566–574. [CrossRef]
139. Velasco, M.; Russo, B.; Monjo, R.; Paradinas, C.; Djordjević, S.; Evans, B.; Martínez-Gomariz, E.; Guerrero-Hidalga, M.; Cardoso, M.; Brito, R.; et al. Increased Urban Resilience to Climate Change—Key Outputs from the RESCCUE Project. *Sustainability* **2020**, *12*, 9881. [CrossRef]
140. Subiza-Pérez, M.; Hauru, K.; Korpela, K.; Haapala, A.; Lehvävirta, S. Perceived Environmental Aesthetic Qualities Scale (PEAQS)—A self-report tool for the evaluation of green-blue spaces. *Urban For. Urban Green.* **2019**, *43*, 126383. [CrossRef]
141. Casal-Campos, A.; Jefferies, C.; Momparler, S.P. Selecting SUDS in the Valencia Region of Spain. *Water Pract. Technol.* **2012**, *7*, 1–9. [CrossRef]
142. Perales-Momparler, S.; Andrés-Doménech, I.; Andreu, J.; Escuder-Bueno, I. A regenerative urban stormwater management methodology: The journey of a Mediterranean city. *J. Clean. Prod.* **2015**, *109*, 174–189. [CrossRef]
143. Sañudo-Fontaneda, L.; Robina, R. Bringing community perceptions into sustainable urban drainage systems: The experience of Extremadura, Spain. *Land Use Policy* **2019**, *89*, 104251. [CrossRef]
144. Carriquiry, A.N.; Sauri, D.; March, H. Community Involvement in the Implementation of Sustainable Urban Drainage Systems (SUDSs): The Case of Bon Pastor, Barcelona. *Sustainability* **2020**, *12*, 510. [CrossRef]
145. Andrés-Doménech, I.; Anta, J.; Perales-Momparler, S.; Rodriguez-Hernandez, J. Sustainable Urban Drainage Systems in Spain: A Diagnosis. *Sustainability* **2021**, *13*, 2791. [CrossRef]
146. Church, S.P. Exploring Green Streets and rain gardens as instances of small scale nature and environmental learning tools. *Landsc. Urban Plan.* **2015**, *134*, 229–240. [CrossRef]
147. McDonough, K.; Moore, T.; Hutchinson, S. Understanding the Relationship between Stormwater Control Measures and Ecosystem Services in an Urban Watershed. *J. Water Resour. Plan. Manag.* **2017**, *143*, 04017008. [CrossRef]
148. Shafique, M.; Kim, R.; Rafiq, M. Green roof benefits, opportunities and challenges—A review. *Renew. Sustain. Energy Rev.* **2018**, *90*, 757–773. [CrossRef]
149. Lovell, S.T.; Johnston, D.M. Creating multifunctional landscapes: How can the field of ecology inform the design of the landscape? *Front. Ecol. Environ.* **2009**, *7*, 212–220. [CrossRef]
150. Ayers, E.M.; Kangas, P. Soil Layer Development and Biota in Bioretention. *Water* **2018**, *10*, 1587. [CrossRef]
151. Payne, E.G.; Pham, T.; Deletic, A.; Hatt, B.E.; Cook, P.; Fletcher, T.D. Which species? A decision-support tool to guide plant selection in stormwater biofilters. *Adv. Water Resour.* **2018**, *113*, 86–99. [CrossRef]
152. Parés, M.; Rivero, M.; Rull, C. BCN: Pla del verd i de la biodiversitat de Barcelona 2020. Ajuntament de Barcelona i Hàbitat Urbà, Barcelona. 2013, p. 113. Available online: https://ajuntament.barcelona.cat/ecologiaurbana/sites/default/files/PladelverdidelabiodiversitatdeBarcelona2020.pdf (accessed on 15 February 2021).
153. Zhou, Q. A Review of Sustainable Urban Drainage Systems Considering the Climate Change and Urbanization Impacts. *Water* **2014**, *6*, 976–992. [CrossRef]

sustainability

MDPI

Article

Characterisation of Impact Funds and Their Potential in the Context of the 2030 Agenda

Juan C. Santamarta [1,*], Mª Dolores Storch de Gracia [2,3], Mª Ángeles Huerta Carrascosa [2,3], Margarita Martínez-Núñez [3], Celia de las Heras García [3] and Noelia Cruz-Pérez [1]

1 Departamento de Ingeniería Agraria, Náutica, Civil y Marítima, Universidad de La Laguna (ULL), 38200 Tenerife, Spain; ncruzper@ull.edu.es

2 Grupo de Investigación Organizaciones Sostenibles (GIOS), Universidad Politécnica de Madrid, 28006 Madrid, Spain; lola.storch@upm.es (M.D.S.d.G.); ma.huerta@upm.es (M.Á.H.C.)

3 Department of Organizational Engineering, Business Administration and Statistics, Escuela Técnica Superior de Ingenieros Industriales, Universidad Politécnica de Madrid, 28006 Madrid, Spain; margarita.martinez@upm.es (M.M.-N.); cee.delash@gmail.com (C.d.l.H.G.)

* Correspondence: jcsanta@ull.es

Abstract: The European Union has incorporated impact investment through two action plans: the Social Entrepreneurship Initiative and the Investment Plan for Europe. These financing tools seek to fund economic growth and promote job creation. Among the different measures carried out, the regulatory framework for impact investment funds stands out, under which the denomination, European Social Entrepreneurship Fund, is established to designate investment funds focused on social enterprises with the objective of generating a positive impact. It is possible to affirm that the creation of a solid impact intermediation infrastructure, by connecting both sides of supply and demand, is a critical aspect for the development and effective functioning of the impact market. Special importance is given to impact funds capable of attracting private capital. In order to categorise the different impact funds according to the most relevant aspects, a proposal form for the characterisation of impact funds has been drawn up and has been applied to a particular case. The presentation of Creas will allow for contextualising the practices that impact funds carry out and facilitate the general understanding of the article through a specific example that is considered successful in Spain.

Keywords: sustainable development; financing; impact fund; 2030 Agenda

Citation: Santamarta, J.C.; Storch de Gracia, M.D.; Carrascosa, M.Á.H.; Martínez-Núñez, M.; García, C.d.l.H.; Cruz-Pérez, N. Characterisation of Impact Funds and Their Potential in the Context of the 2030 Agenda. *Sustainability* **2021**, *13*, 6476. https://doi.org/10.3390/su13116476

Academic Editor: Fernando Almeida

Received: 21 April 2021
Accepted: 2 June 2021
Published: 7 June 2021

1. Introduction

Impact investment is any investment that, in addition to obtaining a financial return, is made with the intention of generating a positive quantifiable social or environmental impact [1]. Investment funds which allocate their capital solely to impact investments, creating social and environmental value, are known mainly as impact funds, but also as responsible funds, philanthropic investment funds or social impact investment funds [2]. They are constituted as venture capital funds, institutions of alternative collective investment (due to their high specificity) and of a closed type. Broadly speaking, they are characterised by investing in unlisted companies which are in the creation or development phase, with temporary and minority involvement [3].

The implementation of Agenda 2030 requires a mobilisation of resources that will be difficult to achieve exclusively through donations. That is why reimbursable instruments are a fundamental tool in financing development [4]. As the United Nations Development Programme rightly states, sustainable and responsible investments represent sources of capital with high potential for achieving the Sustainable Development Goals. In 2016, $18.2 trillion was invested in this asset class. In addition, the green bond market for sustainable businesses is growing, and in 2018, it increased by 78% to $155.5 billion [5]. This line of new opportunities to promote development is where we find impact investing, which has been gaining popularity over the last decade.

The term "sustainable development" is first mentioned in the report "Our common future", published by the United Nations [6] in 1987. It defines the concept as "meeting the needs of the present generation without compromising the ability of future generations to meet their own needs". Sustainable development is based on balanced progress between its three dimensions: economic, social and environmental [7]. The origin of the term is linked to the concern that began to emerge in a widespread way during the 1980s, when progress in development, both economic and social, began to be linked as being solely responsible for serious environmental consequences at the global level. While it is true that unrest was already present in society at that time, the level of alarm was not so high. During those years, the first assessments began to be made which established that, if the trend continued in the same way, the future consequences would be increasingly critical, reaching the point of putting the survival capacity of the human species at risk [8].

Since then, sustainable development has gained increasing importance over the years, today becoming the guiding principle for long-term global development. It is indisputable that it has led to a considerable transformation internationally; however, the world population is still far from reaching a perfect scenario, understood as the balance between the economic, social and environmental dimensions, which would be necessary to guarantee the long-term survival of both the human species and all the other living beings that inhabit the planet today [8]. The way of life that today's society leads is increasingly moving away from this vital final objective and is reaching a critical state in the three dimensions that make up sustainable development [9]. If we continue at the same pace, one planet is not going to be enough to sustain life as we know it today. It is estimated that by 2050, more than nine billion people will inhabit the earth. If the desire for prosperity is associated with a consumption similar to the current one, and proportional to population growth, the consequences that this entails will lead our planet to destruction [10].

The United Nations proposed the Millennium Development Goals (MDG) at the beginning of the century, in September 2000, after a whole decade of conferences, through the Millennium Declaration. Through it, the 192 member countries established the commitment of a global alliance to reduce extreme poverty through eight objectives, with the year 2015 being the deadline for their fulfilment. The MDG were a historic event in terms of global resource mobilisation [11]. However, progress was not sufficient. The incomplete results required a more ambitious project, based on the errors and successes of the Millennium Agenda, with which to continue the path towards a sustainable and egalitarian future. This gave rise to the Sustainable Development Goals [12].

The global contextual framework under which the Millennium Development Goals were developed is very distinct from the reality at the end of their period of validity. Therefore, a new agenda was needed, the demands of which do not correspond to those of the MDG. For this reason, the agenda to be elaborated had to be: a comprehensive agenda, understanding the disparate global needs; integral, including the three dimensions of development, economic, social and environmental; participatory, where people from different countries, living in diverse situations, present their heterogeneous realities; and universal, since the benefits of development are for all, the responsibilities should also be for all, but with a distribution of obligations proportional to the resources and competencies of each actor, where special importance is given to global public goods, promoting the spirit of a shared mission where, without the commitment of the entire community, the objectives are unattainable [13]. The aim was to promote a spirit of a shared mission, in which objectives cannot be achieved without the commitment of the entire community; to promote a multilevel approach based on the principle of subsidiarity; and, last but not least, to be transformative and creative, since a change in the method used to address problems that have not yet been resolved is essential [14]. The result of these efforts has been Agenda 2030 [15], which is a plan of action embodied in the 17 Sustainable Development Goals.

As for the funding required, it is difficult to accurately estimate the capital and resources that are necessary for the successful completion of Agenda 2030. However, all experts agree that the need for economic mobilisation is truly high [4]. The capacities and characteristics of the different financial sources are very diverse, and therefore it should be emphasised that there is no single formula to the key for success for financing such a broad agenda as Agenda 2030. The aim is to set out the different financial alternatives available, choosing the one that best suits each case [16]. The specific features of a financing mechanism may be valid for promoting certain initiatives, but not others. They may also be valid for application in one country, but not in another, for financing the same activity [4]. However, it should be stressed that it would be wrong to think that one source of financing can replace another because all are necessary.

One of these forms of funding is impact funds. According to one of the world's leading nonprofit organisations in impact investing, the Global Impact Investing Network (GIIN), impact investing can be defined as any investment made with the intention of generating a positive and measurable social or environmental impact, while obtaining a financial return. This new form of investment has been gaining popularity in recent years around the world and towards all types of assets, both in emerging and developed markets. The European Union, in its Regulation 346/2013 [17], calls for social entrepreneurship, elaborating on what it had previously outlined in a report that put the focus on social enterprises by placing them at the centre of the social economy and innovation [18].

The overall objective of this paper is to present a document that can serve as a reference guide on the most relevant aspects related to impact funds. The aim is to address the following two research questions:

- What common elements allow us to characterise and compare different impact funds in such a way that we can assess their suitability as an SDG financing tool?
- Are impact funds not only an objective in themselves, but also an appropriate financing tool to achieve the Sustainable Development Goals under the 2030 Agenda?

To this end, this research has been structured as follows:

(a) Considering the 4 principles of The Global Impact Investing Network, a proposal for the classification of impact funds has been developed.
(b) This classification considers aspects such as: size of investment, investor profile, target area of impact and resulting impact.
(c) In the Results section, the Spanish impact funds have been selected, and the proposed classification has been applied to obtain data on how they operate in terms of investment, region, etc.
(d) On the other hand, a subsection focused on the Creas Impacto case study has been added to the results. In addition, the funds have been related to the sustainable development goals of the 2030 Agenda.
(e) Finally, the discussion and conclusions obtained are added.

2. Materials and Methods

The Global Impact Investing Network (GIIN) has defined the four principles that impact investing should have. These four principles are.

2.1. *Intentionality of the Investment in Its Positive Social and Environmental Contribution Together with a Financial Return*

This first principal encompasses two of the key elements in the definition of impact investment, adding transparency as a primary property when setting the target financial market and the impact to be addressed.

2.2. *Use of Evidence and Impact Data When Designing the Investment*

In order to drive and increase the contribution towards creating a positive impact, evidence-based data, both qualitative and quantitative, must be rigorously included in the early stages of investment. In this way, it can be justified by empirical facts that the social or environmental needs being addressed are real. This principle encourages the use of evidence that is credible and accessible to the investor to: define the investment strategies that are essential to address the needs identified; set the impact indicators, as well as the results, whether numerical or qualitative, that are expected to be achieved; and increase the rigour of practices by improving the analytical capacity of impact.

2.3. Management of Impact Development

Throughout the process of implementation and development of the investment, obstacles may arise that divert investors from their final objectives. Therefore, the control and monitoring of performance data, throughout the entire life cycle of the investment, takes on a key role if the social and/or environmental achievements outlined above are to be reached. In this way, identifying possible risks and establishing the corresponding mitigation plans to alleviate the negative consequences, as well as creating an iterative process where the feedback collected is taken into consideration, guarantees the successful achievement of the proposed milestones. It should be noted that informing and sharing data, both with investors and with the entities in which the investment is being made, and comparing the data with previous stages of the life cycle are very useful in order to study the impact and the financial trends of the investment.

2.4. Contribution to the Growth of Impact Investment

Under this last statement, the aim is to expand and promote the effective execution of impact investment through: transparency in the development of impact practices; commitment to share the approaches and standards used to describe the objectives, strategies and performance of impact practices; consideration of the performance and quality of the impact management of other investors in one's own; and, finally, the sharing of both positive and negative learning, evidence and data collected.

To carry out this study of impact funds, a bibliographic and documentary review has been conducted by compiling a wide variety of reports published over the past 10 years in relation to projects' central themes and impact investment, as well as the Sustainable Development Goals and their funding needs. Meetings have also been held with experts in order to focus the projects in the best possible way and to complete and verify the information collected in the project, with the help of experts in both the field of sustainable development and impact investment. All this was complemented with a documentary analysis of the various impact funds operating in Spain to complete the classification proposal established through public data disclosed and other information published on websites, as well as data from other studies carried out by third parties.

With all this information on impact investment, a characterisation sheet of impact funds has been made (Appendix A). The objective is to categorise the different impact funds according to the aspects that have been considered most relevant in this project.

The classification proposal allows the different funds belonging to the impact ecosystem to be categorised according to different characteristics grouped into six main categories. This classification is useful both for analysing and comparing the impact fund ecosystem, and for use by organisations with impact objectives seeking funding according to their needs and the availability of funds.

The first of the categories refers to the "Fund data" in order to identify: the name of the fund, managing entity, city where the headquarters are established and the year in which the first closing took place. Funds with predetermined impact objectives may make all or part of their investments in organisations that generate a positive impact. Knowing the degree of commitment to these practices, in terms of the percentage of capital managed, is extremely important. Finally, the EuSEF (European Social Entrepreneurship Fund) "label" ensures that a fund meets the criteria set by the European Commission to be considered an impact fund. These criteria include aspects relating to the composition of the portfolio, which must be made up of at least 70% impact, the financial instruments available, the recipients of the investments and the eligible categories of investment. These are included in Regulation (EU) No. 346/2013 of the European Parliament and of the Council of 17 April 2013 on European Social Entrepreneurship Funds, which unifies and facilitates the identification of these funds throughout the European Union.

Moving on to the second of the categories, we find "Size of investment". Although it is true that there is no consensus on how to categorise the size of a fund according to the assets under management that it holds (small, medium, large), experts agree that, depending on the total capital, the fund will suffer from some disadvantages or others. A smaller fund will have higher fixed costs per unit and a smaller portfolio, while a larger fund will lose flexibility [19]. It is estimated that the minimum portfolio size for a mutual fund to be sustainable should be at least EUR 20 million.

The third category has the characteristics that define the "Investor profile", which will determine the type of investment products most appropriate according to investors' motivations in terms of their financial return priorities or social and/or environmental impact; the returns funds can provide (performance measured in percentage); the associated risk aversion, on a scale of 1 to 7 as recommended by the CNMV for the preparation of the key investor information document (1 being the least risk averse and 7 the most); and the investment time horizon, which is usually divided indicatively into three categories: short, for periods of less than 1 year; medium, between 1 and 5 years; and long for investments lasting more than 5 years [20].

The fourth category, "Impact target area", attempts to analyse two fundamental aspects: on the one hand, whether there is a main purpose of the impact (environmental, social or both), and, on the other hand, whether the funds are focused on a single sector, or whether, on the contrary, they have a wider range of investments covering both social and environmental projects of different categories. In addition, in order to relate impact investment to Agenda 2030, it is interesting to identify whether the funds themselves identify with their contribution to the SDGs through specific objectives and targets [21].

"Resulting impact" refers to the geographical place where the change is sought to be made, as well as the scale at which it takes place. This last aspect plays a relevant role when analysing whether impact investment, in this case through impact funds, can be an optimal source of funding for the Agenda, which needs a systemic transformation applicable on a large scale. Equally relevant is the study of the indices used to measure impacts in order to identify some of the impacts proposed as being more common, or if, on the contrary, there is no consensus between funds.

Additionally, the last category, "Other features", includes the state of development of the target company (depending on this, the company will have some financial needs or others, which will determine the optimal financial instruments to be used by the fund and the return on investment). In addition, the analysis of the origin of the fund's capital is considered relevant, as are its main investors.

3. Results

To validate the classification proposal, it has been decided to apply it to the Spanish impact funds listed in Table 1. The table also presents the bibliography used for the documentary analysis of each of them, which has been complemented by Urriolagoitia et al. [22].

Table 2 shows the application of the classification sheet through which all the available information of the impact funds is collected.

The results of the four aspects analysed (Size of the investment, Investor profile, Target area of impact, Impact generated) are shown below.

Size of the investment: The size of the investment portfolios of Spanish impact funds, referring to total capital, is very diverse. The smallest fund is Creas Inicia with only EUR 125,000, and the largest is Global Financial Inclusion with a current total of EUR 32.3 million and a target size of EUR 50 million (Figure 1).

Table 1. Spanish impact funds (author prepared based on Urriolagoitia & Hehenberger, 2019).

Impact Fund	Bibliography
GAWA Microfinance Fund	GAWA Capital, 2019
Global Financial Inclusion Fund	GAWA Capital, 2019
Magallanes Impacto FIL	Magallanes Value, 2019
Q-Impact I	Qualitas Equity, 2019
Fondo de Emprendimiento e Innovación Social	Seed Capital Bizkaia, 2019
Creas Impacto	Creas, 2019
Creas Desarrolla	Creas, 2019
Creas Inicia	Creas, 2019
Impact Equity S.L.	Ship2B, 2019
Equity4Good S.L.	Ship2B, 2019
Next Venture Capital	-
Rezinkers	-

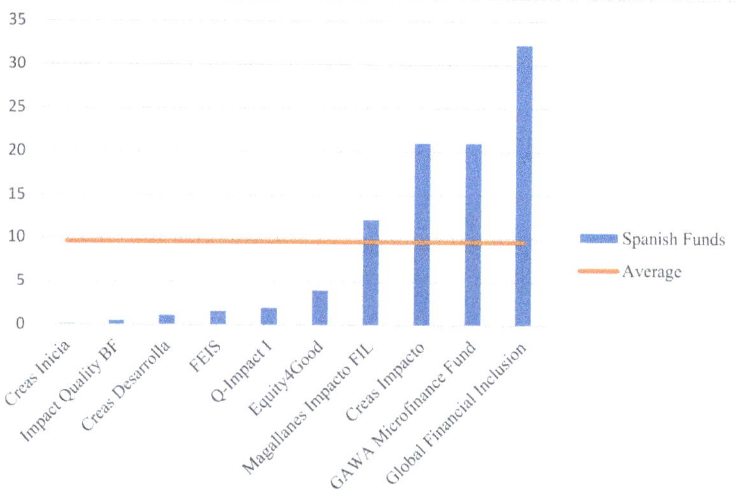

Figure 1. Size of Spanish impact portfolios in millions of euros.

Investor profile: Except for Creas Inicia, which clearly states that it prioritises social and/or environmental impact over economic return, the rest of the funds do not refer to this aspect. However, it can be deduced from their narratives that they are somewhere between profit and impact. As far as the return on investments is concerned, the data from the different funds have been obtained from the Foro Impacto report, since, except for Creas Inicia, the other funds do not provide information on this subject. Expected returns range from 2% (Creas Desarrolla) to 15% (Fondo de Emprendimiento e Innovación Social), apart from the two funds managed by the Ship2B Foundation, whose expectations have a multiplier factor of 1.5, i.e., a return of 50%. The average return, excluding Impact Equity I and Equity4Good, whose values are not considered representative of the rest of the Spanish landscape, is 6.34%. As regards the time horizon of the investments, the data collected show that they are in an interval of between 3 and 7 years, opting for the medium and long term, without going so far as to make patient capital investments, which require longer than 10 years.

Table 2. Summary of the application of the sheet to different funds.

			GAWA Microfinance Fund	Creas Desarrolla	Creas Inicia	Global Financial Inclusion	Fondo de Emprendimiento e Innovación Social	Impact Equity BF	Creas Impacto	Magallanes Impacto FIL	Equity4Good	Q-Impact I
Fund Data	1	Fund Name	GAWA Microfinance Fund	Creas Desarrolla	Creas Inicia	Global Financial Inclusion	Fondo de Emprendimiento e Innovación Social	Impact Equity BF	Creas Impacto	Magallanes Impacto FIL	Equity4Good	Q-Impact I
	2	Management Company Name	GAWA Capital (Madrid)	Fundación Creas Valor Social	Fundación Creas Valor Social	GAWA Capital (Madrid)	Seed Capital Bizkaia	Fundación Ship2B	Self-managed	Magallanes Value Investors	Fundación Ship2B	Qualitas Equity
	3	Headquearters city	Luxemburgo	Zaragoza/Madrid	Zaragoza/Madrid	Luxemburgo	Bilbao	Barcelona	Madrid	Madrid	Barcelona	Madrid
	4	Year of First Closure	2010	2012	2013	2014	2014	2016	2018	2018	2018	2019
	5	Portfolio formed only by impact investment projects	100%	100%	100%	100%	100%	100%	100%	100%	100%	Minimum 70%
	6	Owns FESE Label	No	No	No	No	No	No	Yes	No	No	Yes
Investment Size	7	Size of the fund as a function of its total capital	21 million	750 thousand–1.5 million	125 thousand	32.3 million	1.6 million	0.5 million	221 million 30 million (objective)	12.2 million	4 million	2 million 30 million (objective)
	8	Size of each investment	1–3.5 million	25–250 thousand	10–25 thousand	1–3.5 million	Maximum 600 thousand	50–100 thousand	500 thousand–3 million	n/a	40–400 thousand	500 thousand–3 million
	9	Financing instruments	Capital 35% and Debt 65%	Capital and Debt	Capital and Debt	Capital 35% and Debt 65%	Capital and Debt	Venture Philantropy	Capital, debt and quasi-capital	Unquoted debt	Venture Philantropy	Capital and Debt
Investor Profile	10	Investor Motivation	n/a	Medium	Prioritizes impact	n/a	n/a	Medium	Medium	n/a	Medium	Medium
	11	Profitability	6.40%	2.00%	No	7.0–8.0%	10.0–15.0%	50.0% (x1.5)	7.00%	2.0–4.0%	50.0% (x1.5)	6.00%
	12	Risk aversion (scale from 1 to 7)	n/a	n/a	n/a	n/a	n/a	n/a	n/a	Big risk	n/a	6
	13	Investment timeline	3–7 years	n/a	n/a	3–7 years	Long	n/a	3–7 years	5 years	n/a	Long term
Target Area of Impact	14	Impact with a main purpose	Social	Both	Both	Social	Both	Both	Both	Social	Both	Both
	15	Impact target area	Unique	Multiple	Multiple	Unique	Mutiple	Mutiple	Mutiple	Unique	Mutiple	Mutiple
	16	Which are they?	Financial inclusion	n/a	n/a	Financial inclusion	Renewable energies, organic farming, bioconstruction, people at risk of social exclusion, development cooperation, fair trade	Environment, climate, health and social	Health and welfare, environmental sustainability, education, social innovation	Financial inclusion	Environment, climate, health and social	Sustainability, education, social inclusion
	17	Identified as SGSs	No	No	No	No	No	No	No	No	No	No
Resulting Impact	18	Impact scale	n/a	n/a	n/a	n/a	Local: Bizkaia	n/a	n/a	n/a	n/a	n/a
	19	Place where the impact is generated	Latin America, Asia, Sub-Saharan Africa	Spain	Spain	Latin America, Asia, Sub-Saharan Africa	Spain	Spain	Europe and Spain	Developing countries	Spain	Europe and Spain
	20	Impact measurement	IRIS (GIIN), The Social Performance Task Force, ODS	Theory of change	Theory of change	IRIS (GIIN), The Social Performance Task Force, ODS	n/a	Theory of change	EVPA and Theory of change	IRIS (GIIN), ODS	Theory of change	IRIS (GIIN)
Other features	21	State of development of the target company	Growth	Start	Seed	Growth	All	Seed and start	Growth	Growth	Seed and start	Growth
	22	Origin of the fund's capital	Private	Private	Private	Private and Public	Public	Private	Public	Private	Private and Public	Private
	23	The fund's main investors	Family offices, development institutions, individuals	Family offices and private investors	Donor partners	Family offices, development institutions, individuals, AECID	Provincial Council of Bizkaia	36 investors	FEI, AXIS-ICO	n/a	European Investment Fund, Impact Equity	Investors of the Qualitas Equity funds

Target area of impact: Apart from the two funds managed by GAWA Capital and the Magallanes Impacto FIL fund, which only invest in social impact in a single target area, financial inclusion, the remaining Spanish funds do not prioritise social over environmental impact, and both are at the same level of importance. All of them have multiple target areas, focusing on health and wellbeing, environmental sustainability and climate, education, social innovation, social inclusion, renewable energies, ecological agriculture, bioconstruction and fair trade. However, none of these focal areas are identified by the funds within the framework of the Sustainable Development Goals, or at least that is what they show on their respective websites. Despite this, the issues addressed are closely related to those proposed by the United Nations.

Resulting impact: Only the Social Entrepreneurship and Innovation Fund establishes the scale at which they seek to generate impact, which is at the local level in the province of Biscay. The rest of the funds do not specify this aspect. However, all of them disclose the geographical dimension in which they wish to generate it. Apart from GAWA Microfinance Fund, Global Financial Inclusion and Magallanes Impacto FIL, which establish developing countries belonging to Latin America, Sub-Saharan Africa and Asia as their geographical target, the rest of the funds seek to generate impact within the borders of Spain.

3.1. Study Case: Creas Impacto

Among the different impact funds in Spain, the case of Creas Impacto should be highlighted. The Creas fund has been chosen, considered the pioneer in impact investing in Spain, as it is the first to have the FESE label. The presentation of Creas will allow for contextualising the practices that impact funds carry out and will facilitate the general understanding of the article through a specific example that is considered successful in Spain.

Creas Impacto is the first institutional impact fund in Spain, with a total capital of EUR 21 million. The first closing of EUR 16 million was carried out in October 2018 by the European Investment Fund (EIF) and the second of EUR 5 million in April 2019 by the Official Credit Institute (ICO) through AXIS. Creas Impacto has been set up under European regulations as a European Social Entrepreneurship Fund (EuSEF), as it is a venture capital company under the supervision of the Spanish National Securities Market Commission (CNMV). The financial capacity of each investment is between EUR 500,000 and 3 million. These are carried out through capital increases and purchase and sale operations of shares or convertible participative loans, as well as hybrid financing instruments.

Collective investment institutions in the area of impact have evolved notably in Spain since the creation in 2011 of the Creas fund, a pioneer in impact investing in Spain.

In 2013, the regulatory framework (Regulation No. 346/2013) for impact investment funds was approved, under which the name of European Social Entrepreneurship Fund (EuSEF) was established to designate investment funds focused on social enterprises with the aim of generating a positive impact. In this way, investors are made easier to identify, and funds whose purpose goes beyond obtaining financial profitability are made easier to access, allowing for enterprises to attract a greater number of investments and promoting impact investment.

To obtain this denomination, the fund must have an investment portfolio where at least 70% is allocated to social companies and must also provide the relative information (social objectives of the fund, the social companies where it invests and how it evaluates the achievement of the objectives by the companies) in a standardised way. Another requirement refers to the role of the fund manager, who must demonstrate good business governance and effective control systems and avoid any conflict of interest. In addition, FESE denomination funds must be supervised by the authorities of the country where they are established. In case of not fulfilling any of the obligations, the FESE figure can be withdrawn. In Spain, there are only two funds established under European regulations and therefore with the label FESE, Q-Impact I and Creas Impacto.

Target companies must have an innovative approach and a sustainable profitability model that can respond to social problems, and they must either be in the early stages of growth or, if operating on a larger scale, have a differentiating business model. Their business model must be consolidated, with sales traction and positive performance. In addition, they must belong to one of Creas Impacto's priority impact areas—education, health and welfare, social innovation and environmental sustainability—and generate their impact mainly in Spain, although 20% of investments may be made in social enterprises whose impact is made in Europe.

Not only do these impact areas coincide with the goals of the SDGs (education (SDG 4), health and welfare (SDG 3), social innovation (SDG 1,2,5) and environmental sustainability (SDG 13,14,15), but the global approach is aligned with the triple core of the SDGs (economic, social and environmental).

Creas Impacto goes beyond simply making the investment and participates on the board of directors of the target companies, defining itself as a hands-on investor. Thanks to the diverse and multidisciplinary profile of the experts, who belong to the business, social and financial worlds, Creas Impacto actively provides support for financial, management and strategic decisions. It has a long history of investments and disinvestments, which allows it to have in-depth knowledge of the different stages of the investment life cycle. Likewise, as an impact fund, it collaborates in the process of measuring impact based on the theory of social change and the international evaluation models proposed by the European Venture Philanthropy Association (EVPA). Creas Impacto's experience comes from its other two funding instruments belonging to the impact ecosystem, both managed through its Creas Valor Social Foundation: Creas Inicia and Creas Desarrollo (Table 3).

Table 3. Creas Impacto and Creas Desarrolla as funding instruments of Creas Impacto.

	Creas Inicia	
Social Enterprise	**Description**	**Impact Area**
iWOPI	Platform through which the user "donates" the km of sport he or she does to social projects that collaborate with the application	Health Funding
Civiclub	Encourage society to take sustainable actions through a points and rewards system	Social and environmental awareness
Disjob	Employment page for people with disabilities	Inclusion Work
Sensovida	Telecare system for the elderly	Health
	Creas Desarrolla	
Social Enterprise	Description	Impact Area
Koiki	"Last mile" delivery system by people in social centres	Inclusion Work Environment
Emzingo	Training program on innovation, responsible leadership and the connection between business performance and social and environmental impact	Education
Sadako	Robots with artificial intelligence that separate and recycle garbage in landfills	Technology
Jump Math	Mathematics education programme for primary and secondary school children	Education
Whats cine	Innovative audio-visual platform adapted for people with visual and hearing disabilities	Inclusion
Smileat	Locally produced organic baby food products and healthy recipes	Environment Nutrition Health

Below (Table 4) is the proposed classification directly applied to the Creas Impacto case:

Table 4. Results of the analysis for Creas Impacto Fund.

<table>
<tr><th colspan="2">Classification of Impact Funds</th></tr>
<tr><th colspan="2">Fund data</th></tr>
<tr><td>Fund name: Creas Impacto</td><td>Management company name: Self-management</td></tr>
<tr><td>Headquarter city: Madrid</td><td>Year of first closure: 2018</td></tr>
<tr><td>The portfolio consists only of impact investment projects:</td><td>x Yes ☐ No</td></tr>
<tr><td>Is labelled ESEF:</td><td>x Yes ☐ No</td></tr>
<tr><th colspan="2">Size of investment</th></tr>
<tr><td>Size of the fund as a function of its total capital (EUR):
21 million (objective 30 million)</td><td>Size of each investment (EUR): 0.5–3 million</td></tr>
<tr><td rowspan="2">Financing instruments:</td><td>x Debt/Loan x Capital ☐ Donation x Capital Hybrid Model</td></tr>
<tr><td>☐ Venture Philanthropy Hybrid Model ☐ Other</td></tr>
<tr><th colspan="2">Investor profile</th></tr>
<tr><td>Investor motivation:</td><td>x Financial return ☐ Social and environmental impact</td></tr>
<tr><td>Financial return expectations (profitability): 7%</td><td>Risk aversion (scale from 1 to 7): -</td></tr>
<tr><td>Investment timeline:</td><td>☐ Short (< 1 year) ☐ Medium (1–5 years) x Long (> 5 years)</td></tr>
<tr><th colspan="2">Impact target area</th></tr>
<tr><td>Main aim impact:</td><td>☐ Environmental ☐ Social x Both</td></tr>
<tr><td>Impact target:</td><td>☐ One x Multiple</td></tr>
<tr><td>What are they? Education, health and welfare, social innovation and environmental sustainability</td><td>Are the impact areas identified by the fund through the sustainable development objectives? ☐ Yes x No</td></tr>
<tr><th colspan="2">Resulting impact</th></tr>
<tr><td>Scale on which the impact is generated:</td><td>☐ Local ☐ Regional ☐ National ☐ International x Undetermined</td></tr>
<tr><td>Geographic location of impact:</td><td>x Europe ☐ Asia ☐ Africa ☐ North America ☐ South America</td></tr>
<tr><td rowspan="2">The measurement of the impact is done by:</td><td>☐ Fund-specific rates ☐ Objective rates of sustainable development</td></tr>
<tr><td>x Rates established by another organisation. Which one? EVPA</td></tr>
<tr><th colspan="2">Other features</th></tr>
<tr><td>State of development of the target company:</td><td>☐ Seed ☐ Start x Growth ☐ Maturity</td></tr>
<tr><td>Source of capital for the fund:</td><td>x Public ☐ Private ☐ Both</td></tr>
<tr><td>Fund's main investors:</td><td>FEI and ICO</td></tr>
</table>

3.2. Impact Investing as a Financing Tool for Agenda 2030

After the implementation of the Sustainable Development Goals in 2015, the impact community began to study how it could join the global effort that the Agenda entails, and some impact funds use the SDG as a framework under which to develop their investments. Investors with extensive experience in the impact arena say that aligning impact practices with the Sustainable Development Goals can bring advantages in fund development in three key areas: communication, impact strategy and objectives and attracting new sources of capital.

Triodos Investment Management, one of the world's largest impact fund managers with total assets under management of EUR 4.2 billion, strongly believes that investment funds that align their practices with the SDG 17 will be able to attract larger amounts of capital, thereby helping to solve the problems their investments were intended to address [23]. Moreover, Foro Impacto establishes as one of its main objectives to promote impact investment in Spain within the framework of Agenda 2030, which may lead to progress towards alignment with the SDG by the funds that carry out their practices in our country [24].

3.3. Investment and Impact Funds: SDG 17

Within the nineteen specific goals that make up SDG 17, the following five goals can be highlighted as those where impact funds have high potential to contribute. Within the field of "Finance", impact investment would participate in:

- 17.3 Mobilise additional financial resources from multiple sources for developing countries.
- 17.5 Adopt and implement investment promotion systems in favour of least developed countries.

It seems sensible to think that impact investment could count as a mobilised financial resource if it aligns its objectives specifically with the SDG and focuses on disadvantaged geographical areas and specifically on developing countries. In fact, impact funds could go a step further, since, as mentioned in the first subsection of this section, not only do they seek to mobilise financial resources to developing areas, they also cover a wider spectrum by addressing the impact dimension in any demographic area that requires it.

In the field of "Systemic Issues", which is divided into three sections, two of them, "Multi-stakeholder partnerships" and "Data, Monitoring and Accountability", are reflected:

- 17.16 Enhance the Global Partnership for Sustainable Development, complemented by multi-stakeholder partnerships that mobilise and exchange knowledge, expertise, technology, and financial resources to support the achievement of the Sustainable Development Goals in all countries, particularly developing countries.
- 17.17 Develop and promote effective partnerships in the public, public-private and civil society spheres, building on partnership experience and resource mobilisation strategies.

Additionally, in "Data, Monitoring and Accountability":

- 17.19 By 2030, build on existing initiatives to develop indicators to measure progress towards sustainable development and to complement gross domestic product, and support statistical capacity-building in countries in development.

Regarding the first of the blocks, "Alliances between multiple stakeholders", impact funds, like other impact intermediaries, establish agreements between various agents belonging to different sectors with a common goal: to generate a positive outcome for society or the environment and an economic benefit. They contribute their knowledge and experience in order to achieve the double objective through an action plan. Impact funds connect both sides of the market, the supply of impact capital and the demand for it from partner companies. Impact entrepreneurs are strong allies, whose efforts must be supported [24].

In the second of the blocks, "Data, Monitoring and Accountability", which refers to the process of measuring results, impact funds have extensive experience since one of the intrinsic characteristics of impact investment itself is the measurement of results. In fact, it is possible that the alignment of impact funds with the SDG can be an advantage in overcoming one of the major pitfalls in this type of fund: the impact measurement fund. It may be easier to find formulas for impact assessment if the SDG indicators themselves are taken as a reference.

It should be noted that financial intermediaries of great importance in promoting these practices, such as Foro Impacto and GIIN, are identified as contributors to achieving SDG 17.

4. Discussion

In addressing the need for this present generation to develop sustainability, we must create a balance between the evolution of the environment, the economy and society. This point is crucial for the success of a harmonious and long-term coexistence between humans and nature. Currently, the achievement of that equilibrium is more important than ever, since the combination of an increasing human population, unawareness of overconsumption and environmental exploitation has led to the opposite effect. Indeed, after many decades of conferences and debates, the United Nations the Millennium Development Goals (MDG) were finally settled, focusing on the reduction of extreme poverty and setting a deadline for this goal. Nevertheless, many errors were made, and the results were not as expected. Thus, again, new goals were fixed and a new deadline set, focusing on global public good and promoting these goals through the spirit of a common mission that needs to be present as a catalyst to achieve all the above.

As regards the validation of the classification proposal, as mentioned above, the information published by the funds themselves is not sufficient, and the resources available for carrying out this work have not been sufficient to undertake more exhaustive work by means of interviews with those responsible for all funds. Available information has been used as a reliable source, from which the following conclusions can be drawn regarding the profile of the impact funds operating in Spain:

i. Foundations are positioned as entities with a high potential to promote the creation of impact funds in Spain, which still continue to be a new product limited to operating in the most important cities.
ii. Spanish impact funds focus only on impact investment with portfolios formed solely by social organisations with a positive and quantifiable impact on society/environment. However, only two of the funds have the "European Social Entrepreneurship Fund" label, which they recently obtained, confirming the novelty of the impact product. Perhaps this could act as a catalyst if other funds take them as a reference.
iii. The investment portfolios are very small and, proportionate to this, are the sizes of each investment. However, the financing tools they use offer diversity. Most of their capital comes from various private sector actors, although some public entities have been key to development.
iv. The funds seek a balance between impact and profitability, but the data show rather low returns compared to those generally generated by traditional investment. The risk of investments is high, which is linked to being established as venture capital companies, where high risk is an intrinsic feature. The investment period ranges from 3 to 7 years. However, impact investment requires patient capital, with longer terms and continuous support. As identified in the Impact Forum report, patient capital is scarce because there are not enough players willing to assume the high risk involved in investing in social enterprises in their early stages: "*companies with objectives of generating significant social impact need patient capital (more than ten years) with return expectations that reflect the additional costs and risks they face in achieving their objectives*" [22].
v. Moving on to the impact that Spanish funds seek to generate, these funds are not specialised in a specific topic, but rather the areas of impact are multiple with both social and environmental implications. These areas are not identified through the Sustainable Development Goals, nor are the results measured with the indices proposed by the Agenda but are carried out through those proposed by third parties. Therefore, based on the previous information, it is proposed as a main recommendation that the different funds unify and align their efforts in the same direction of work as the Agenda by including in the impact considerations, integrated throughout each of the stages that make up the investment process and the Sustainable Development Goals: objectives, targets and indicators.
vi. Finally, the development phase of the target companies, in which a greater number of funds are concentrated, is the growth phase. As concluded in the report of Foro Impacto, new funds need to be created that focus on the pregrowth stages. There is a funding gap between the initial phase, when social enterprises outgrow the requirements to receive grants, but are nevertheless too small and too high risk for investors.

The European Union calls for social entrepreneurship, focusing on social enterprises by placing them at the centre of the social economy and innovation [25]. The objective of social entrepreneurship is on long-term results. From a broad view, social entrepreneurship can be considered as a holistic concept that encompasses many diverse perspectives [26–28]. The umbrella construct of social entrepreneurship covers key phenomena: community entrepreneurship, social change agents, institutional entrepreneurs, social ventures, entrepreneurial nonprofit organisations, social enterprise and social innovation [29].

As mentioned above, some authors have argued that *"social entrepreneurs cannot reasonably be expected to solve social problems on a large-scale"*. One is the moral argument suggesting that moral egoism and social atomisation govern societies and thus inhibit any businesses, including social entrepreneurial ventures, to become 'moral' leaders. Another assertion is the political argument proposing that social entrepreneurs are often driven by a preconceived mental model to prioritise one's values and beliefs over the political and social desirability of particular social ends [30]. Other arguments include forces of institutional isomorphism and legitimacy pressure [29], suggesting that dominant institutions will inevitably force social entrepreneurs to fit within the existing and prevailing systems of rules, norms and cultural scripts, thus inhibiting societal change. To these arguments, the proposition is that institutional complexity can trigger the social venture to develop innovative and creative responses, which in turn can amplify, extend, bridge or even transform the social value proposition [26].

In general, future research on this ethically sensitive area is needed to determine how managers make decisions, especially given the potential for charges of exploitation of vulnerable populations. Due to the inherently ethical nature of such decisions, a foray into more normative territory may also be justified [31]. Business requires that enough people share a problem that can be addressed, however imperfectly, by a single solution. Only then is the problem likely to justify the production, exchange and delivery costs borne by business. Deciding to supply this entrepreneurial solution, therefore, requires some sense of demand for the product. These social entrepreneurs have transformed the charitable work of many nonprofit organisations and nongovernmental organisations around the world by acknowledging the gap between what customers must pay for a business solution to be operationally sustainable and what individuals in a particular market might actually be able to pay and seeking charitable donations to bridge this divide [32].

5. Conclusions

An exhaustive work of bibliographic compilation from diverse and relevant sources has been carried out, establishing a general vision of the impact panorama and serving as a fundamental basis for the achievement of the rest of the objectives. Based on the analysis carried out, a classification proposal was drawn up which included the main characteristics in a simplified form.

With regard to the objective of studying the potential of impact funds as an economic and support resource, which will drive Agenda 2030, it can be concluded that the Sustainable Development Goals present a great opportunity for impact investors to support this global agenda through capital investment in projects that address these critical challenges we face. However, the effects that impact investment are having today are made in an isolated way and not in an interconnected way as the Agenda needs. This problem has a double origin; on the one hand, it is the state of development in which the impact ecosystem and the investment products are still found. On the other hand, a large part of the funds, at least those operating in Spain, do not align their practices with the objectives of the Agenda, despite the fact that they seek to impact the same areas under which the latter operates and that such alignment has been shown to bring about multiple and mutual benefits, both for the funds themselves and for the Agenda.

Despite these two considerations, from what has been explained previously, it can be concluded that impact funds should operate in line with the principles of Agenda 2030 since, in addition to being aligned with the achievement of SDG 17, they have a high potential for mobilising not only public capital but, above all, private capital and for financing the sustainable development proposed in the SDG. In turn, the Sustainable

Development Goals provide the context for the funds to see how their strategies and objectives are part of an even larger project towards a better future for all.

Impact funds contribute to the achievement of the Agenda. However, despite the multiple benefits that alignment causes, a large part of the impact funds does not use the Sustainable Development Goals to identify its objectives and its specific goals, nor does it use the metrics that the United Nations presents to evaluate achievements.

It should be noted once again that this paper does not propose impact funds as an exclusive remedy to the Agenda's disparate funding problems, but rather as a tool or instrument that complements and helps in raising and mobilising funds. As Antonio Guterres, UN Secretary General, said: "There is no single solution for financing the SDG. The financial needs of Agenda 2030 are truly high, which is why any income, however insignificant it may seem, must be considered. The simple fact of eliminating any of these small sources of funding would entail a very high symbolic cost."

It is proposed as the main recommendation that different funds unify and align their efforts in the same direction of work as the Agenda, including impact considerations, integrated throughout each of the stages that make up the investment process and the Sustainable Development Goals: objectives, goals and indicators.

In this way, it is much easier, both for organisations seeking financing and for investors who wish to deposit their savings, entities that proactively seek change and economic benefit, to identify if the fund is suitable for them. In addition, as mentioned previously, it allows many other investors who, despite not knowing the impact market, are attracted by the alignment with the 2030 Agenda, which is world renowned. Additionally, on the other hand, it allows the funds themselves to identify with an ambitious movement of universal change.

Author Contributions: Conceptualisation, M.D.S.d.G.; methodology, C.d.l.H.G.; validation, M.Á.H.C. and M.M.-N.; formal analysis, J.C.S.; writing—review and editing, N.C.-P. All authors have read and agreed to the published version of the manuscript.

Funding: This research received no external funding.

Institutional Review Board Statement: Not applicable.

Informed Consent Statement: Not applicable.

Data Availability Statement: The data presented in this study are available on request from the corresponding author.

Conflicts of Interest: The authors declare no conflict of interest.

Appendix A

Table A1. Proposed classification of impact funds.

Classification of Impact Funds	
Fund data	
Fund name:	Management company name:
Headquarter city:	Year of first closure:
The portfolio consists only of impact investment projects:	☐ Yes ☐ No
Is labelled ESEF:	☐ Yes ☐ No
Size of investment	
Size of the fund as a function of its total capital (EUR):	Size of each investment (EUR):
Financing instruments:	☐ Debt/Loan ☐ Capital ☐ Donation ☐ Capital Hybrid Model
	☐ Venture Philanthropy Hybrid Model ☐ Other
Investor profile	
Investor motivation:	☐ Financial return ☐ Social and environmental impact
Financial return expectations (profitability): ____ %	Risk aversion (scale from 1 to 7): ____
Investment timeline:	☐ Short (< 1 year) ☐ Medium (1–5 years) ☐ Long (> 5 years)
Impact target area	
Main aim impact:	☐ Environmental ☐ Social ☐ Both
Impact target:	☐ One ☐ Multiple
What are they?	Are the impact areas identified by the fund through the sustainable development objectives? ☐ Yes ☐ No
Resulting impact	
Scale on which the impact is generated:	☐ Local ☐ Regional ☐ National ☐ International ☐ Undetermined
Geographic location of impact:	☐ Europe ☐ Asia ☐ Africa ☐ North America ☐ South America
The measurement of the impact is done by:	☐ Fund-specific rates ☐ Objective rates of sustainable development
	☐ Rates established by another organisation. Which one?
Other features	
State of development of the target company:	☐ Seed ☐ Start ☐ Growth ☐ Maturity
Source of capital for the fund:	☐ Public ☐ Private ☐ Both
Fund's main investors:	

References

1. Global Impact Investment Network. Financing the Sustainable Development Goals: Impact Investing in Action. 2018. pp. 1–28. Available online: https://thegiin.org/assets/Financing%20the%20SDGs_Impact%20Investing%20in%20Action_Final%20Webfile.pdf (accessed on 18 August 2019).
2. Impact Hub Madrid. *Transformar la Ciudad a Través de la Inversión de Impacto*; Impact Hub Madrid: Madrid, Spain, 2018.
3. Unirisco. Available online: http://www.unirisco.com/ (accessed on 25 May 2019).
4. Alonso, J.A. Movilizando los recursos y los medios de apoyo para hacer realidad la agenda de desarrollo post-2015. *Coop. Esp. Madr.* **2015**, *53*, 1689–1699. [CrossRef]
5. PDNU (Programa de Desarrollo de las Naciones Unidas). *Financing the 2030 Agenda*; Programa de Desarrollo las Naciones Unidas: New York, NY, USA, 2018.
6. Keeble, B.R. The Brundtland Report: "Our Common Future". *Med. War* **1988**, *4*, 17–25. [CrossRef]
7. Egelston, A.E. *Sustainable Development: A History*; Springer Science & Business Media: Berlin, Germany, 2013. [CrossRef]
8. Gómez Gutiérrez, C.; Díaz Duque, J.A. *Origen del Concepto de Desarrollo Sostenible*; Universidad de Alcalá: Madrid, Spain, 2013; pp. 7–16.
9. Sierra Montoya, J.E. Los Tres Pilares del Desarrollo Sostenible. 2019. Available online: http://www.i-ambiente.es/?q=blogs/los-tres-pilares-del-desarrollo-sostenible (accessed on 25 July 2019).

10. Luengo Montero, M. El Futuro Será Sostenible o no Será. 2018. Available online: https://elpais.com/elpais/2018/09/11/eps/1536681073_060705.html (accessed on 25 March 2019).
11. Razavi, S. *The 2030 Agenda: Challenges of Implementation to Attain Gender Equality and Women's Rights*; Gender & Development: London, England, 2016; pp. 24–41.
12. Sachs, J.D. From millennium development goals to sustainable development goals. *Lancet* **2012**, *379*, 2206–2211. [CrossRef]
13. Lomazzi, M.; Borisch, B.; Laaser, U. The Millennium Development Goals: Experiences, achievements and what's next. *Glob. Health Act.* **2014**, *7* (Suppl. S1). [CrossRef] [PubMed]
14. CEPAL, B. Agenda 2030 para el Desarrollo Sostenible. 2019. Available online: https://biblioguias.cepal.org/c.php?g=447204&p=6366258 (accessed on 15 July 2019).
15. Clark, H. *Openness for All: The Role for OGP in the 2030 Development Agenda*; Open Government Partnership Summit: Washington, DC, USA, 2015.
16. Colglazier, B.W.; Na, U. Sustainable development agenda: 2030. *Science* **2015**, *349*, 1048–1050. [CrossRef] [PubMed]
17. UE. *Reglamento UE No346/2013 del Parlamento Europeo y del Consejo de 17 de Abril de 2013 Sobre los Fondos de Emprendimiento Social Europeos*; European Parliament: Strasbourg, France, 2013; pp. 18–38.
18. Comisión Europea. *Construir un Ecosistema para Promover las Empresas Sociales en el Centro de la Economía y la Innovación Sociales. Comunicación de La Comisión Al Parlamento Europeo Al Consejo, Al Comité Económico y Social Europeo y Al Comité de Las Regiones*; European Parliament: Strasbourg, France, 2011.
19. Rial, L. ¿Cómo Afecta el Tamaño de un Fondo de Inversión a la Rentabilidad? 2019. Available online: https://www.rankiapro.com/como-afecta-tamano-fondo-inversion-rentabilidad/ (accessed on 30 July 2019).
20. BBVA. ¿Qué es el Horizonte Temporal en Una Inversión? 2018. Available online: https://www.bbvaassetmanagement.com/am/am/es/es/particular/actualidad/noticias/tcm:864-769645/horizonte-temporal-inversion (accessed on 25 July 2019).
21. Sims, D.W.; Queiroz, N. Fish stocks: Unlimited by-catch limits recovery. *Nature* **2016**, *531*, 448. [CrossRef] [PubMed]
22. Urriolagoitia, L.; Hehenberger, L. *La Inversión de Impacto en España: Intermediación de Capital Una llamada a la Acción Para Impulsar el Mercado de las Inversiones de Impacto en España en el Marco del GSG*; Foro Impacto: Madrid, Spain, 2019.
23. Triodos Investment Management. Available online: https://www.triodos-im.com/ (accessed on 29 August 2019).
24. Impact Forum. In Proceedings of the VI S2B Impact Forum, Barcelona, Spain, 25 October 2019.
25. Amin, A.; Cameron, A.; Hudson, R. The social economy in context. *Placing Soc. Econ.* **2010**, 1–15. [CrossRef]
26. Cherrier, H.; Goswami, P.; Ray, S. Social entrepreneurship: Creating value in the context of institutional complexity. *J. Bus. Res.* **2018**, *86*, 245–258. [CrossRef]
27. Dwivedi, A.; Weerawardena, J. Conceptualizing and operationalizing the social entrepreneurship construct. *J. Bus. Res.* **2018**, *86*, 32–40. [CrossRef]
28. Hervieux, C.; Voltan, A. Framing Social Problems in Social Entrepreneurship. *J. Bus. Ethics* **2018**, *151*, 279–293. [CrossRef]
29. Macke, J.; Sarate, J.A.R.; Domeneghini, J.; da Silva, K.A. Where do we go from now? Research framework for social entrepreneurship. *J. Clean. Prod.* **2018**, *183*, 677–685. [CrossRef]
30. Sud, M.; Vansandt, C.V.; Baugous, A.M. Social entrepreneurship: The role of institutions. *J. Bus. Ethics* **2009**, *85* (Suppl. S1), 201–216. [CrossRef]
31. Félix González, M.; Husted, B.W.; Aigner, D.J. Opportunity discovery and creation in social entrepreneurship: An exploratory study in Mexico. *J. Bus. Res.* **2017**, *81*, 212–220. [CrossRef]
32. McMullen, J.S.; Bergman, B.J. The promise and problems of price subsidization in social entrepreneurship. *Bus. Horiz.* **2018**, *61*, 609–621. [CrossRef]

sustainability

MDPI

Communication

Trilemma of Nordic–Baltic Forestry—How to Implement UN Sustainable Development Goals

Lars Högbom [1,2,*], Dalia Abbas [3], Kęstutis Armolaitis [4], Endijs Baders [5], Martyn Futter [6], Aris Jansons [5], Kalev Jõgiste [7], Andis Lazdins [5], Diana Lukminė [4], Mika Mustonen [8], Knut Øistad [9], Anneli Poska [10], Pasi Rautio [11], Johan Svensson [12], Floor Vodde [7], Iveta Varnagirytė-Kabašinskienė [4], Jan Weslien [1], Lars Wilhelmsson [1] and Daiga Zute [5]

1 The Forestry Research Institute of Sweden–Skogforsk, 751 83 Uppsala, Sweden; jan-olov.weslien@skogforsk.se (J.W.); Lars.Wilhelmsson@skogforsk.se (L.W.)
2 Department of Forest Ecology and Management, Swedish University of Agricultural Sciences (SLU), 901 03 Umeå, Sweden
3 Department of Environmental Science, American University, Washington, DC 20016, USA; saleh@american.edu
4 LAMMC, Institute of Forestry, Girionys, Kaunas District, 501 27 Kaunas, Lithuania; kestutis.armolaitis@lammc.lt (K.A.); diana.lukmine@lammc.lt (D.L.); iveta.kabasinskiene@lammc.lt (I.V.-K.)
5 SILAVA, Latvian State Forest Institute, LV-2169 Salaspils, Latvia; endijs.baders@silava.lv (E.B.); aris.jansons@silava.lv (A.J.); andis.lazdins@silava.lv (A.L.); daiga.zute@silava.lv (D.Z.)
6 Department of Aquatic Sciences and Assessment, Swedish University of Agricultural Sciences (SLU), P.O. Box 7010, 750 07 Uppsala, Sweden; martyn.futter@slu.se
7 Institute of Forestry and Rural Engineering, Estonian University of Life Sciences, Kreutzwaldi 5, 51006 Tartu, Estonia; kalev.jogiste@emu.ee (K.J.); floortje.vodde@emu.ee (F.V.)
8 Natural Resources Institute Finland (Luke), P.O. Box 2, FI-00791 Helsinki, Finland; mika.mustonen@luke.fi
9 NiBio, Norwegian Institute of Bioeconomy Research, 1431 Ås, Norway; knut.oistad@nibio.no
10 Department of Geology, Tallinn University of Technology, Ehitajate tee 5, 19086 Tallinn, Estonia; anneli.poska@taltech.ee
11 Natural Resources Institute Finland (Luke), Ounasjoentie 6, FI-96200 Rovaniemi, Finland; pasi.rautio@luke.fi
12 Department of Wildlife, Fish and Environmental Studies, Swedish University of Agricultural Sciences (SLU), 901 03 Umeå, Sweden; johan.svensson@slu.se
* Correspondence: lars.hogbom@skogforsk.se; Tel.: +46-(0)-18188549

Citation: Högbom, L.; Abbas, D.; Armolaitis, K.; Baders, E.; Futter, M.; Jansons, A.; Jõgiste, K.; Lazdins, A.; Lukminė, D.; Mustonen, M.; et al. Trilemma of Nordic–Baltic Forestry—How to Implement UN Sustainable Development Goals. *Sustainability* **2021**, *13*, 5643. https://doi.org/10.3390/su13105643

Academic Editors: Margarita Martinez-Nuñez and Mª Pilar Latorre-Martínez

Received: 22 March 2021
Accepted: 14 May 2021
Published: 18 May 2021

Abstract: Forests are the dominant land cover in Nordic–Baltic countries, and forestry, the management of forests for improved ecosystem-service (ES) delivery, is an important contributor to sustainability. Forests and forestry support multiple United Nations Sustainability Goals (UN SDGs) and a number of EU policies, and can address conflicting environmental goals. Forests provide multiple ecosystem services and natural solutions, including wood and fibre production, food, clear and clean water and air, animal and plant habitats, soil formation, aesthetics, and cultural and social services. Carbon sequestered by growing trees is a key factor in the envisaged transition from a fossil-based to a biobased economy. Here, we highlight the possibilities of forest-based solutions to mitigate current and emerging societal challenges. We discuss forestry effects on forest ecosystems, focusing on the optimisation of ES delivery and the fulfilment of UN SDGs while counteracting unwanted effects. In particular, we highlight the trilemma of (i) increasing wood production to substitute raw fossil materials, (ii) increasing forest carbon storage capacity, and (iii) improving forest biodiversity and other ES delivery.

Keywords: United Nations Sustainable Development Goals; ecosystem services; forest; forestry; management practices; European Green Deal

1. Introduction

The Past and Today—Various Dimensions of the Northern Forest

The forest landscape in Nordic–Baltic countries, comprising Estonia, Finland, Latvia, Lithuania, Norway, and Sweden, is largely a 20th century creation of active management and governance aimed at optimising the economic value of forest production by increasing standing forest biomass. Today's forest landscape consists of a mosaic of stands of different ages, dominant tree species, productivity, and sizes (<1 to >100 ha) intermixed with semiopen and open land cover with low or no forest productivity (e.g., mires and bogs). There is also large diversity of forest owners ranging from the state to private individuals. Nordic–Baltic countries and their inhabitants derive economic, social, and environmental benefits from forests and forestry. Forestry has historically had a significant contribution to national and rural economies, and is still an important sector today. The development of a Nordic–Baltic forestry system mirrors the economic, environmental, and social development in the region, and forests play a significant role in the development of societal welfare. This societal welfare stems from activities ranging from global bioproduct mills to local sawmills and thermal power plants, and to local communities hunting, and picking berries and mushrooms.

Guiding principles for forestry are quite similar within the region, but with differences in the size, structure, and sectorial share of the national GDP. These differences are explained by the forest cover, terrain, and other sectors' development and share of the economy, among other factors. In addition to their economic importance, forests and woodlands provide other essential forest-ecosystem services [1–3]. Forest products are extracted and used in construction, paper and paperboard products, and in replacing fossil fuel and materials. However, production forestry interferes with biogeochemical cycles [4–6], changes hydrology [7,8], and affects biodiversity [9].

The Nordic–Baltic region encompasses a variety of climates, ranging from sub-Arctic in the north to nemoral in the south, with a more maritime climate along the Atlantic and the Baltic Sea coast, and a more continental climate in the east (Figure 1). Scots pine (*Pinus sylvestris* L.), Norway spruce (*Picea abies* (L.) Karst.), hardwood birches (*Betula pendula* (Roth.) and *Betula pubescens (Ehrh.)),* and aspen (*Populus tremula* L.), are the dominant tree species in the boreal forest zone in Finland, Norway, and northern Sweden. A range of broadleaf species, including beech (*Fagus sylvatica* L.), oak (*Quercus robur* L.), ash (*Fraxinus excelsior* L.), and alder (*Alnus glutinosa* L.) are common in the nemoral zone.

European cultural inheritance is largely related to forests, especially in Northern Europe, where forests are deeply rooted in cultural and belief systems. Many fairy tales and sagas are set in forests that are described as both dark and frightful places, and as a refuge for asylum and tranquillity [10]. Even today, when most of us live in cities and may have only limited contact with forests and rural areas, forests and trees make many essential and highly valued contributions to our well-being, e.g., recreational and outdoor activities.

During the last millennium, the ever-growing agrarian and industrial expansion, and the need for building material, firewood, and charcoal for mining and metal smelters led to a severe reduction in woodland coverage in Europe [11]. Forests historically provided construction material for housing, heating, and products such as charcoal, tar, resins, and potash, which are important material for industrial processes and shipbuilding. Humans actively use forests for food production, including livestock grazing and crop production (e.g., slash-and-burn agricultural farming). While the deforestation of midlatitude Europe accelerated over time, and woodland cover had reached its minimum prior to the Industrial Revolution, deforestation peaked in northern Europe during the 19th century [12]. Concerns about deforestation led to political actions supporting afforestation and intensifying forest growth, including the development of the world's first National Forest Inventories (NFIs) in Nordic countries about 100 years ago. Other restoration efforts that aimed to increase forest growth included applying different silvicultural practices, tree-breeding programmes, and peatland drainage. Approximately 1.5 million ha in Sweden and 4.7 million ha in Finland were drained, which increased available forest production areas, but also substantially impacted other land-cover types, habitats, biodiversity, ecosystem services, carbon loss, and ecosystem function [13–15].

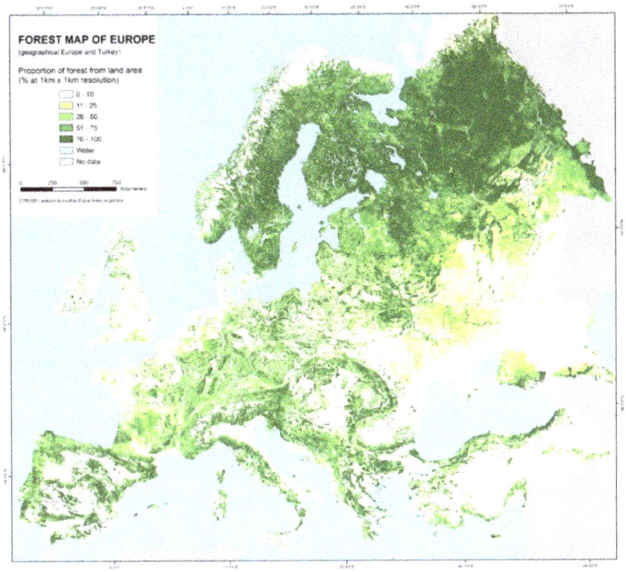

Figure 1. Forest map of Europe showing high proportion of forested area in Nordic and Baltic countries. With permission of the European Forest Institute [16–18].

During the past 100 years, standing volume increased by over 50% in the Nordic–Baltic region due to active forest management and favourable climatic conditions, and despite substantial logging during the same period. In boreal Sweden, for example, the total standing volume in the 1920s was 1.6×10^9 m^3; today, it is 3.6×10^9 m^3 [19], equivalent to a 111% increase. In Finland, the standing volume in 1990 was 1.9×10^9 m^3, and this increased to 2.5×10^9 m^3 (2018). At the same time, the total drain (removals and natural losses) of roundwood was 2.1×10^9 m^3, more than the whole standing volume in 1990 [20,21].

Forests regulate climate, buffering the effects of extreme temperatures and precipitation, and sustaining the water cycle [22,23]. They filter and purify both surface- and groundwater, and can mitigate floods and droughts. In cities, growing trees can contribute to cooling both with their shade and by transpiring water, in addition to improving air quality [24–26].

These multiple benefits and multiple use objectives are not intrinsically considered in forest planning. Forests can produce multiple and frequent benefits of goods and services, concurrently and from the same piece of land [27]. A recent study of northern Sweden showed that the demand for conservation, sociocultural, and economic land-use interests exceeds the available land area by 2 to 4 times [28].

Forestry and reindeer husbandry largely utilise overlapping areas; for example, in the reindeer-herding area in Sweden (160,000 km^2 around 55% of the Swedish land area), forestry and reindeer herding have had several confrontations [29]. In Finnish Lapland, forestry, reindeer herding, and tourism coexist in the same area that partly overlaps with Sápmi as the cultural region of the Sámi people, which creates difficulties between different land-use modes. As an example, reindeer herders and tourism entrepreneurs prefer continuous cover forest management, but the prevailing methods are based on stand-based even-aged forestry with final cuttings.

Biodiversity has decreased in many Nordic–Baltic forests due to a combination of historical habitat fragmentation and the logging of old growth stands. Replacing old growth stands, some of which are many centuries old, by conventional production forests fails to offer the same habitat complexity and diversity due to considerably shorter rotation lengths. Measures have been taken to help to reduce habitat fragmentation, and to maintain and create habitats by a combination of planned management and conservation. Intact forest landscapes were defined, e.g., in a recent mapping by Watson et al. [30], as larger (>500 km^2) mosaics of forests and natural open ecosystems. Intact forest landscapes include so-called primary forests, which show no or low influence of human activities or habitat fragmentation, but may contain the historical human influence of, e.g., preindustrial selective logging [31]. In northern Europe, large intact forest landscapes are only found on the Swedish side of the Scandinavian mountain range, the forest border area between Finland and Northwest Russia, and the Kola peninsula forest belt [32]. Elsewhere in the region, most forests were systematically transformed into plantation forests [33]. Intact forest landscapes and primary forests were lost, and the remaining patches are highly fragmented, with the effects of fragmentation even further pronounced when edge effects are considered [34].

Forest biodiversity is also threatened by the so-called "green shift". A global transition from an economy based on petrochemicals towards an economy that is sustainable, biobased, and circular is now underway with, e.g., the Paris Agreement, the European Green Deal [35], and the Renewable Energy Resources Act and Renewable Energy Directive. The economically, environmentally, and socially sustainable use of natural resources is the fundamental principle. Biobased innovation could, in this context, reduce dependence on fossil fuels, thereby making a positive contribution to meeting EU and UN climate goals. As a renewable natural resource, woody biomass can be valorised in innovative bioproducts alongside a range of conventional forest-industry products [36].

However, this also highlights what could best be described as a trilemma, with conflicting goals concerning biodiversity, carbon storage, and wood supply.

In this paper, we address and highlight some of the most critical aspects for a sustainable and advanced Nordic–Baltic forestry into the future in the context of some of the most relevant of the 17 UN SDGs, particularly SDG 12: responsible consumption and production, SDG 13: climate actions, and SDG 15: life on land. Further, we discuss the SDGs in the context of the European Green Deal. Our key message is that avenues into the future for sustainable forestry and a forest sector that contribute to societal development have to be sought out by addressing the conflicts risks, and integration and synergy opportunities among multiple, sometimes diverging, and even disparate high-level targets. Thus, a balancing act lies ahead. In this paper, we highlight how these three SDGs provide a perspective on a trilemma that needs to be explored and resolved to promote forestry, and forest and forest-landscape sustainability in a growing bioeconomy.

2. Sustainable Forest Management—SDG 12

Responsible production and consumption implies the need for sustainable forestry. However, the definitions of sustainable forestry include multiple subjective assessments and are consequently a moving target. Although forestry is generally orientated towards sustainability, sustainable forestry per se is as much of a matter of public perceptions of forest biology, ecology, hydrology, and woody-biomass production. Contreras-Hermosilla [27] argued that it is hard to agree on sustainability objectives and their relative importance, and thereby be able to assess whether a forest is sustainably managed or not.

Sustainable forest-production management is about understanding, planning, and balancing different goals and actions to achieve optimal ecosystem services to legitimate stakeholders, and avoid risks of negative environmental effects. This includes both combinable goals and spatially distributed functional landscape mosaics to achieve goals not possible to combine within acceptable limits. The complexity and lack of detailed knowledge concerning the side effects of management and actions show a common need for continuous monitoring to warn of both predicted and unpredicted environmental effects.

The forest-based bioeconomy can be defined as all economic activities that affect forest ecosystem services, ranging from forest biomass production to tourism, recreation, and nonwood products [37]. It is viewed as a way to mitigate climate change, which is best achieved within a balanced combination of ecosystems services [37,38]. This definition could be established on the pan-European or EU level, which would provide the basis for policy measures supporting economic activities and innovations related to the entire spectrum of forest ecosystem services.

Most forestry-derived bioenergy is currently produced as a secondary source, i.e., bark, black liquor, sawdust, and other byproducts from pulp- and sawmills. A minor fraction (e.g., in Sweden, 20%) of forestry-derived biofuel originates from branches and top residues from timber harvest. The extraction of forest biofuel from logging residues such as branches, twigs, tops, and needles (or leaves) has caused a fierce debate about sustainability from the perspectives forest production and forest protection, and for carbon balances [39,40].

Raising awareness of the need to manage forests for multiple potentially competing goals among landowners, foresters, and other decision makers is critical. Planning measures and natural solutions provide positive examples of ecological and spatial networks supporting endangered species, and improving resistance and recovery capacity for (i.e., response diversity) future ecosystem challenges [41].

3. Climate Actions—SDG 13

In addition to their potential for fossil-fuel substitution, forests and forestry have important roles to play in both climate mitigation and adaptation. Growing trees and other vegetation sequesters carbon. As a stand matures, more carbon is stored in the stems, stumps, and coarse roots. At the same time, decomposition increases, and there is a balance between growth and decomposition, meaning effectively no net CO_2 sequestration on the stand level [42,43]. As forests mature and age, carbon accumulates in understory vegetation, humus, and the upper soil layers. Net losses of sequestered carbon on the stand level can be caused by various kinds of calamities, e.g., insects, forest fires, and extreme weather. Old-growth stands with high levels of accumulated carbon thus also release high amounts of carbon if devastated by forest fires or storms [44,45].

Over the short term, harvesting returns carbon to the atmosphere through the use of extracted biomass and the decomposition of roots, branches, and soil organic matter. Site preparation, drainage, and soil scarification can exacerbate this problem. Initially, regenerated young forests may not accumulate carbon at rates equivalent to site capacity [46–48]. Middle-aged stands, typically 20–60 years old, in the region have fast growth and a high carbon uptake. As a stand matures, net carbon uptake decreases, although carbon could still accumulate in the soil. The total ecosystem carbon storage in forests varies by stand age, structure, soil type, species, site, bioclimatic conditions, and stand history. Many of these factors are included in integrated landscape-scale monitoring based on eddy-flux measurements [49,50].

Over the years, there has been a fierce debate of whether forestry and forest products are carbon-neutral [51]. Carbon calculations are largely dependent on how system boundaries are defined. Results based on single-stand or even single-tree calculations differ from landscape-scale calculations [52]. However, with balanced age distribution across larger geographic scales and a sufficiently large proportion of forest land set aside, forest function as a carbon sink can be maintained over the longer term. On the short term, given the needs to immediately reduce greenhouse-gas emissions and increase forest carbon storage, longer rotation periods could be considered [53,54]. On the stand level, total biomass production (and hence carbon storage) is highest if precommercial and commercial thinning is not performed, up to the stage when carbon release through natural decline and tree mortality exceeds the carbon uptake by living trees [55]. In upland forests, roughly half the carbon is stored above ground in tree trunks, branches, twigs, and needles or leaves [56–59]. The rest is stored below ground, which implies that forest soils are at least as important for carbon storage as above-ground vegetation. In addition, soil carbon is more recalcitrant toward disturbances [60].

Postharvest forestry activities also influence climate impact. Following the final felling, stand-level CO_2 emissions increase due to the decomposition of stumps, roots, and logging residues [61]. Carbon is also lost by the leaching of dissolved organic carbon (DOC) [62,63]. Soil scarification can increase the decomposition of soil organic matter, in particular in the humus layer [64]. On the other hand, soil scarification can increase carbon uptake by enhanced tree growth [65]. For carbon balance, an important factor is how fast the growth of the new stand counteracts soil CO_2 emissions. However, we have too many knowledge gaps to predict forest-management activities on soil carbon stocks with sufficient certainty to guide foresters, e.g., on soil preparation or nitrogen addition [64]. As these activities significantly improve tree growth, targeted research on the effects of forest-management actions on soil carbon storage is urgently needed.

4. Life on Land—SDG 15

Forestry and Habitat Maintenance

Nordic–Baltic forests harbour more than 25,000 known species of plants, animals, and fungi. There are about 2000 living forest species on average on the national red list of each Scandinavian country [66]. While forestry focuses on the active management of a handful of commercially important tree species, many other species are also affected. For example, in the Swedish Red List, as many as 1400 species are considered to be affected by clear felling [67].

Habitat loss and fragmentation are probably the most negative effects of forestry on flora and fauna. These processes usually involve several elements including habitat loss (decrease in total habitat area in a landscape), habitat isolation (e.g., increase in mean distance between habitat patches in a landscape), decrease in patch size, and increased edge effects [68,69]. The relative impact of these elements on biodiversity loss is still a matter of controversy [70–73].

Evaluating the impact of habitat loss and fragmentation on biodiversity is hugely more difficult when soil biodiversity is also considered, where we have numerous blind spots [74]. Planning, e.g., afforestation efforts to decrease habitat fragmentation, is not a simple issue, and local knowledge is needed, as the best methods might differ depending on landscape composition and location across elevation and geographical gradients in the region [75].

Stability, Resilience and Functional Diversity

Forestry practices that primarily focus on economically valuable species may, in combination with climate change, worsen forest ecosystem resilience. Forest management alters ecosystems and patterns of forest disturbances, and thereby natural ecosystem functions. Diversity in tree species, within and between stands, provides a higher range of options to respond to stresses and new environmental hazards. Furthermore, understanding the role of plant diversity, microorganisms, and other decomposers in successional trajectories, and the stability of secondary forests demands analysis of how these changes differ from changes in soil-organism diversity caused by more severe fires, wind throws, and pest outbreaks [76,77].

Ensuring species and habitat diversity is important for maintaining ecosystem resilience, especially under uncertainties associated with climate change [78]. It is, however, a paramount challenge for the forestry sector to offer enough suitable habitats to support viable populations of all species. Nature reserves, national parks, and voluntary set-aside forests offer such habitats and benefits. Forest certification standards and national legislations are partly directed towards creating and retaining habitats within the managed areas.

5. Discussion

Ecosystem resilience can be defined as the capacity over time for an ecosystem to resist external stress and disturbances or restore ecosystem structures and functioning to a predisturbance stable state [79]. Ecosystem resilience can have an antagonistic relationship with economic resilience, as the latter benefits from forestry-related forest disturbance (e.g., harvesting) as a source of income. This highlights the challenges in balancing multiple dimensions of sustainability in the forest landscape.

The bioeconomy, an economy that is based on renewable biological resources, such as forests and relies on sustainable biobased solutions, could, together with increased circularity, offer a way forward in building a more sustainable future [80,81]. This is particularly urgent in the societal transition to a fossil-free future. Circularity requires a new look at the economic model that we created, and rethinking the way we produce and consume. There is a plethora of new innovative biobased products that could replace fossil-based products, but these cannot currently compete in the marketplace due to the low price of fossil-based materials. Hence, new measures are needed to promote the use of more climate-friendly biobased materials.

Biological diversity defines the capacity of an ecosystem to adapt and evolve in a changing environment, and is therefore a prerequisite for a viable circular bioeconomy. By promoting a more holistic view on economic development, the circular bioeconomy could also contribute to protecting biodiversity and other important forest ecosystem services by replacing fossil products and contributing to climate mitigation. Given the uncertainty triggered by climate change, ensuring the diversity of species and conditions seems to be the most effective means to improve the resilience of the ecosystems [78].

Maintaining sustainable, climate-smart forestry is a balancing act among management strategies to meet conflicting goals. There is a challenge for the forestry sector to offer enough suitable habitats to support sufficient biodiversity to promote ecosystem resilience. Considering factors such as key habitats and biological legacies can create a conceptual frame to address forest regeneration, afforestation, and restoration efforts [82].

The magnitude and composition of Nordic–Baltic forests is the result of long-term investments in forest resources and economic benefits, and national priorities and governance that are generated by these investments. Ecosystem goods and services derived from healthy and functional forest ecosystem processes can contribute to social welfare [80] and economic wellbeing. Forests in Nordic–Baltic countries have provided ecosystem services and benefits throughout human history. These services have varied over time, but for more than a century, the benefits that contribute to income, employment, and social development have increased and developed.

Under the European Green Deal [35], a growth strategy that aims to transform the EU into a fair and prosperous society with a greener competitive economy, the European Commission adopted the EU Biodiversity Strategy 2030 in May 2020 [83]. To address biodiversity loss in the EU, the strategy aims to widen the network of protected areas and promote ecosystem restoration. In addition to the strict protection of primary and old-growth forests, forests and forest landscapes need to be ameliorated both qualitatively and quantitatively. Building on the biodiversity strategy, the Commission is preparing a new EU forest strategy during 2021. The key objectives include measures to increase absorption of CO_2, to reduce the incidence of forest fires, and to promote the bioeconomy in full respect for ecological principles favourable for biodiversity.

To achieve the objectives of the Green Deal policies, the measures must focus on both the protection and the use of forests. Management practices improving the quality and resilience in (all) multiple-use forests are key to both these objectives, and in providing products to circular and bioeconomy services (recreation, healthy products) and new business opportunities in line with the Green Deal. The sustainable re- and afforestation and restoration of degraded forests can contribute to carbon sequestration while also improving forest resilience and promoting the circular bioeconomy. A forest-based circular bioeconomy has great potential to contribute to the European Green Deal as part of a bundle of solutions.

6. Prospects

Promoting forestry, forest and forest landscape sustainability in a growing bioeconomy is arduous given the many actual, potentially diverging, and sometimes disparate high-level targets reflected in the UN Sustainable Development Goals [84]. Routes towards integration and synergy between different land-use modes and interests have to be mapped and promoted, whereas potential and real conflicts have to be acknowledged and avoided or mitigated [28]. Given the national and regional importance of the Nordic–Baltic forest sector, multiple challenges lie ahead to pave out future strategic and operational avenues.

In this paper, we explored and discussed a trilemma that originates from trying to simultaneously achieve SDGs 12, 13, and 15, and that encompasses some of the most challenging risks for conflict and opportunities for synergy and integration. Clearly, the current overlap of different present-day demands on forests and forest landscapes, as well as future expected and not yet defined demands, requires a wider acceptance of governance and management that acknowledges multifunctionality and is supported by evidence-based policies [85]. One example of this is the expanding renewable-energy sector. With the production and consumption of clean energy as a high-level global policy ambition [86], as reflected in SDG 13, the wind-power footprint on landscapes is becoming increasingly manifested [87–89]. The recent strategy for the sustainable development of wind power in Sweden [90] clearly defines the inland of northern Sweden as the focal area for new development, where production forestry is a dominating land use. Therefore, a substantial share of forest land for forest production and for meeting SDG 12 targets would transform to forest land used for energy production. Moreover, as wind power is commonly established on higher elevations where good wind conditions occur, the impact on the few remaining natural and near-natural forests with rich pools of biodiversity values (SDG 15) in such hinterland areas [32] must be expected, as well as pronounced visual, vibration, auditory, and light impact on landscape values, including human benefits and values [89].

Forests undeniably have great potential to substitute concrete or steel building materials and fossil-based raw materials or energy [90,91]. Wood used as construction material both offers long-term carbon storage and replaces materials with a much larger carbon footprint (e.g., concrete and steel). The sustainability of wood as an energy source is, however, debated. Here, the definitions and system boundaries for sustainability assessments affect the outcome. There is variation in national policies related to perceptions of the role of bioenergy in national climate strategies, but there have generally not been strong driving forces in forest policy to utilise forest biomass for energy. Where there were efforts for supporting the use of forest biomass as energy, such utilisation is generally recognised and supported for environmental and social reasons, as economic driving forces are considered to be weak, and profits minimal (see [92] and references therein). While the environmental sustainability of wood as an energy source is debated, there might be carbon benefits when replacing fossil fuels, but at the same time disadvantaging biodiversity [93].

Forests are the dominant land cover in the Nordic–Baltic countries and forestry. The management of forests for improved ecosystem-service (ES) delivery is an important contributor to sustainability. Forests provide multiple ecosystem services and natural solutions, including wood and fibre production, food, clear and clean water and air, animal and plant habitats, soil formation, aesthetics, and cultural and social services. A precursor to a successful transition to a biobased, circular economy is recognising the trilemma of (i) increasing wood production to substitute raw fossil materials, (ii) increasing forest carbon storage capacity, and (iii) improving forest biodiversity and ES delivery.

Author Contributions: The manuscript is a summary of a series of work-shops held by the SNS-founded project PROFOR. L.H. has been responsible for editing but all authors have contributed to discussions, writing and comment on numerous versions on the manuscript. All authors have read and agreed to the published version of the manuscript.

Funding: This paper is based on workshops organized by the Nordic–Baltic network PROFOR founded by SNS Nordic Forest Research (grant number N2020-05).

Institutional Review Board Statement: Not applicable.

Informed Consent Statement: Not applicable.

Data Availability Statement: Not applicable.

Conflicts of Interest: The authors declare no conflict of interest.

References

1. Hansen, K.; Malmaeus, M. Ecosystem services in Swedish forests. *Scand. J. For. Res.* **2016**, *31*, 626–640. [CrossRef]
2. Lawlor, K.; Sills, E.; Atmadja, S.; Lin, L.; Songwathana, K.; Sunderland, T.C.H.; O'Connor, A.; Muir, G.; Nerfa, L.; Nodari, G.R.; et al. Sustainable Development Goals: Their Impacts on Forests and People. 2019. Available online: http://www.bosquesandinos.org/wp-content/uploads/2020/01/Sustainable_Development_Goals__Their_Impacts_on_Forests_and_People_compressed.pdf (accessed on 20 January 2021).
3. Díaz, S.; Demissew, S.; Carabias, J.; Joly, C.; Lonsdale, M.; Ash, N.; Larigauderie, A.; Adhikari, J.R.; Arico, S.; Báldi, A.; et al. The IPBES Conceptual Framework—connecting nature and people. *Curr. Opin. Environ. Sustain.* **2015**, *14*, 1–16. [CrossRef]
4. Abbas, D.; Current, D.; Phillips, M.; Rossman, R.; Hoganson, H.; Brooks, K.N. Guidelines for harvesting forest biomass for energy: A synthesis of environmental considerations. *Biomass Bioenergy* **2011**, *35*, 4538–4546. [CrossRef]
5. Strandberg, G.; Kjellström, E.; Poska, A.; Wagner, S.; Gaillard, M.-J.; Trondman, A.-K.; Mauri, A.; Davis, B.A.S.; Kaplan, J.O.; Birks, H.J.B.; et al. Regional climate model simulations for Europe at 6 and 0.2 k BP: Sensitivity to changes in anthropogenic deforestation. *Clim. Past* **2014**, *10*, 661–680. [CrossRef]
6. Futter, M.N.; Högbom, L.; Valinia, S.; Sponseller, R.A.; Laudon, H. Conceptualizing and communicating management effects on forest water quality. *AMBIO* **2016**, *45*, 188–202. [CrossRef] [PubMed]
7. Cui, X.; Liu, S.; Wei, X. Impacts of forest changes on hydrology: A case study of large watersheds in the upper reaches of Minjiang River watershed in China. *Hydrol. Earth Syst. Sci.* **2012**, *16*, 4279–4290. [CrossRef]
8. Zhang, M.; Liu, N.; Harper, R.; Li, Q.; Liu, K.; Wei, X.; Ning, D.; Hou, Y.; Liu, S. A global review on hydrological responses to forest change across multiple spatial scales: Importance of scale, climate, forest type and hydrological regime. *J. Hydrol.* **2017**, *546*, 44–59. [CrossRef]
9. Bremer, L.L.; Farley, K.A. Does plantation forestry restore biodiversity or create green deserts? A synthesis of the effects of land-use transitions on plant species richness. *Biodivers. Conserv.* **2010**, *19*, 3893–3915. [CrossRef]
10. Lindhjem, H.; Reinvang, R.; Zandersen, M. Landscape experiences as a cultural ecosystem service in a Nordic context. *Conceptsvalues Decis. Mak.* **2015**, *2015*, 549. [CrossRef]
11. Roberts, N.; Fyfe, R.M.; Woodbridge, J.; Gaillard, M.-J.; Davis, B.A.S.; Kaplan, J.O.; Marquer, L.; Mazier, F.; Nielsen, A.B.; Sugita, S.; et al. Europe's lost forests: A pollen-based synthesis for the last 11,000 years. *Sci. Rep.* **2018**, *8*, 1–8. [CrossRef] [PubMed]
12. Poska, A.; Väli, V.; Tomson, P.; Vassiljev, J.; Kihno, K.; Alliksaar, T.; Villoslada, M.; Saarse, L.; Sepp, K. Reading past landscapes: Combining modern and historical records, maps, pollen-based vegetation reconstructions, and the socioeconomic background. *Landsc. Ecol.* **2018**, *33*, 529–546. [CrossRef]
13. Päivänen, J.; Hånell, B. *Peatland Ecology and Forestry—A Sound Approach. Helsingin Yliopiston Metsätieteiden Laitoksen Julkaisuja*; Department of Forest Sciences, University of Helsinki: Helsinki, Finland, 2012; Volume 3, 267p.
14. Nieminen, M.; Sarkkola, S.; Laurén, A. Impacts of forest harvesting on nutrient, sediment and dissolved organic carbon exports from drained peatlands: A literature review, synthesis and suggestions for the future. *For. Ecol. Manag.* **2017**, *392*, 13–20. [CrossRef]
15. Nieminen, M.; Piirainen, S.; Sikström, U.; Löfgren, S.; Marttila, H.; Sarkkola, S.; Laurén, A.; Finér, L. Ditch network maintenance in peat-dominated boreal forests: Review and analysis of water quality management options. *AMBIO* **2018**, *47*, 535–545. [CrossRef]
16. Päivinen, R.; Päivinen, R.; Lehikoinen, M.; Lehikoinen, M.; Schuck, A.; Schuck, A.; Häme, T.; Häme, T.; Väätäinen, S.; Väätäinen, S.; et al. *Mapping Forest in Europe by Combining Earth Observation Data and Forest Statistics*; Springer: Berlin/Heidelberg, Germany, 2003; Volume 76, pp. 279–294.
17. Schuck, A.; Van Brusselen, J.; Päivinen, R.; Häme, T.; Folving, S. Compilation of a calibrated European forest map derived from NOAA-AVHRR data. European Forest Institute. *EFI Intern. Rep.* **2002**, *13*, 44.
18. Kempeneers, P.; Sedano, F.; Seebach, L.; Strobl, P.; San-Miguel-Ayanz, J. Data Fusion of Different Spatial Resolution Remote Sensing Images Applied to Forest-Type Mapping. *IEEE Trans. Geosci. Remote Sens.* **2011**, *49*, 4977–4986. [CrossRef]
19. Anonymous. *Forest Statistics, Official Statistics of Sweden*; Swedish University of Agricultural Sciences: Umeå, Sweden, 2019; 144p. Available online: www.slu.se/en/Collaborative-Centres-and-Projects/the-swedish-national-forest-inventory/forest-statistics/forest-statistics/ (accessed on 20 January 2021).
20. Anonymous. *Forest Resources by Region [Web Publication]*; Natural Resources Institute: Helsinki, Finland, 2019. Available online: Stat.luke.fi/en/forest-resources-region_en-2 (accessed on 31 August 2020).
21. Anonymous. *Total Roundwood Removals and Drain [Web Publication]*; Natural Resources Institute: Helsinki, Finland, 2020. Available online: Stat.luke.fi/en/roundwood-removals-and-drain (accessed on 31 August 2020).
22. Sheil, D. How plants water our planet: Advances and imperatives. *Trends Plant Sci.* **2014**, *19*, 209–211. [CrossRef]
23. Makarieva, A.M.; Gorshkov, V.G. Biotic pump of atmospheric moisture as driver of the hydrological cycle on land. *Hydrol. Earth Syst. Sci.* **2007**, *11*, 1013–1033. Available online: www.hydrol-earth-syst-sci.net/11/1013/2007/1 (accessed on 31 August 2020). [CrossRef]

24. McDonald, A.G.; Bealey, W.J.; Fowler, D.; Dragosits, U.; Skiba, U.; Smith, R.; Donovan, R.G.; Brett, H.E.; Hewitt, C.N.; Nemitz, E. Quantifying the effect of urban tree planting on concentrations and depositions of PM10 in two UK conurbations. *Atmos. Environ.* **2007**, *41*, 8455–8467. [CrossRef]
25. Roy, S.; Byrne, J.; Pickering, C. A systematic quantitative review of urban tree benefits, costs, and assessment methods across cities in different climatic zones. *Urban For. Urban Green.* **2012**, *11*, 351–363. [CrossRef]
26. Livesley, S.J.; McPherson, E.G.; Calfapietra, C. The Urban Forest and Ecosystem Services: Impacts on Urban Water, Heat, and Pollution Cycles at the Tree, Street, and City Scale. *J. Environ. Qual.* **2016**, *45*, 119–124. [CrossRef]
27. Contreras-Hermosilla, A. *Towards Sustainable Forest Management: An Examination of the Technical, Economic and Institutional Feasibility of Improving Management of the Global Forest Estate*; FAO Forestry Policy and Planning Division: Rome, Italy, 1999. Available online: Agris.fao.org/agris-search/search.do?recordID=XF2000394043 (accessed on 20 January 2021).
28. Svensson, J.; Neumann, W.; Bjärstig, T.; Zachrisson, A.; Thellbro, C. Landscape Approaches to Sustainability—Aspects of Conflict, Integration, and Synergy in National Public Land-Use Interests. *Sustainability* **2020**, *12*, 5113. [CrossRef]
29. Sandström, C.; Moen, J.; Widmark, C.; Danell, Ö. Progressing toward co-management through collaborative learning: Forestry and reindeer husbandry in dialogue. *Int. J. Biodivers. Sci. Manag.* **2006**, *2*, 326–333. [CrossRef]
30. Watson, J.E.M.; Evans, T.; Venter, O.; Williams, B.; Tulloch, A.; Stewart, C.; Thompson, I.; Ray, J.C.; Murray, K.; Salazar, A.; et al. The exceptional value of intact forest ecosystems. *Nat. Ecol. Evol.* **2018**, *2*, 599–610. [CrossRef] [PubMed]
31. Potapov, P.; Hansen, M.C.; Laestadius, L.; Turubanova, S.; Yaroshenko, A.; Thies, C.; Smith, W.; Zhuravleva, I.; Komarova, A.; Minnemeyer, S.; et al. The last frontiers of wilderness: Tracking loss of intact forest landscapes from 2000 to 2013. *Sci. Adv.* **2017**, *3*, e1600821. [CrossRef]
32. Jonsson, B.G.; Svensson, J.; Mikusiński, G.; Manton, M.; Angelstam, P. European Union's Last Intact Forest Landscapes are at A Value Chain Crossroad between Multiple Use and Intensified Wood Production. *Forests* **2019**, *10*, 564. [CrossRef]
33. Svensson, J.; Bubnicki, J.W.; Jonsson, B.G.; Andersson, J.; Mikusiński, G. Conservation significance of intact forest landscapes in the Scandinavian Mountains Green Belt. *Landsc. Ecol.* **2020**, *35*, 2113–2131. [CrossRef]
34. Svensson, J.; Andersson, J.; Sandström, P.; Mikusiński, G.; Jonsson, B.G. Landscape trajectory of natural boreal forest loss as an impediment to green infrastructure. *Conserv. Biol.* **2019**, *33*, 152–163. [CrossRef] [PubMed]
35. Anonymous. The European Green Deal. COM/2019/640 Final. 2019. Available online: Eur-lex.europa.eu/legalcontent/EN/TXT/?qid=1588580774040&uri=CELEX:52019DC0640 (accessed on 20 January 2021).
36. Lilja, K.; Loukola-Ruskeeniemi, K. *Wood-Based Bioeconomy Solving Global Challenges*; Finnish Ministry of Economic Affairs and Employment, Enterprise and Innovation Department: Helsinki, Finland, 2017; ISSN 2342-7922.
37. Winkler, G. Towards a Sustainable European Forest-Based Bioeconomy—Assessment and the Way Forward. European Forest Institute. Available online: Efi.int/sites/default/files/files/publication-bank/2018/efi_wsctu8_2017.pdf (accessed on 20 January 2021).
38. Kauppi, P.E.; Ausubel, J.H.; Fang, J.; Mather, A.S.; Sedjo, R.A.; Waggoner, P.E. Returning forests analyzed with the forest identity. *Proc. Natl. Acad. Sci. USA* **2006**, *103*, 17574–17579. [CrossRef]
39. Kumar, A.; Adamopoulos, S.; Jones, D.; Amiandamhen, S.O. Forest Biomass Availability and Utilization Potential in Sweden: A Review. *Waste Biomass Valorization* **2021**, *12*, 65–80. [CrossRef]
40. IRENA. *Bioenergy from Boreal Forests: Swedish Approach to Sustainable Wood Use*; International Renewable Energy Agency: Abu Dhabi, United Arab Emirates, 2019; ISBN 978-92-9260-119-5.
41. Kolström, M.; Lindner, M.; Vilén, T.; Maroschek, M.; Seidl, R.; Lexer, M.J.; Netherer, S.; Kremer, A.; Delzon, S.; Barbati, A.; et al. Reviewing the Science and Implementation of Climate Change Adaptation Measures in European Forestry. *Forests* **2011**, *2*, 961–982. [CrossRef]
42. Seedre, M.; Kopáček, J.; Janda, P.; Bače, R.; Svoboda, M. Carbon pools in a montane old-growth Norway spruce ecosystem in Bohemian Forest: Effects of stand age and elevation. *For. Ecol. Manag.* **2015**, *346*, 106–113. [CrossRef]
43. Ķēniņa, L.; Jaunslaviete, I.; Liepa, L.; Zute, D.; Jansons, Ā. Carbon Pools in Old-Growth Scots Pine Stands in Hemiboreal Latvia. *Forests* **2019**, *10*, 911. [CrossRef]
44. Hanewinkel, M.; Cullmann, D.A.; Schelhaas, M.-J.; Nabuurs, G.-J.; Zimmermann, N.E. Climate change may cause severe loss in the economic value of European forest land. *Nat. Clim. Chang.* **2012**, *3*, 203–207. [CrossRef]
45. Seidl, R.; Schelhaas, M.-J.; Rammer, W.; Verkerk, P.J. Increasing forest disturbances in Europe and their impact on carbon storage. *Nat. Clim. Chang.* **2014**, *4*, 806–810. [CrossRef] [PubMed]
46. Jonsson, M.; Wardle, D.A. Structural equation modelling reveals plant-community drivers of carbon storage in boreal forest ecosystems. *Biol. Lett.* **2009**, *6*, 116–119. [CrossRef]
47. Chaudhary, A.; Burivalova, Z.; Koh, L.P.; Hellweg, S. Impact of Forest Management on Species Richness: Global Meta-Analysis and Economic Trade-Offs. *Sci. Rep.* **2016**, *6*, 23954. [CrossRef] [PubMed]
48. Brockerhoff, E.G.; Barbaro, L.; Castagneyrol, B.; Forrester, D.I.; Gardiner, B.; González-Olabarria, J.R.; Lyver, P.O.; Meurisse, N.; Oxbrough, A.; Taki, H.; et al. Forest biodiversity, ecosystem functioning and the provision of ecosystem services. *Biodivers. Conserv.* **2017**, *26*, 3005–3035. [CrossRef]
49. Chi, J.; Nilsson, M.B.; Kljun, N.; Wallerman, J.; Fransson, J.E.; Laudon, H.; Lundmark, T.; Peichl, M. The carbon balance of a managed boreal landscape measured from a tall tower in northern Sweden. *Agric. For. Meteorol.* **2019**, *274*, 29–41. [CrossRef]

50. Rebane, S.; Jõgiste, K.; Kiviste, A.; Stanturf, J.A.; Kangur, A.; Metslaid, M. C-exchange and balance following clear-cutting in hemiboreal forest ecosystem under summer drought. *For. Ecol. Manag.* **2020**, *472*, 118249. [CrossRef]
51. Berndes, G.; Abt, B.; Asikainen, A.; Cowie, A.; Dale, V.; Egnell, G.; Lindner, M.; Marelli, L.; Paré, D.; Pingoud, K.; et al. Forest biomass, carbon neutrality and climate change mitigation. *Sci. Policy* **2016**, *3*, 3–27. [CrossRef]
52. Cintas, O.; Berndes, G.; Cowie, A.L.; Egnell, G.; Holmström, H.; Ågren, G.I. The climate effect of increased forest bioenergy use in Sweden: Evaluation at different spatial and temporal scales. *Wiley Interdiscip. Rev. Energy Environ.* **2016**, *5*, 351–369. [CrossRef]
53. Lilja, S.; Wallenius, T.; Kuuluvainen, T. Structure and development of oldPicea abiesforests in northern boreal Fennoscandia. *Écoscience* **2006**, *13*, 181–192. [CrossRef]
54. Luyssaert, S.; Schulze, E.-D.; Börner, A.; Knohl, A.; Hessenmöller, D.; Law, B.E.; Ciais, P.; Grace, J. Old-growth forests as global carbon sinks. *Nat. Cell Biol.* **2008**, *455*, 213–215. [CrossRef] [PubMed]
55. Nabuurs, G.-J.; Lindner, M.; Verkerk, P.J.; Gunia, K.; Deda, P.; Michalak, R.; Grassi, G. First signs of carbon sink saturation in European forest biomass. *Nat. Clim. Chang.* **2013**, *3*, 792–796. [CrossRef]
56. Marklund, L.G. Biomassafunktioner för gran i Sverige. *Sver. Lantbr. Inst. För Skogstaxeringrapport* **1987**, *43*, 127. (In Swedish)
57. Hakkila, P. *Utilization of Residual Forest Biomass*; Springer: Berlin/Heidelberg, Germany, 1989; p. 568.
58. Petersson, H.; Ståhl, G. Functions for below-ground biomass of Pinus sylvestris, Picea abies, Betula pendula and Betula pubescens in Sweden. *Scand. J. For. Res.* **2006**, *21*, 84–93. [CrossRef]
59. Armolaitis, K.; Varnagirytė-Kabašinskienė, I.; Stupak, I.; Kukkola, M.; Mikšys, V.; Wójcik, J. Carbon and nutrients of Scots pine stands on sandy soils in Lithuania in relation to bioenergy sustainability. *Biomass Bioenergy* **2013**, *54*, 250–259. [CrossRef]
60. Liski, J.; Ilvesniemi, H.; Mäkelä, A.; Starr, M. Model analysis of the effects of soil age, fires and harvesting on the carbon storage of boreal forest soils. *Eur. J. Soil Sci.* **1998**, *49*, 407–416. [CrossRef]
61. Armolaitis, K.; Stakėnas, V.; Varnagirytė-Kabašinskienė, I.; Gudauskienė, A.; Žemaitis, P. Leaching of organic carbon and plant nutrients at clear cutting of Scots pine stand on arenosol. *Baltic For.* **2018**, *24*, 50–59.
62. Oni, S.K.; Tiwari, T.; Ledesma, J.L.J.; Ågren, A.M.; Teutschbein, C.; Schelker, J.; Laudon, H.; Futter, M.N. Local- and landscape-scale impacts of clear-cuts and climate change on surface water dissolved organic carbon in boreal forests. *J. Geophys. Res. Biogeosciences* **2015**, *120*, 2402–2426. [CrossRef]
63. Campeau, A.; Bishop, K.; Amvrosiadi, N.; Billett, M.F.; Garnett, M.H.; Laudon, H.; Öquist, M.G.; Wallin, M.B. Current forest carbon fixation fuels stream CO2 emissions. *Nat. Commun.* **2019**, *10*, 1–9. [CrossRef]
64. Mayer, M.; Prescott, C.E.; Abaker, W.E.; Augusto, L.; Cécillon, L.; Ferreira, G.W.; James, J.; Jandl, R.; Katzensteiner, K.; Laclau, J.-P.; et al. Tamm Review: Influence of forest management activities on soil organic carbon stocks: A knowledge synthesis. *For. Ecol. Manag.* **2020**, *466*, 118127. [CrossRef]
65. Mjöfors, K.; Strömgren, M.; Nohrstedt, H.Ö.; Johansson, M.-B.; Gärdenäs, A.I. Indications that site preparation increases forest ecosystem carbon stocks in the long term. *Scand. J. For. Res.* **2017**, *32*, 717–725. [CrossRef]
66. Larsson, A.; Bjelke, U.; Dahlberg, A.; Sandström, J. Tillståndet i skogen–rödlistade arter i ett nordiskt perspektiv. *ArtDatabanken Rapporterar* **2011**, *9*, 4–13.
67. Anonymous. Swedish University of Agricultural Sciences, Species Information Centre. 2020. Available online: www.artdatabanken.se/en/ (accessed on 20 January 2021).
68. Lindenmeyer, D.; Franklin, J.F. *Conserving Forest Biodiversity: A Comprehensive Multiscaled Approach*; Island Press: Washington, DC, USA, 2002; ISBN 978-1-59726-853-0.
69. Fahrig, L. Effects of Habitat Fragmentation on Biodiversity. *Annu. Rev. Ecol. Evol. Syst.* **2003**, *34*, 487–515. [CrossRef]
70. Fahrig, L. Rethinking patch size and isolation effects: The habitat amount hypothesis. *J. Biogeogr.* **2013**, *40*, 1649–1663. [CrossRef]
71. Hanski, I. Habitat fragmentation and species richness. *J. Biogeogr.* **2015**, *42*, 989–993. [CrossRef]
72. Fahrig, L. Ecological Responses to Habitat Fragmentation Per Se. *Annu. Rev. Ecol. Evol. Syst.* **2017**, *48*, 1–23. [CrossRef]
73. Rybicki, J.; Abrego, N.; Ovaskainen, O. Habitat fragmentation and species diversity in competitive communities. *Ecol. Lett.* **2019**, *23*, 506–517. [CrossRef]
74. Guerra, C.A.; Heintz-Buschart, A.; Sikorski, J.; Chatzinotas, A.; Guerrero-Ramírez, N.; Cesarz, S.; Beaumelle, L.; Rillig, M.C.; Maestre, F.T.; Delgado-Baquerizo, M.; et al. Blind spots in global soil biodiversity and ecosystem function research. *Nat. Commun.* **2020**, *11*, 1–13. [CrossRef]
75. Palmero-Iniesta, M.; Espelta, J.M.; Gordillo, J.; Pino, J. Changes in forest landscape patterns resulting from recent afforestation in Europe (1990–2012): Defragmentation of pre-existing forest versus new patch proliferation. *Ann. For. Sci.* **2020**, *77*, 1–15. [CrossRef]
76. Trumbore, S.E.; Brando, P.M.; Hartmann, H. Forest health and global change. *Science* **2015**, *349*, 814–818. [CrossRef]
77. Frelich, L.E.; Jõgiste, K.; Stanturf, J.; Jansons, A.; Vodde, F. Are Secondary Forests Ready for Climate Change? It Depends on Magnitude of Climate Change, Landscape Diversity and Ecosystem Legacies. *Forests* **2020**, *11*, 965. [CrossRef]
78. Seidl, R.; Thom, D.; Kautz, M.; Martin-Benito, D.; Peltoniemi, M.; Vacchiano, G.; Wild, J.; Ascoli, D.; Petr, M.; Honkaniemi, J.; et al. Forest disturbances under climate change. *Nat. Clim. Chang.* **2017**, *7*, 395–402. [CrossRef] [PubMed]
79. Palahí, M.; Pantsar, M.; Costanza, R.; Kubiszewski, I.; Potočnik, J.; Stuchtey, M.; Nasi, R.; Lovins, H.; Giovannini, E.; Fioramonti, L.; et al. Investing in Nature as the true engine of our economy: A 10-point Action Plan for a Circular Bioeconomy of Wellbeing. *Knowl. Action* **2020**, *2*, 58. [CrossRef]

80. Widmark, C.; Heräjärvi, H.; Kurttila, M.; Lier, K.; Mutanen, A.A.; Øistad, K.; Routa, J.; Saranpää, P.; Tolvanen, A.; Viitanen, J. The Forest in Northern Europe's Emerging Bioeconomy—Reflections on the Forest's Role in the Bioeconomy. 2020. Available online: https://forbioeconomy.com/app/uploads/2021/01/The-Forest-in-Northern-Europe\T1\textquoterights-Emerging-Bioeconomy.pdf (accessed on 20 January 2021).
81. Jõgiste, K.; Korjus, H.; Stanturf, J.A.; Frelich, L.; Baders, E.; Donis, J.; Jansons, A.; Kangur, A.; Köster, K.; Laarmann, D.; et al. Hemiboreal forest: Natural disturbances and the importance of ecosystem legacies to management. *Ecosphere* **2017**, *8*, e01706. [CrossRef]
82. Summers, J.K.; Smith, L.M.; Fulford, R.S.; Crespo, R.D.J. The Role of Ecosystem Services in Community Well-Being. *Ecosyst. Serv. Glob. Ecol.* **2018**, *145*, 13.
83. Anonymous. EU Biodiversity Strategy for COM/2020/380 Final. 2020. Available online: Eu/legal-content/EN/TXT/?uri=CELEX:52020DC0380 (accessed on 20 January 2021).
84. Nilsson, M.; Griggs, D.; Visbeck, M. Policy: Map the interactions between Sustainable Development Goals. *Nat. Cell Biol.* **2016**, *534*, 320–322. [CrossRef] [PubMed]
85. Hetemäki, L. The role of science in forest policy–Experiences by EFI. *For. Policy Econ.* **2019**, *105*, 10–16. [CrossRef]
86. IRENA International Renewable Energy Agency. Available online: https://irena.org/wind (accessed on 19 March 2021).
87. Northrup, J.M.; Wittemyer, G. Characterising the impacts of emerging energy development on wildlife, with an eye towards mitigation. *Ecol. Lett.* **2012**, *16*, 112–125. [CrossRef] [PubMed]
88. Eichhorn, M.; Tafarte, P.; Thrän, D. Towards energy landscapes—"Pathfinder for sustainable wind power locations". *Energy* **2017**, *134*, 611–621. [CrossRef]
89. Diógenes, J.R.F.; Claro, J.; Rodrigues, J.C.; Loureiro, M.V. Barriers to onshore wind energy implementation: A systematic review. *Energy Res. Soc. Sci.* **2020**, *60*, 101337. [CrossRef]
90. Oliver, C.D.; Nassar, N.T.; Lippke, B.R.; McCarter, J.B. Carbon, Fossil Fuel, and Biodiversity Mitigation With Wood and Forests. *J. Sustain. For.* **2014**, *33*, 248–275. [CrossRef]
91. Gustavsson, L.; Nguyen, T.; Sathre, R.; Tettey, U. Climate effects of forestry and substitution of concrete buildings and fossil energy. *Renew. Sustain. Energy Rev.* **2021**, *136*, 110435. [CrossRef]
92. Stupak, I.; Asikainen, A.; Jonsell, M.; Karltun, E.; Lunnan, A.; Mizaraite, D.; Pasanen, K.; Parn, H.; Raulundrasmussen, K.; Roser, D. Sustainable utilisation of forest biomass for energy—Possibilities and problems: Policy, legislation, certification, and recommendations and guidelines in the Nordic, Baltic, and other European countries. *Biomass Bioenergy* **2007**, *31*, 666–684. [CrossRef]
93. Baumgartner, R.J. Sustainable Development Goals and the Forest Sector—A Complex Relationship. *Forests* **2019**, *10*, 152. [CrossRef]

sustainability

MDPI

Article

Sustainability Goals and Firm Behaviours: A Multi-Criteria Approach on Italian Agro-Food Sector

Lucia Briamonte [1], Raffaella Pergamo [1], Brunella Arru [2,*], Roberto Furesi [2], Pietro Pulina [2] and Fabio A. Madau [2]

1 CREA-PB, 00198 Rome, Italy; lucia.briamonte@crea.gov.it (L.B.); raffaella.pergamo@crea.gov.it (R.P.)
2 Department of Agriculture, University of Sassari, 07100 Sassari, Italy; rfuresi@uniss.it (R.F.); ppulina@uniss.it (P.P.); famadau@uniss.it (F.A.M.)
* Correspondence: brarru@uniss.it

Abstract: Today, the transition to a more sustainable model of the agro-food system is increasingly impellent, requiring all actors' commitment. In particular, small and medium agro-food business (SMABs) play a decisive and central role in the food and economies of national and underdeveloped areas. Our study aims to identify, through desk research, the level of commitment and communication to the sustainability of SMABs operating in southern Italy. In this study, we followed the Food and Agriculture Organization's (FAO) approach to implementing such a transition, using their principles as a diagnostic tool to interpret business operations. The data were analysed using two approaches: a regime analysis to assess which FAO principles are commonly followed to make the above transition possible, and an extension of the Abraham and Pingali (2020) framework to describe the commitment of SMABs to the Agenda 2030 goals with respect to the behaviour of small and medium enterprises (SMEs). We found that the SMABs' behaviours are more oriented towards some FAO principles: those that explain their commitment to improving natural resources and livelihoods, fostering inclusive economic growth, and achieving sustainable development goal 7 of Agenda 2030 than towards others. The contribution of our study lies in providing detailed insights into sustainable actions taken by SMABs while testing the FAO's principles as a new model to evaluate business operations.

Keywords: agro-food business; small and medium enterprise (SME); Food and Agriculture Organization (FAO); Agenda 2030; regime analysis

check for updates

Citation: Briamonte, L.; Pergamo, R.; Arru, B.; Furesi, R.; Pulina, P.; Madau, F.A. Sustainability Goals and Firm Behaviours: A Multi-Criteria Approach on Italian Agro-Food Sector. *Sustainability* **2021**, *13*, 5589. https://doi.org/10.3390/su13105589

Academic Editors: Margarita Martinez-Nuñez and Mª Pilar Latorre-Martínez

Received: 22 March 2021
Accepted: 14 May 2021
Published: 17 May 2021

1. Introduction

Global food challenges in the 21st century require substantive changes in agriculture and the food system. These challenges are accelerating the transition to sustainable food and agriculture (SFA) to enable world food security and healthier diets, societal well-being, and environmental safeguards [1].

SFA is at the centre of the 2030 Agenda, which, in shifting the debate from 'willingness' to 'the ability to act', aims to spur people and institutions to an urgent rethinking of the global development model. In this vein, the Food and Agriculture Organisation (FAO) [2] has also developed a vision for SFA based on five principles which are aimed at providing a basis for developing policies, strategies, regulations, and incentives that enable SFA and rural development: (1) increasing the productivity, employment, and value addition in food systems; (2) protecting and enhancing natural ecosystems; (3) improving livelihoods and fostering inclusive economic growth; (4) enhancing the resilience of people, communities, and ecosystems, and (5) adapting governance to new challenges. Twenty interconnected actions, which countries together with key stakeholders should take to accelerate the transition to SFA, are derived from these principles. Moreover, these 20 practical and interconnected FAO actions, in addiction to aiming to transform food and agriculture, intend to drive achievement across the sustainable development goals (SDGs) of Agenda 2030 [2]. In effect, FAO plays a critical role in the 2030 Agenda [3] and FAO's strategic framework is explicitly aligned with the SDGs [4]. This is why FAO punctually indicates for each action which the SDGs are on which a contribution is made [2]. Evidence on business behaviours undertaken along the interlinks among FAO's actions and SDGs allow one to display whether and how the agri-food sector contributes to achieving sustainable development according to the Agenda 2030 [4].

Among the agents of SFA transition, the agriculture and agro-food business (hereinafter AFBs) play an essential role in improving and revitalising rural contexts [5], responding to a growing world population's food demand, and fostering inclusive growth; these are the cornerstones of the success of Agenda 2030 [6]. In particular, the role of SMABs, as well as of families and smallholders, has become increasingly more decisive, being the backbone of many rural societies, promoting innovation, and playing a central role in national food and economies [7–11].

Against this background, the transition to SFA from conventional unsustainable food practices requires the consideration of pivotal interlinks among the incomes of family and smallholder farms, sustainable practices, improved productivity, and technological innovation and efficiency across the sector [12–17].

This is also the approach of the European Green Deal, which calls on the European farmers and AFBs to re-adjust their work practices to the new green objectives (EU COM/2019/640 final; EU COM/2020/381). In effect, the European Green Deal, with the 'Farm to Fork' and 'Biodiversity' strategies and a reformed Common Agricultural Policy (CAP), aims to switch Europe's agricultural sector towards a more sustainable model, ensuring food security and the preservation of environmental and human health and making the EU the first climate-neutral continent in the world. The CAP and European Green Deal objectives' achievement requires that the entire agro-food sector (farms, agro-food industry, and organisations), which is known to be one of the main drivers of the EU economy [18], is involved in the changing process.

However, wanting to get upstream of the speech, although the European Green Deal is Europe's new growth strategy to transform the EU economy for a sustainable future, we need to refer first of all to the Agenda 2030 program, which also includes some of the strategic objectives of the CAP, and that is "the cornerstone of defining EU policies and interventions" [19] (p. 9). This is why Agenda 2030 is the ideal strategic framework for addressing the issue of sustainability for investigating AFBs.

In Italy, within a few years, many AFBs have been initiated that pursue social, environmental, and economic sustainability goals; Italy ranks third (preceded by Israel and Spain) at the international level in terms of sustainable agro-food start-ups [20]. These start-ups aim for innovative solutions for implementing more efficient use of resources, introducing the 'short supply chain', and using natural materials in production. The Italian agro-food system has shown progressive advancement and a good capacity for the sustainable management of agricultural resources [21]. Moreover, Italy's civil, social, and political traditions have contributed over time to an orientation towards sustainable entrepreneurship, fostered by a positive bond with the territory and the environment, along with a high level of social cohesion and stakeholders' proximity [22]. Within this context, SMEs, considered the backbone of the Italian economy, epitomise the entrepreneurs' ethical values that lead to adopting sustainability practices and strategies and contributing to sustainable development [22,23].

However, Italy has prominent differences between its north and south. South Italy is one of the EU's most underdeveloped areas and has a lower gross domestic product (GDP) and industrialisation rate than the peninsula. According to ISMEA, "in the South, the agri-food sector assumes greater economic importance than the Italian average and the rest of the country [. . . , and] the agri-food chain is an important production pillar in the South" [24] (p. 37). Previous studies have shown that commitment to sustainability provides SMEs with competitive advantages, creating new market access, aligning activities with shifting customer preferences, capitalising on innovative solutions, filling market gaps due to market failure, and addressing economic disequilibria [8,25–27]. In this sense, the recovery of competitiveness of the agro-food sector of southern Italy must take place through the transition towards a sustainable AFB model to address this region's economic backwardness.

Based on these considerations, investigating the level of commitment and communication to sustainability among southern Italian SMABs is an important research topic.

Our study contributes to addressing this topic by diagnosing the current sustainability objectives mainly promoted by the southern Italian SMABS as well as analysing the issues that need to be addressed to achieve a higher level of sustainability. Specifically, a qualitative analysis was carried out to assess the FAO's principles and actions [2] towards which the sustainability practices of southern Italy's SMABs tend to be the most commonly oriented. Another contribution of this study is to present the FAO's principles as a diagnostic tool for evaluating business operations.

Furthermore, since FAO [2] links each action to several SDGs of the 2030 Agenda, we analysed the sustainable behaviours of SMABs also from the SDG point of view, adapting the Abraham and Pingali framework [28].

To our knowledge, this is the first study that has highlighted SMABs' behaviour regarding their compliance with the sustainability principles of the FAO. Moreover, as far as we know, this is the first study that uses the Abraham and Pingali framework [28] to look into the compliance of SMABs' behaviours with the SDGs in light of the specificities of the agro-food sector.

Our results can be a good starting point for future discussions on actions to be taken to improve SMABs competitiveness in this macro area.

2. Methodology

2.1. Data Source

The analysis presented in this study was carried out using a qualitative research approach, focusing on the SMABs operating in southern Italy. The firms' data were selected from the AIDA database (the Bureau van Dijk), which offers financial, demographic, and commercial information on Italian firms. Before selection, the firms were screened for the following criteria:

- Employees: a minimum of 10 (micro-businesses excluded);
- Legal status: active and viable;
- Availability of an official website.

A total of 720 Italian firms were selected. The research was carried out in July 2019 through a two-phase analysis of each firm's website.

First, we only selected the AFBs that have their own active websites, reducing the sample to 650 firms, of which 616 were SMABs. Then, companies that communicated one or more concrete actions (and not a simple declaration of intent) in line with the FAO's principles of sustainable food and agriculture actions were identified. This criterion helped to identify firms that implement and communicate sustainable actions in the field and not just state a commitment to sustainability. After applying these criteria, a total of 193 southern Italian firms were selected, of which 180 were SMABSs.

Websites were used as valid sources of data for several reasons. Communication plays a fundamental role and is an integral part of every sustainability plan or strategy [29]. In the case of AFBs, entered in the database consulted, the website, in addition to carrying out a communication function, also could refer to an idea of the web reputation of the company itself that can trace the picture of the main sustainable behaviours assumed over time and highlight the path taken.

Due to the growing demand from stakeholders for greater transparency, social and environmental responsibility, and dialogue, companies' awareness of the need to not only adopt sustainability activities but also inform their stakeholders about sustainability performance has grown, leading to constant growth in the size and complexity of communication on social and environmental issues [30].

Websites are the main communication channels for sustainable initiatives. They enable the communication of relevant information about the firms' commitment towards sustainable practice to a wider range of stakeholders than traditional media [31–33]. This is why many researchers chose companies' websites to measure their sustainability practices [34].

Our research focuses on SMABs' commitment towards sustainability. According to Hasim et al. (2018), 'commitment' can be understood as the extent of information provided by firms on their website, with demonstrated actions towards achieving sustainable development [35]. Moreover, due to the advantages offered by the Internet [36–38], websites are particularly appealing to SMEs because they mitigate traditional burdens related to firm size [39].

Therefore, this study considers website analyses as a legitimate research tool.

2.2. Research Model

The analysis focuses on assessing the ongoing sustainability behaviours mainly promoted by southern Italy's SMABs according to the five FAO principles stated in 'transforming food and agriculture to achieve the SDGs' [2] and the Agenda 2030 goals, bearing in mind the interlinks between the former and the latter. In effect, the FAO document [2] explains how each FAO action links multiple SDGs and integrates the three dimensions of sustainable development (economic, social, and environmental).

Regarding the five principles of FAO, although they are complementary, by highlighting the importance assigned to each one, it is possible to deepen which aspect is favoured by AFBs and offer considerations on the process of integrating sustainability targets into their practices. Moreover, we think that the analysis of the AFBs' behaviours according to the FAO principles and actions [2] is a suitable tool to overcome the limitations of the sustainability assessment of SMABs [40,41].

In reference to the Agenda 2030—keeping in mind the pivotal role of the private sector in entrepreneurs' engagement and know-how transfer; job creation; and alternative revenue streams, and in particular, of the role of SMEs for the implementation of the 2030 agenda—we examined the agro-food SMEs' behaviours in light of the 17 sustainable development goals (SDGs) of Agenda 2030 according to the nature of the promoted actions of firms.

To this end, the Abraham and Pingali framework [28] was adapted. The authors offer a seminal paper that makes clear the link between the SDGs and the agri-food sector, with a particular focus on smallholder farming, identified which explicitly depend on firm growth for their achievement, and classified the SDGs that specifically pertain to the agricultural sector into four clusters: poverty goals, nutrition goals, social goals, and environmental goals. To look at the goals in the context of agro-food SMEs' behaviours, and considering the need to broaden the horizons from the agricultural sector to the agro-industrial one, the Abraham and Pingali clustering has been revisited, offering its evolved version, which includes all SDGs. The relevance of the SDGs not previously included in the Abraham and Pingali framework with the four clusters identified by the authors was assessed based on the content of the Agenda 2030, leading to the attribution of SDG 4 to the social cluster and of SDG7 and SDg14 to the environmental cluster. The SDGs 11, 16 and 17, not being directly attributable to one of the 4 clusters of the framework, were included in a new cluster named "residual cluster". This "extended framework" can produce a picture of SDGs' categories on which to act by stimulating, accelerating and supporting SMEs' behaviours to foster their transition to sustainable food and agriculture.

Figure 1 shows the existence of the link between the five FAO principles and the SDGs (which we have decided to analyse in light of the extended classification proposed by Abraham and Pingali [28]) and that SMABs' sustainable behaviour can be read in light of these two frameworks.

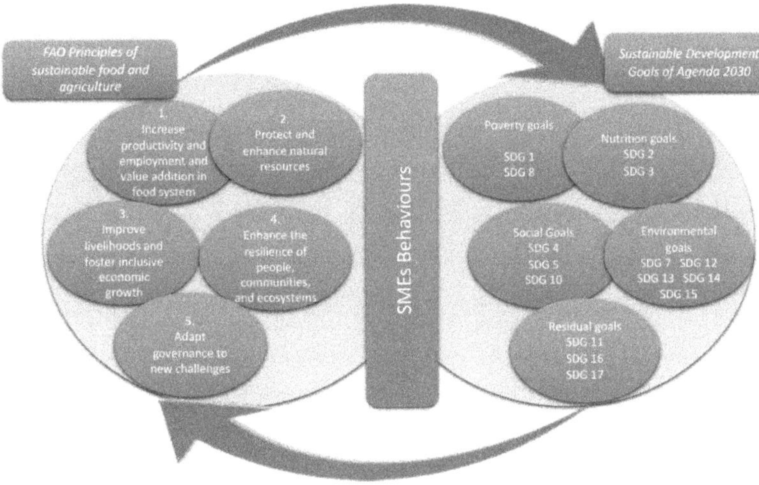

Figure 1. Research model.

2.3. Method

Given the exploratory character of this study and our research objectives, qualitative research was considered the most appropriate technique for analysing AFB behaviours' diversity. Qualitative research is gaining popularity in the small business and entrepreneurship research community [42]. It deals with non-numerical information, allowing the interpretation of a phenomenon and building a meaningful picture without compromising its richness and dimensionality [43].

Our qualitative analysis aimed at describing the diversity of sustainable behaviours of AFBs. Since the description of the diversity of characteristics of some topics of interest within a given population can be carried out using coding, the unidimensional description with downward coding [44] was deemed suitable for the purpose of our analysis. In effect, it involves organizing data into three logical levels of diversity; that is, objects, dimensions for each object and categories for each dimension, allowing us to move towards a lower level of abstraction [44]. Therefore, based on the unidimensional description of diversity proposed by Jansen [44], we analysed the diversity of sustainable behaviours of AFBs starting from the higher level of abstraction; that is, the sustainability behaviour of firms (the main object), following the middle level of abstraction; namely the FAO's principles and actions (the dimensions of objects), down to the lower abstraction level analysis consisting of the SDGs to which each FAO's action contributes (the categories of objects). It is important to highlight that the identification of SDGs relevant to each principle and actions FAO was made based on the explicit links indicated by FAO [2]). This articulation is indicated in Figure 2.

A chart with double entries was used to grasp each company's behaviour according to the three logical levels of diversity (Figure 3).

The vertical reading of the blue columns of the chart makes it possible to highlight each company's behaviour concerning the 5 FAO principles and their respective actions (the last two lines of the blue cells). It concerns the analysis of the "dimensions of the object" that has been extended to all AFBs to offer an overview of the overall scenario of the agro-food sector in southern Italy.

To analyse the categories of dimensions (each company's behaviours concerning each FAO principles and actions respects to SDGs), based on the FAO document [2], the SDGs on which each FAO action has an impact were first identified. These interlinks are signalled with green cells in the chart. Each green cell allows a dichotomous response variable (yes/no). The sum of the "yes" is shown in the cells of the totals.

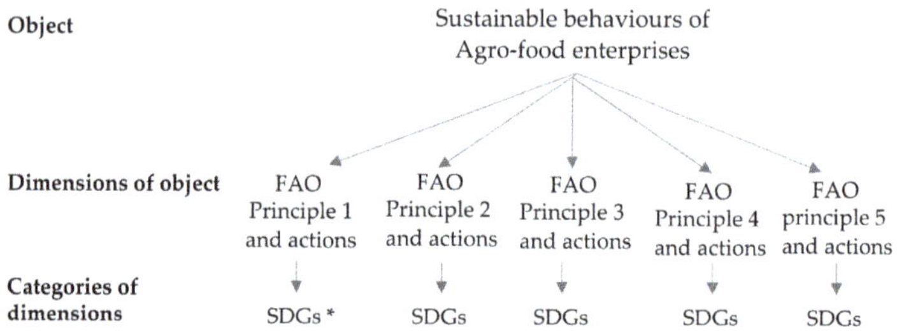

Figure 2. Articulation of object, dimensions, and categories. * Each SDG on each FAO's action gives a contribution according to the FAO framework [2].

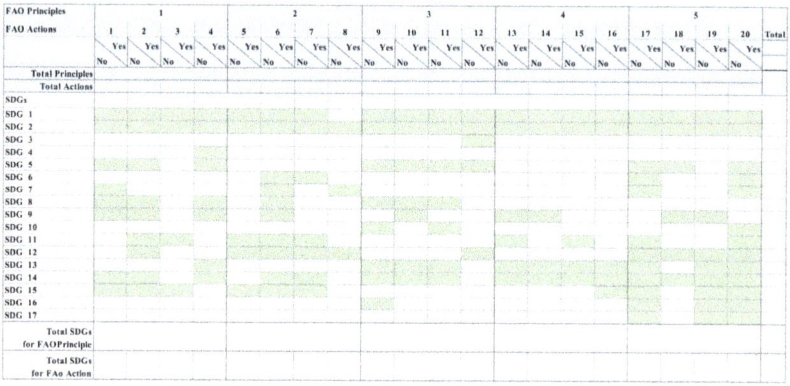

Figure 3. Contribution of FAOs to SDGs, according to FAO's framework [2].

Information on the AFBs' behaviours was collected based their descriptions on their websites. More precisely, we focused on information on sustainability real actions, tools and the performance of each company.

The drawing-up of the chart required a preliminary step and two subsequent steps.

In the preliminary step, for each firm's website, the presence or absence of a sustainability report and a section of the website dedicated to the sustainability field was noted. This was interpreted as the first signal of an AFBs' awareness of the importance of sustainability.

The next step aimed to describe the dimensions of the object. The behaviours (in conformity with the FAO's principles) of each selected AFB were individuated and related to each action indicated by the FAO guidelines [2]. The output was a dichotomous response variable (yes/no). Two researchers independently analysed the firms' websites. They looked for the presence or absence of a firm's behaviours according to the FAO's actions ('yes' for the presence of each behaviour and 'no' for its absence).

The sum of the positive responses was reported in the blue cell totals.

In the final step, for each selected AFB, we tried to relate each behaviour catalogued as previously described to the 17 SDGs of Agenda 2030 (analysis of categories of dimensions). In short, each FAO's action of the firm was ascribed to the SDG on which it has had an impact (i.e., by vertically selecting a cell among the green ones). Multiple behaviours of each firm related to one FAO action and the same SDG hold a value of one. Multiple behaviours for the same FAO's action and SDG were not considered.

Care was taken to ensure that all behaviours represented were real behaviours and not a simple declaration of intent. The presence of SA800 certification was interpreted as acting in line with action 9 (third principle) and SDG 8. Similarities among the independently generated data were noted, and after several iterations, a consensus was reached on the final coding of the major and minor themes. Finally, a third researcher checked for problems and inconsistencies; discrepancies were resolved through discussion.

The values attained by each firm in each cell of the chart shown in Figure 3 were added in the summary chart, shown in Figure 4.

FAO Principles	1				2				3				4				5				
SDGs / Actions	1	2	3	4	5	6	7	8	9	10	11	12	13	14	15	16	17	18	19	20	Total
SDG 1																					
SDG 2																					
SDG 3																					
SDG 4																					
SDG 5																					
SDG 6																					
SDG 7																					
SDG 8																					
SDG 9																					
SDG 10																					
SDG 11																					
SDG 12																					
SDG 13																					
SDG 14																					
SDG 15																					
SDG 16																					
SDG 17																					
Total Principles																					
Total Actions																					

Figure 4. Summary chart of SMABs' behaviours.

It must be underlined that the analysis of the categories of dimensions concerns only SMABs under the role they play in the southern Italian economy.

The SMABs' behaviours were analysed based on the extended Abraham and Pingali framework [28], providing important insights worthy of further thought.

To better grasp the behaviour of the SMABs, a multi-criteria analysis (MCA) was applied to the sample.

MCA allows one to compare alternative courses of action based on multiple factors. Among various MCA methods, regime analysis (RA) was chosen to rank the five FAO principles based on how the selected firms promote sustainability. RA is an evaluation method suitable for handling sustainability problems owing to its applicability to complex scenarios [45]. This method allows the management of quantitative and qualitative information, which is why it was previously used to rank different sustainable development attributes [45–47]. RA requires defining a priori a distinct set of ith alternatives, evaluating each one's impact on a plurality of jth criteria for all criteria together [48,49].

The first phase of RA concerns building an 'impact matrix' by assignment of the 'behaviour indices' (p_{ij}) of each alternative with respect to each criterion, thereby adopting an appropriate judgement scale.

The second phase is devoted to constructing a 'regime matrix' through a pairwise comparison based on the 'behaviour indices' attributed to the 'impact matrix'.

The elements of the impact matrix are composed as follows:

$$a_{i\,i',j} = +1 \text{ if } p_{ij} > p_{i\,i'j}, a_{i\,i',j} = -1 \text{ if } p_{ij} < p_{i\,i'j}, a_{i\,i',j} = 0 \text{ if } p_{ij} = p_{i\,i'j}, \quad (1)$$

where for each comparison, j is the value arising from comparing the two alternatives i and i' according to the j criterion.

The final phase concerns obtaining the ranking of the alternatives. The aggregate priority of each alternative, i.e., the preference of option i respect to alternative i' (considering all the criteria adopted) is expressed by the C_i value:

$$C_i = \frac{\sum_{i'=1}^{n-1} c_{ii'} \cdot j}{n-1} \quad (2)$$

where $c_{ii'}$ is the weight attributed to the criteria. C_i is conveniently normalised so that it can be included between -1 and +1. The alternative that reports the highest final value is the most attractive one according to the criteria set adopted.

In our study, the impact matrix is represented by the relationship between the 180 observed SMABs—i.e., the alternatives in the RA—and the five FAO principles. The behaviour indices reflect (for each firm) the frequency of 'yes' reported in the scheme shown in Figure 3. Since the highest frequency assessed was 8, we adopted a 1–9 judgement scale, i.e., p_{ij} could assume a value from 1 (any presence of firm action related to the FAO principle) to 9 on the basis (8 actions).

The regime matrix was built based on Formula (1); it means that, in each cell, the value is equal to +1, -1, or 0 in cases of positive, negative, and null difference, respectively, between the behaviour indexes (from time to time) considered in the pairwise comparisons; finally, Formula 2 was applied to estimate the ranking among the FAO principles.

It is important to stress that in the MCA, for analytical purposes, the variables should be independent [48]. For this reason, the principles and actions of FAO were considered methodologically independent variables and interlinks among principles and actions were not investigated.

3. Results

A preliminary descriptive analysis was carried out to offer a snapshot of what is occurring in southern Italy regardless of the agricultural firms' size. This will lead to an understanding of the role of SMABs in achieving the objectives of sustainable development.

Findings showed that among the AFBs located in South Italy that had a website (n = 650), only 30% (n = 193) implemented at least one action as defined by the FAO [2]. What emerges is a picture of the different attentions paid by firms to the issues of sustainable development. Only 38% of large AFBs (13 out of 34) implement these objectives, and the percentage is even lower in the case of SMABs (of which only 29% (180 out of 616) declared on the website to have put actions that positively impact sustainable development). Of these, 26% (n = 8 large firms and 42 SMABs) had a section of their respective websites dedicated to sustainability, although only seven in all (of which there were three large firms) had a sustainability report. These data can be interpreted as a sign of how many sustainability values are adopted by the SMABs.

Speaking about the actions taken in detail, Table 1 shows that 20% of AFBs take at least one action related to the protection and enhancement of natural resources (second principle), followed by 13% of firms being committed to improve livelihoods and foster inclusive economic growth (third principle). These overall rates take different values according to firm size, but the principles' positions do not change.

Looking at the actions taken by firms for each FAO principle, the findings show that the most reported action concerns the reduction in losses, encouraging 'reuse and recycling', and promoting sustainable consumption (action 8), followed by action 12 (improving nutrition and promoting balanced diets).

Table 1. Number of AFBs taking at least one action per principle and action.

FAO Principles	1. Increase Productivity and Employment and Value Addition in Food Systems					2. Protect and Enhance Natural Resources					3. Improve Livelihoods and Foster Inclusive Economic Growth					4. Enhance Resilience of People, Communities, and Ecosystems					5. Adapt Governance to New Challenges				
FAO Actions	**1**	**2**	**3**	**4**	**TP1**	**5**	**6**	**7**	**8**	**TP2**	**9**	**10**	**11**	**12**	**TP3**	**13**	**14**	**15**	**16**	**TP4**	**17**	**18**	**19**	**20**	**TP5**
N. AFBs % (on 650)	3	23	6	35	**58** *8.9*	31	21	33	90	**130** *20.0*	38	4	8	51	*84* *12.9*	1	0	25	7	*33* *5.1*	4	4	0	0	*7* *1.1*
N. large AFBs % (on 34)	0	3	0	3	**5** *14.7*	3	5	5	9	**13** *38.2*	7	2	1	5	*8* *23.5*	0	0	3	1	*3* *8.8*	2	1	0	0	*2* *5.9*
N. SMABs % (on 616)	3	20	6	32	**53** *8.6*	28	16	28	81	**117** *19.0*	31	2	7	46	*76* *12.3*	1	0	22	6	*30* *4.9*	2	3	0	0	*5* *0.8*

The total number of firms that take at least one action per principle is reported in the TP columns (TP = total for a given principle). This number can be lower than the sum of firms per single action because some firms can appear in more actions.

In order to describe the dimensions of the object and categories of dimensions with regard to SMABs, each of their behaviour was assessed according to the FAO's actions that impact the SDGs that have been listed. Table 2 summarizes the actions taken by the SMABs.

Table 2. Summary of actions performed by SMABs according to FAO's actions and SDGs.

Principles / SDGs Actions	Principle 1				Principle 2				Principle 3				Principle 1				Principle 5				Total
	1	2	3	4	5	6	7	8	9	10	11	12	13	14	15	16	17	18	19	20	
1	2	0	0	0	0	0	0		1	1	3	0	0	0	0	2	0	0	0	0	9
2	0	3	4	16	3	0	7	6	10	0	2	3	1	0	0	0	1	0	0	0	56
3												34									34
4				5																	5
5	0	0		2					2	0	0	0					0	0		0	4
6						4	0										0			0	4
7	1					1		30									0			0	32
8	0	13		9		1			19	0	5										47
9	0	0		0		0				1			0	0				0	0		1
10									1		0									0	1
11		0	0		0	0	0						0		0		0			0	0
12		5			22	11	18	54				11					1	3	0	0	125
13				1					0	0	0		0	0	22	4	0		0	0	27
14	0	0		0		0	1		0	0	0		0	0	0	0	0	0	0	0	1
15	0	0	2		4	1	2									0	0		0	0	9
16									0								0		0	0	0
17																	0		0	0	0
Total	3	21	6	33	29	18	28	90	33	2	10	48	1	0	22	6	2	3	0	0	355
	63				165				93				29				5				

Using the Abraham and Pingali's [28] extended framework, we analysed the categories of dimensions, classifying the SMABs' behaviour from SDGs' viewpoint (Table 3).

Table 3. Number of actions reported for each sustainable development goal.

Goals	Poverty		Nutrition		Social		Environmental		Residual	
	Goal 1	9	Goal 2	56	Goal 4	5	Goal 6	4	Goal 11	-
	Goal 8	47	Goal 3	34	Goal 5	4	Goal 7	32	Goal 16	-
					Goal 10	1	Goal 12	125	Goal 17	-
							Goal 13	27	Goal 9	1
							Goal 14	1		
							Goal 15	9		
Total	56		90		10		198		1	

The findings show that the environmental goals receive, without doubt, the greatest attention from the SMABs. This indicates that they understood that natural resources are the material basis of human society and made this the foundation of their primary sector activities. The maximum attention given to SDG 12 (responsible consumption and production) shows that the actions of the SMABs are inspired by the aims of reducing their environmental impact, promoting the use of renewable energy sources, and making responsible purchases. In particular, 43% of the behaviours concern the FAO action 8. This action is also the most important in SDG 7, which is affordable clean energy (17% of businesses). These data demonstrate companies' commitment to promoting their transition towards a sustainable energy system through technological investment in renewable energy resources.

The second priority was nutrition goals. The results show that companies' actions aimed at reducing hunger were mainly aimed at favouring their transformation to the SFA system by sharing knowledge, building capacities, and fostering participation in modern value chains (action FAO 4).

Finally, the poverty goals, and in particular the SDG 8 (decent work and economic growth), recorded a commitment from businesses for FAO action 9, which is 'empower people and fighting inequalities'.

The extent of the actions attributable to the social and residual SDGs are negligible.

The application of regime analysis allowed us to rank the FAO principles according to the behaviour of each SMAB that is summarised in Table 2. Using a 180 (firms) × 5 (principles) matrix and developing the method illustrated in the methodological section, we assessed that the second FAO principle, i.e., 'protect and enhance natural resources' was preferred over other principles (Table 4). This principle shows a score of about 0.7, implying that, as a whole, SMABs that are expressly sustainability-oriented complied with it with a probability of approximately 70%.

Table 4. Final ranking of Food and Agriculture Organization principles.

FAO Principles	Probability
1. Increase productivity and employment and value addition in food systems	0.490
2. Protect and enhance natural resources	0.682
3. Improve livelihoods and foster inclusive economic growth	0.548
4. Enhance resilience of people, communities and ecosystems	0.381
5. Adapt governance to the new challenges	0.351

'Improve livelihoods and foster inclusive economic growth' ranks second, with an estimated score of 0.548.

The other principles showed probabilities of less than 50%. This is particularly surprising for the first FAO principle because increasing productivity and added value in the food system should represent the primary objectives of the SMABs. These firms probably tended not to declare this effort on the website or elsewhere because they considered it implicit.

4. Discussion

SMEs are considered to be the 'major engine' of economic growth and socio-economic development [50] and they play (now and in the near future) a leading role in SFA [51]. Therefore, this study aimed to investigate which sustainability actions are most important for the SMABs and which are neglected so as to provide policymakers with the basis for planning appropriate strategies to achieve all the SDGs of the 2030 Agenda.

Despite the variety of sustainability practices pursued by SMABs, the propensity for actions that lead to protect and improve natural resources prevails (65%), disavowing previous studies that perceived SMEs as failures in relation to environmental sustainability due to their low take-up rates of sustainable business practices [52] (p. 172). This datum fits well with the CAP's aim of protecting natural resources, which, in the SMABs, is a worthy ally to push for SFA.

A possible explanation of this result suggests that SMABs, due to the rapid transformation of the agro-food system and the pressure on them to tackle environmental, health, and food safety problems [53], are more aware of both their environmental responsibility

(ER) and competitive advantage derived from voluntary ER practices and their disclosure [54,55]. Indeed, organisations driven by sustainability competitiveness are prone to improve their performance related to energy and waste management, increase production and decrease sources of input, introduce eco-products, and implement ecological labelling and green marketing [56]. This interpretation of the results contrasts with previous studies in the Italian context, according to which SMEs understood the environmental responsibility as an added cost rather than a market opportunity, while not considering the market that is highly responsive to their environmental practices [57].

Within the second FAO principle, more actions were recorded for FAO action 8 (reduce losses, encourage reuse and recycle, and promote sustainable consumption). This indicates that SMABs have begun to understand that even if they are small, they contribute to pollution worldwide [52], and as producers, they will be central stakeholders in achieving an optimised, zero-waste production and distribution system [58] because they are likely to design business systems that reduce environmental impacts [59]. This propensity of SMABs bodes well for the transition towards the circular economy paradigm and meets the aim of the 'Circular Economy Package', which is the new action plan of Europe's new agenda for sustainable growth (EC 2018) and the Green Deal.

The second principle mainly considered is 'improve livelihoods and foster inclusive economic growth' (24% of large AFBs and 42% of SMABs), which is primarily directed to reduce poverty and food insecurity in rural areas. This principle shares several aims with the CAP that supports farmers' income and adopts market measures, and seeks to ensure sustainable and inclusive rural development. The fact that SMABs are engaged in behaviours related to this FAO principle is of considerable importance.

First, because SMEs dominate the agro-industry sector and are a core component of any rural development strategy [60,61], our results can be read in the light of previous studies focused on social sustainability in agriculture according to which firms producing more social outputs are considered of great value [62]. Therefore, investments in the social dimension [63] in SMABs can find a perfect breeding ground for the success of policies aimed at promoting sustainable rural development, especially rural vitality and food security, which are considered among the most significant public goods from agriculture [64].

Second, we have noticed that many businesses have the tab 'territory' site menu, in several cases positioned before the 'about us' tab. This datum can be read according to two opposing interpretations. On the one hand, SMABs, especially Italian ones and producers of traditional food products, use 'territory' as a strategic resource in a fiercely competitive market, leveraging the synergistic link between authentic agro-food products and their region of origin [65,66]. On the other hand, it can be the expression of sincere interest in promoting its territory as a lever to increase the attention towards all local products and promote food and wine tourism. In this context, the producer feels part of a community and aims to contribute to its survival and growth.

Looking at the single FAO actions, the most important appears in the FAO action 12 (improve nutrition and promote balanced diets). The overwhelming majority of firms that have implemented policies/steps under this action have carried out nutrition education and awareness programs, promoting the consumption of locally grown nutritious food.

However, the commitment shown in action 9 (empower people and fight inequalities) is no less important. This action aims to provide rural firms with the tools and capacity to build resilient livelihoods. In this respect, many firms have activated programs to help small producers and young people enter networks that allow them to enter the market. This behaviour generates benefits that fall under social indicators related to the society as a whole (such as the quality of rural areas and contribution to local employment) [67] and can complement the concept of supply chain responsibility (SCR). Within the two-way relationship between supply chains (SCs) (that depend on community resources such as entrepreneurs) and the social well-being of communities (that build and maintain prosperity, thanks to the opportunity offered by SCs), the firms that participate in the agro-food supply chain reduce the producers' disadvantages and enhance the rural community's development. It is not just actions aimed at meeting external pressures, disclosing their "status of responsibility", maintaining their reputation and obtaining legitimacy to their operations and presence in the market [68–70]. SMABs also feel that they "have the responsibility to promote functional communities or community sustainability proactively" [71].

By focusing on the third most-widely reported principle (increase productivity and employment and value addition in food systems), on the one hand, SMABs are committed to creating the conditions for the producers' skills and knowledge so that they can participate in modern value chains (FAO action 4). On the other hand, just over 10% of the adopted behaviour related to FAO action 2 (connecting smallholders to markets). To make resilient and stable agro-food business, this system must encourage all the actors (and therefore, also smallholders) to cooperate, since, through cooperation, the agro-food system creates a new development process centred on sustainability, thereby creating value for territories and agro-food districts, promoting their uniqueness, and enhancing environmental protection and social cohesion [72]. In this sense, the policymakers' role is crucial in fostering such collaboration, particularly improving specific assistance to local AFBs for collective projects and spreading knowledge amongst other rural stakeholders [73].

Concerning the last two indicators reported, the lack of attention to principle 5 is not surprising (adapt governance to new challenges). Only four companies reported the FAO action 18 (strengthen innovation systems), having made investments in agricultural R&D with collaboration between firms and universities. The least attention to principle 4 (enhancing the resilience of people, communities, and ecosystems) raises some questions.

On the one hand, the FAO actions 13, 14, and 16 may appear to be out of the reach of SMABs. On the other hand, only three large companies, but as many as 25 SMABs, have implemented actions aimed explicitly at reducing greenhouse gas emissions (FAO action 15—address and adapt to climate change). It is a sign that climate change is becoming a primary issue not only within the political agenda but also in businesses. In this context, education programs and outreach plans for SMABs can find fertile ground.

Turning our gaze towards SMABS' behaviours recorded based on the SDGs in which each FAO's action impacts, it appears that SMABs are more likely to adopt behaviours aimed at protecting and improving natural resources. The fact that there is only about 50% of probability that companies invest in actions aimed at 'increasing productivity and employment and adding value to food systems' raises doubts about the willingness or ability of companies to create networks in which skills and knowledge can be shared with small producers to develop their skills so as to connect small farms to markets.

Finally, by moving the analysis towards the SDGs, a significant finding was made: many SMABs are committed to achieving SDG 7, especially adopting innovations in the field of renewable energy. Its importance stems from the fact that the European Commission states that the clean energy supply for food and agriculture is crucial to deliver the European Green Deal. Once again, SMABs are proving to be agents of sustainable development.

5. Final Remarks

The aim of this study was to understand the behaviours of SMABs that promote the transition towards the SFA. In effect, the agro-food sector has always performed essential functions on the food, environmental, and social levels. Still, today, it is called and engaged in an even more complex effort if we consider that the future economy cannot ignore the social, economic, environmental, food, water equity, and energy issues and preserve biodiversity.

Our analysis has contributed to the literature by using the FAO approach outlined in 'transforming food and agriculture to achieve the SDGs' (2018), aiming to support and accelerate the transition to more sustainable agro-food systems. The previous literature investigates the agro-food system's actors' behaviours about single or multiple dimensions of sustainability or SDGs. However, as far as we know, no studies investigate specific behaviours according to FAO's actions.

Another relevant contribution lies in investigating which categories of SDGs are greatly promoted by SMABs, bringing to light the important signs of their pivotal role in achieving effective Green Deal targets.

Finally, a further contribution lies in having carried out an analysis of sustainability actions of SMABs through the lens of digital disclosure, i.e., of firms that use online tools to communicate their commitment to sustainability. In fact, among the different types and tools available for external communication of corporate sustainability, online has experienced rapid growth in recent years [74] because it is "a privileged means of communication towards sustainable development, where information disclosed knows no border" [75] (p. 253). In this vein, the corporate website represents an important medium for voluntary sustainable disclosure, and their dialogical communication capabilities may be considered an indication of demonstrated willingness and preparedness to promote and support communication with external actors [76]. For that reason, the choice of the website as a tool for our investigation arises from the consideration that it is no longer a simple 'channel' among many, but often the strategic lever of dialogue and confrontation with the outside world.

This study had some limitations. First, it was based on a sample of southern Italian firms; thus, the results need to be tested in different areas and countries. Second, the study did not consider multiple behaviours relating to the same FAO's action and SDGs. Future research can account for such multiple behaviours as well as other common standards used by SMABs. Future research can also expand the study with direct interviews with companies to see if they take more actions than those posted on their website, with an undoubted advantage in terms of verifying our results. Finally, we used the FAO's principles as a diagnostic tool to interpret business operations. In this perspective, our findings do not allow one to give information about possible management implications on the prescriptive side. In effect, while providing a detailed overview of corporate behaviour, our findings are not directly able to give indications on the intentions that led to the adoption of observed behaviours by firms. However, we are conscious that the analysis of how pursuing the formal sustainability objectives can affect the company's management can be a valid next step of our research.

Author Contributions: Conceptualization, B.A., F.A.M., L.B., R.P., P.P. and R.F.; methodology, B.A., F.A.M., L.B., R.P., P.P. and R.F.; formal analysis, B.A., F.A.M.; investigation, B.A., L.B., R.P.; writing—original draft preparation, B.A., F.A.M., L.B., R.P.; writing—review and editing, B.A., F.A.M., L.B., R.P., P.P. and R.F.; supervision, R.F., P.P.; project administration, P.P. All authors have read and agreed to the published version of the manuscript.

Funding: This research received no external funding.

Data Availability Statement: The Aida database—Bureau Van Dijk was used to select the sample of firms.

Conflicts of Interest: The authors declare no conflict of interest.

References

1. Campagnolla, C.; Rametsteiner, E.; Gutierrez, D. Sustainable Agriculture and Food Systems: Towards a Third Agricultural Revolution. *Fome Zero Zero Hunger* **2019**, 140–157. [CrossRef]
2. FAO. *Transforming Food and Agriculture to Achieve the SDGs: 20 Interconnected Actions to Guide Decision-Makers*; FAO: Roma, Italy, 2018; ISBN 978-92-5-130626-0.
3. FAO. Sustainable Development Goals. Available online: http://www.fao.org/sustainable-development-goals/indicators/en/ (accessed on 27 April 2021).
4. FAO. Food and Agiculture in the 2030 Agenda 2016. Available online: http://www.fao.org/3/i6105e/i6105e.pdf (accessed on 27 April 2021).
5. Ngenoh, E.; Kurgat, B.K.; Bett, H.K.; Kebede, S.W.; Bokelmann, W. Determinants of the Competitiveness of Smallholder African Indigenous Vegetable Farmers in High-Value Agro-Food Chains in Kenya: A Multivariate Probit Regression Analysis. *Agric. Food Econ.* **2019**, *7*, 2. [CrossRef]
6. Röös, E.; Fischer, K.; Tidåker, P.; Källström, H.N. How Well Is Farmers' Social Situation Captured by Sustainability Assessment Tools? A Swedish Case Study. *Int. J. Sustain. Dev. World Ecol.* **2019**, *26*, 268–281. [CrossRef]

7. Terlau, W.; Hirsch, D.; Blanke, M. Smallholder Farmers as a Backbone for the Implementation of the Sustainable Development Goals. *Sustain. Dev.* **2019**, *27*, 523–529. [CrossRef]
8. Westman, L.; Luederitz, C.; Kundurpi, A.; Mercado, A.J.; Weber, O.; Burch, S.L. Conceptualizing Businesses as Social Actors: A Framework for Understanding Sustainability Actions in Small-and Medium-sized Enterprises. *Bus. Strategy Environ.* **2019**, *28*, 388–402. [CrossRef]
9. Zarbà, C.; Chinnici, G.; D'Amico, M. Novel Food: The Impact of Innovation on the Paths of the Traditional Food Chain. *Sustainability* **2020**, *12*, 555. [CrossRef]
10. Colman, P.; Harwell, J.; Found, P. Value Creation through Innovation in the Primary Sector. *Int. J. Qual. Serv. Sci.* **2020**, *12*, 475–487.
11. De Bernardi, P.; Azucar, D. A European Food Ecosystem: The EIT Food Case Study. In *Innovation in Food Ecosystems*; Springer: Cham, Switzerland, 2020; pp. 245–280.
12. Anríquez, G.; Stamoulis, K. Rural Development and Poverty Reduction: Is Agriculture Still the Key? 2007. Available online: https://ageconsearch.umn.edu/record/289048/ (accessed on 27 April 2021).
13. Asfaw, S.; Shiferaw, B.; Simtowe, F.; Lipper, L. Impact of Modern Agricultural Technologies on Smallholder Welfare: Evidence from Tanzania and Ethiopia. *Food Policy* **2012**, *37*, 283–295. [CrossRef]
14. Irz, X.; Lin, L.; Thirtle, C.; Wiggins, S. Agricultural Productivity Growth and Poverty Alleviation. *Dev. Policy Rev.* **2001**, *19*, 449–466. [CrossRef]
15. Kato, E.; Mekonnen, D.K.; Tiruneh, S.; Ringler, C. *Sustainable Land Management and Its Effects on Water Security and Poverty: Evidence from a Watershed Intervention Program in Ethiopia*; Intl Food Policy Res Inst: Washington, DC, USA, 2019; Volume 1811.
16. Wezel, A.; Casagrande, M.; Celette, F.; Vian, J.-F.; Ferrer, A.; Peigné, J. Agroecological Practices for Sustainable Agriculture. A Review. *Agron. Sustain. Dev.* **2014**, *34*, 1–20. [CrossRef]
17. Zeweld, W.; Van Huylenbroeck, G.; Tesfay, G.; Speelman, S. Smallholder Farmers' Behavioural Intentions towards Sustainable Agricultural Practices. *J. Environ. Manag.* **2017**, *187*, 71–81. [CrossRef]
18. Finco, A.; Bentivoglio, D.; Bucci, G. Lessons of Innovation in the Agrifood Sector: Drivers of Innovativeness Performances. *Econ. Agro Aliment.* **2018**. [CrossRef]
19. Di Marco, L.; Obiettivi Di Sviluppo Sostenibile e Politiche Europee. Dal Green Deal al Next Generation EU. 2020. Available online: https://asvis.it/public/asvis2/files/Pubblicazioni/Quaderno_Obiettivi_di_sviluppo_sostenibile_e_politiche_europee.pdf (accessed on 27 April 2021).
20. Osservatorio Food Sustainability Cresce l'innovazione in Risposta Alle Sfide Di Sostenibilità Agroalimentare. 2019. Available online: https://www.osservatori.net/it/prodotti/formato/report/crescita-innovazione-sfide-sostenibilita-agroalimentare (accessed on 10 March 2021).
21. Coluccia, B.; Valente, D.; Fusco, G.; De Leo, F.; Porrini, D. Assessing Agricultural Eco-Efficiency in Italian Regions. *Ecol. Indic.* **2020**, *116*, 106483. [CrossRef]
22. Arru, B. An Integrative Model for Understanding the Sustainable Entrepreneurs' Behavioural Intentions: An Empirical Study of the Italian Context. *Environ. Dev. Sustain.* **2019**, *22*, 3519–3576. [CrossRef]
23. Del Baldo, M. Corporate Social Responsibility, Entrepreneurial Values and Transcendental Virtues in Italian SMEs. *Int. J. Bus. Soc. Sci.* **2014**, *5*, 25–51.
24. ISMEA. Rapporto Sulla Competitività Dell'agroalimentare Nel Mezzogiorno. 2019. Available online: http://www.ismeamercati.it/flex/cm/pages/ServeBLOB.php/L/IT/IDPagina/9942 (accessed on 27 April 2021).
25. Cohen, B.; Winn, M.I. Market Imperfections, Opportunity and Sustainable Entrepreneurship. *J. Bus. Ventur.* **2007**, *22*, 29–49. [CrossRef]
26. Jansson, J.; Nilsson, J.; Modig, F.; Hed Vall, G. Commitment to Sustainability in Small and Medium-sized Enterprises: The Influence of Strategic Orientations and Management Values. *Bus. Strategy Environ.* **2017**, *26*, 69–83. [CrossRef]
27. Moore, S.B.; Manring, S.L. Strategy Development in Small and Medium Sized Enterprises for Sustainability and Increased Value Creation. *J. Clean. Prod.* **2009**, *17*, 276–282. [CrossRef]
28. Abraham, M.; Pingali, P. Transforming smallholder agriculture to achieve the SDGs. In *The Role of Smallholder Farms in Food and Nutrition Security*; Springer: Cham, Switzerland, 2020; pp. 173–209.
29. Genç, R. The Importance of Communication in Sustainability & Sustainable Strategies. *Procedia Manuf.* **2017**, *8*, 511–516.
30. Gray, R.; Javad, M.; Power, D.M.; Sinclair, C.D. Social and Environmental Disclosure and Corporate Characteristics: A Research Note and Extension. *J. Bus. Financ. Account.* **2001**, *28*, 327–356. [CrossRef]
31. Adams, C.A.; Frost, G.R. Accessibility and Functionality of the Corporate Web Site: Implications for Sustainability Reporting. *Bus. Strategy Environ.* **2006**, *15*, 275–287. [CrossRef]
32. Da Giau, A.; Macchion, L.; Caniato, F.; Caridi, M.; Danese, P.; Rinaldi, R.; Vinelli, A. Sustainability Practices and Web-Based Communication: An Analysis of the Italian Fashion Industry. *J. Fash. Mark. Manag.* **2016**, *20*, 72–88. [CrossRef]
33. de Villiers, C.; van Staden, C. Shareholder Requirements for Compulsory Environmental Information in Annual Reports and on Websites. *Aust. Account. Rev.* **2011**, *21*, 317–326. [CrossRef]
34. Sharma, S.; Henriques, I. Stakeholder Influences on Sustainability Practices in the Canadian Forest Products Industry. *Strateg. Manag. J.* **2005**, *26*, 159–180. [CrossRef]

35. Hasim, M.; Hashim, A.; Ariff, N.; Sapeciay, Z.; Abdullah, A. Commitment to Sustainability: A Content Analysis of Website for University Organisations. *IOP Conf. Ser. Earth Environ. Sci.* **2018**, *117*, 12046. [CrossRef]
36. Bosetti, L. Web-Based Integrated CSR Reporting: An Empirical Analysis. *Symphonya Emerg. Issues Manag.* **2018**, 18–38. [CrossRef]
37. Capriotti, P. The world wide web and the social media as tools of CSR communication. In *Handbook of Integrated CSR Communication*; Springer: Cham, Switzerland, 2017; pp. 193–210.
38. Lodhia, S. Factors Influencing the Use of the World Wide Web for Sustainability Communication: An Australian Mining Perspective. *J. Clean. Prod.* **2014**, *84*, 142–154. [CrossRef]
39. Cassetta, E.; Monarca, U.; Dileo, I.; Berardino, C.D.; Pini, M. The Relationship between Digital Technologies and Internationalisation. Evidence from Italian SMEs. *Ind. Innov.* **2020**, *27*, 311–339. [CrossRef]
40. Kassem, E.; Trenz, O. Automated Sustainability Assessment System for Small and Medium Enterprises Reporting. *Sustainability* **2020**, *12*, 5687. [CrossRef]
41. Goswami, R.; Saha, S.; Dasgupta, P. Sustainability Assessment of Smallholder Farms in Developing Countries. *Agroecol. Sustain. Food Syst.* **2017**, *41*, 546–569. [CrossRef]
42. Perren, L.; Ram, M. Case-Study Method in Small Business and Entrepreneurial Research: Mapping Boundaries and Perspectives. *Int. Small Bus. J.* **2004**, *22*, 83–101. [CrossRef]
43. Leung, L. Validity, Reliability, and Generalizability in Qualitative Research. *J. Fam. Med. Prim. Care* **2015**, *4*, 324. [CrossRef]
44. Jansen, H. The Logic of Qualitative Survey Research and Its Position in the Field of Social Research Methods. *Forum Qual. Soz. Forum Qual. Soc. Res.* **2010**, *11*. [CrossRef]
45. Nijkamp, P.; Vindigni, G. Integrated Multicriteria Analysis for Sustainable Agricultural Policy Evaluation. *Riv. Econ. Agrar.* **1999**, *53*, 9–40.
46. Akgün, A.A.; van Leeuwen, E.; Nijkamp, P. A Multi-Actor Multi-Criteria Scenario Analysis of Regional Sustainable Resource Policy. *Ecol. Econ.* **2012**, *78*, 19–28. [CrossRef]
47. Idda, L.; Furesi, R.; Madau, F.A.; Rubino, C. The Italian Fishing and Aquaculture System in a Sustainable Development Perspective: A Multicriterial Approach to the Theme. In Proceedings of the 10th Congress European Association of Agricultural Economists (EAAE): Exploring Diversity in the European Agri-Food System, Zaragoza, Spain, 28–31 August 2002.
48. Munda, G.; Nijkamp, P.; Rietveld, P. Qualitative Multicriteria Evaluation for Environmental Management. *Ecol. Econ.* **1994**, *10*, 97–112. [CrossRef]
49. Nijkamp, P.; Rietveld, P.; Voogd, H. *Multiple Criteria Analysis in Physical Planning*; Elsevier: Amsterdam, The Netherlands, 1990.
50. Šebestová, J.; Sroka, W. Sustainable Development Goals and SMEs Decisions: Czech Republic vs. Poland. *J. East. Eur. Cent. Asian Res. JEECAR* **2020**, *7*, 39–50. [CrossRef]
51. SDSN. *Transformative Changes of Agriculture and Food Systems. Prepared by the Thematic Group 7: Sustainable Agriculture and Food Systems*; Sustainable Development Solutions Network: New York, NY, USA, 2013.
52. Battisti, M.; Perry, M. Walking the Talk? Environmental Responsibility from the Perspective of Small-business Owners. *Corp. Soc. Responsib. Environ. Manag.* **2011**, *18*, 172–185. [CrossRef]
53. Borsellino, V.; Schimmenti, E.; El Bilali, H. Agri-Food Markets towards Sustainable Patterns. *Sustainability* **2020**, *12*, 2193. [CrossRef]
54. Gallardo-Vázquez, D.; Sánchez-Hernández, M.I. Structural Analysis of the Strategic Orientation to Environmental Protection in SMEs. *BRQ Bus. Res. Q.* **2014**, *17*, 115–128. [CrossRef]
55. Panwar, R.; Nybakk, E.; Hansen, E.; Pinkse, J. The Effect of Small Firms' Competitive Strategies on Their Community and Environmental Engagement. *J. Clean. Prod.* **2016**, *129*, 578–585. [CrossRef]
56. Bansal, P.; Roth, K. Why Companies Go Green: A Model of Ecological Responsiveness. *Acad. Manag. J.* **2000**, *43*, 717–736.
57. Cantele, S.; Zardini, A. Is Sustainability a Competitive Advantage for Small Businesses? An Empirical Analysis of Possible Mediators in the Sustainability–Financial Performance Relationship. *J. Clean. Prod.* **2018**, *182*, 166–176. [CrossRef]
58. Jabbour, C.J.C.; de Jabbour, A.B.L.S.; Sarkis, J.; Filho, M.G. Unlocking the Circular Economy through New Business Models Based on Large-Scale Data: An Integrative Framework and Research Agenda. *Technol. Forecast. Soc. Chang.* **2019**, *144*, 546–552. [CrossRef]
59. Madau, F.A.; Arru, B.; Furesi, R.; Pulina, P. Insect Farming for Feed and Food Production from a Circular Business Model Perspective. *Sustainability* **2020**, *12*, 5418. [CrossRef]
60. Garcia-Alvarez-Coque, J.-M.; Mas-Verdu, F.; García, M.S. Determinants of Agri-Food Firms' Participation in Public Funded Research and Development. *Agribusiness* **2015**, *31*, 314–329. [CrossRef]
61. Jindřichovská, I.; Kubíčková, D.; Mocanu, M. Case Study Analysis of Sustainability Reporting of an Agri-Food Giant. *Sustainability* **2020**, *12*, 4491. [CrossRef]
62. Ait Sidhoum, A. Valuing Social Sustainability in Agriculture: An Approach Based on Social Outputs' Shadow Prices. *J. Clean. Prod.* **2018**, *203*, 273–286. [CrossRef]
63. Lebacq, T.; Baret, P.V.; Stilmant, D. Sustainability Indicators for Livestock Farming. A Review. *Agron. Sustain. Dev.* **2013**, *33*, 311–327. [CrossRef]
64. Cooper, T.; Hart, K.; Baldock, D. *Provision of Public Goods through Agriculture in the European Union*; Institute for European Environmental Policy London: London, UK, 2009.

65. Festa, G.; Rossi, M.; Kolte, A.; Situm, M. Territory-Based Knowledge Management in International Marketing Processes—The Case of "Made in Italy" SMEs. *Eur. Bus. Rev.* **2020**, *32*, 425–442. [CrossRef]
66. Iaia, L.; Maizza, A.; Fait, M.; Scorrano, P. Origin Based Agro-Food Products: How to Communicate Their Experiential Value Online? *Br. Food J.* **2016**. [CrossRef]
67. Diazabakana, A.; Latruffe, L.; Bockstaller, C.; Desjeux, Y.; Finn, J.; Kelly, E.; Ryan, M.; Uthes, S. A Review of Farm Level Indicators of Sustainability with a Focus on CAP and FADN. Ph.D. Thesis, Agro Paris Tech, Paris, France, 2014.
68. Kramer, M.R.; Porter, M. *Creating Shared Value*; FSG: Boston, MA, USA, 2011.
69. Maas, K.; Schaltegger, S.; Crutzen, N. Integrating Corporate Sustainability Assessment, Management Accounting, Control, and Reporting. *J. Clean. Prod.* **2016**, 1–12. [CrossRef]
70. Stoian, D.; Donovan, J.; Fisk, J.; Muldoon, M.F. Value Chain Development for Rural Poverty Reduction: A Reality Check and a Warning. *Enterp. Dev. Microfinanc. J.* **2012**, *23*, 54. [CrossRef]
71. Liu, L.; Ross, H.; Ariyawardana, A. Community Development through Supply Chain Responsibility: A Case Study of Rice Supply Chains and Connected Rural Communities in Central China. *Sustainability* **2020**, *12*, 927. [CrossRef]
72. Contini, C.; Marotta, G.; Torquati, B. Multi-Actor Approaches to Implement Cooperative Strategies and Value Chains Based on Sustainability. *Agric. Food Econ.* **2020**, *8*, 7. [CrossRef]
73. Cisilino, F.; Vanni, F. Agri-Environmental Collaborative Projects: Challenges and Perspectives in Italy. *Econ. AGRO Aliment.* **2019**, 459–479. [CrossRef]
74. Lundquist Lundquist CSR Online Awards. 2017. Available online: https://lundquist.it/wp-content/uploads/2017/11/CSR_OA_2017_interattivo-1.pdf (accessed on 15 April 2021).
75. Carvalho, F.; Santos, G.; Gonçalves, J. The Disclosure of Information on Sustainable Development on the Corporate Website of the Certified Portuguese Organizations. *Int. J. Qual. Res.* **2018**, *12*, 253–276.
76. Thimm, H.; Rasmussen, K.B. Website Disclosure of Environmental Compliance Management—The Case of European Production Companies. *J. Environ. Stud. Sci.* **2020**. [CrossRef]

sustainability

MDPI

Article

Capability Assessment toward Sustainable Development of Business Incubators: Framework and Experience Sharing

Nathasit Gerdsri [1,*], Boonkiart Iewwongcharoen [2], Kittichai Rajchamaha [1], Nisit Manotungvorapun [3], Jakapong Pongthanaisawan [4] and Watcharin Witthayaweerasak [5]

1 College of Management, Mahidol University, Bangkok 10400, Thailand; kittichai.raj@mahidol.ac.th
2 Graduate School of Management and Innovation (GMI), King Mongkut's University of Technology Thonburi, Bangkok 10400, Thailand; boonkiart.iew@kmutt.ac.th
3 School of Business Administration, Bangkok University, Pathum Thani 12120, Thailand; nisit.m@bu.ac.th
4 Energy Research Institute, Chulalongkorn University, Bangkok 10330, Thailand; jakapong.p@chula.ac.th
5 Thai Business Incubators and Science Parks Association, Pathum Thani 12120, Thailand; dingalhing@gmail.com
* Correspondence: nathasit.ger@mahidol.ac.th; Tel.: +66-02-206-2000

Citation: Gerdsri, N.; Iewwongcharoen, B.; Rajchamaha, K.; Manotungvorapun, N.; Pongthanaisawan, J.; Witthayaweerasak, W. Capability Assessment toward Sustainable Development of Business Incubators: Framework and Experience Sharing. *Sustainability* **2021**, *13*, 4617. https://doi.org/10.3390/su13094617

Academic Editor: Andrea Pérez

Received: 15 March 2021
Accepted: 13 April 2021
Published: 21 April 2021

Abstract: Business incubators have been widely developed to advise, support, promote, and provide a nurturing environment for new business start-ups and entrepreneurs. The development of a framework for capability assessment allows the management of each incubator to understand its strengths and room for further improvement. Moreover, assessment results across a community, such as a nation or state, can provide insights into resource allocation and various management policies so that policymakers can support the development of business incubators under their supervision. This article describes the development of a capability assessment framework for business incubators (BIs) in Thailand. A case study demonstrating how the capability assessment is analyzed is also presented in the article.

Keywords: capability assessment; business incubator; maturity model; sustainability; Thailand

1. Introduction

There are various roles and duties of business incubators depending on the structure of the unit. One of the widely accepted business incubator configurations is the InfoDev configuration developed by World Bank. InfoDev addresses the roles of a business incubator as: To help clients (entrepreneurs) develop a business model, set up a plan, and find a source of funding; to provide access to experts who can give technical advice; and to create the appropriate environment for active engagement. However, business incubators do not have a primary role as an investor. InfoDev's configuration aims to harmonize with the four stages of the entrepreneurial life-cycle: Germination, pre-incubation, incubation, and post-incubation [1]. The UBI Global benchmark 2015/2016 report identifies the three critical success traits of an incubation program as attracting high-potential startups, ensuring enough resources for operations, and creating a supportive entrepreneurial environment among startups [2].

In East Europe and Central Asia (ECA), InfoDev interviewed nine businesses from eight countries with distinct performance incubators. The results showed that the current business incubators in those countries are mainly supported by the federal government, such as the ministry of ICT, ministry of education, and/or ministry of science. The study also showed that the role of business incubators in the private sector is increasing in many countries. Moreover, some business incubators can reach financial self-sustainability after 2–4 years of operation, and experienced experts, such as CEOs and high-level executives, tend to gratefully participate more with incubator activities [1].

In Brazil, ANPROTEC (Brazilian Association of Science Park and Business Incubators) is composed of private and public members that promote innovation to increase Brazil's economic and social welfare value by providing a variety of activities and services to support entrepreneurs and companies. ANPROTEC is currently a member of the International Association of Scientific Parks and also represents Brazil in the Triple Helix Association (THA). The financial income of ANPROTEC comes from membership, training courses, research, and events. A major feature of incubation in Brazil is the degree of private/public coalitions of partners that support incubation efforts. The Brazilian case has strong national incubator associations. For instance, the Federation of Industries for the State of Sao Paulo (FIESP) operates a dozen incubators [3].

The International Business Incubation Association (InBIA) is a global nonprofit organization in the USA. For over 30 years of service, the goal of InBIA has been to enrich the entire ecosystem by providing industry resources, education, events, and global programming to help their members better serve the needs of their unique communities and regions. Currently, InBIA consists of business incubator developers and managers, corporate joint venture partners, venture capital investors, and economic development professionals. There are more than 2200 members in 62 countries. The Services of InBIA include training and education for members, such as the Business Incubation Management (BIM) Certificate, NewCo Academy Courses, Online Courses, and Customized Training. In addition, InBIA also provides information for their members about industry news and resources and hosts International Conferences on Business Incubation (ICBI). InBIA's income is from membership, training courses, research, and events.

In South Korea, the Korea Business Incubation Association (KOBIA) facilitates technology-based start-ups in collaboration with the Korean Intellectual Property Office (KIPO). The organization trains start-up managers (business incubator managers) as well as students through short-cycle start-up schools or start-up competitions for university students [4]. The association plays this role in five areas. First, the policy area aims to support higher education institutions and research institutes to establish incubators for research commercialization. Second, the business area targets the expansion of the role of incubator entities to support marketing programs. Third, the education area puts an emphasis on the development of an entrepreneurship curriculum in higher-education institutions. Fourth, the international networking area focuses on overseas marketing. Fifth, the public area aims to provide online access and develop the communication tools [5].

In Taiwan, the Chinese Business Incubation Association (CBIA) is a non-profit and membership basis organization. CBIA promotes efficient management, the exchange of information and experience, and resource sharing for the incubators in Taiwan. In addition, CBIA creates networks, conducts research, and provides assistance to policy-makers. CBIA also develops appraisal system and related training programs for incubator professionals. The CBIA's mission is the development of incubation centers, assistance to incubated enterprises in diversified fields, arrangement of specialized activities and skill training courses, provision of educational and practical assistance and materials for incubators and their tenants, publications related to business incubation, and contract establishment with relevant domestic and foreign partners for exchange of experiences (Chinese Business Incubation Association) [6].

The discussion above exemplifies how the executives and management team of business incubators attempt to make the better uses of their capabilities and resources to drive the future development rather than rely on external supports, particularly from government. This attempt has been considered as the pathway toward sustainable development of future business incubators.

Thus, the current capabilities of each incubator need to be assessed in various dimensions. The management team can use the assessment results to guide the future development of each incubator toward sustainable operations.

This leads to two major research questions. The first one is how the capabilities of business incubators can be assessed. The second one is what dimensions of capabilities and their measuring parameters should be used for assessment. Responding to these two research questions, this study reviews different assessment frameworks and presents the development of a capability assessment model for incubators in Thailand. The later section of this paper provides managerial implications on how the assessment framework and model can be strategically implemented.

2. Incubator Development and Evolution in Thailand

Thailand's business incubator was first initiated in 2002 at the country's National Science and Technology Development Agency (NSTDA) where the Business Incubator Centre was established to support startup companies and firms with innovative and technologically driven products. Later in 2004, the Office of the Higher Education Commission (OHEC) under the Ministry of Education initiated the University Business Incubator (UBI) program to reinforce the country's technological commercialization from both public and private higher education institutions [7].

There are currently three platforms of business incubators in Thailand: (1) Business Incubation Center (BIC), (2) University Business Incubator (UBI), and (3) incubators in the private sector. The first two platforms are operated by governmental agencies. The third one is managed by different private firms. BIC is under the supervision of NSTDA. UBI is run by OHEC. For the third platform, there are various companies that support the country's business incubator.

BIC has incubated 74 start-ups and supported established companies with a total of 320-million-baht annual revenue, such as Flexoresearch (an R&D service provider for the pulp, paper, printing, and packaging industries) and KEEN (a bio-remedial firm) [8]. For UBI, the country's fifty-six higher-education institutions have participated in the UBI platform (http://www.mua.go.th/users/bphe/bs/ubi.html (accessed on 20 March 2020)). The third platform of non-governmental agencies is run by private companies/firms in different industrial sectors, for instance, telecommunication service providers and real estate companies. These are companies such as AIS The Startup, The FinLab Accelerator Program, Digital Ventures Accelerator, AddVentures, and Ananda Urban Tech.

A critical problem of the country's university business incubator is a lack of strategic support and insufficient, fragmented, and uncertain financial resources. This is due to the lack of understanding of the risky nature and financial support of start-ups, particularly the technologically based ones. Financial resource support provided by UBI has been spread too thin (due to program rigidities) and has been spent inefficiently (such as duplicate trainings) [7,8].

3. Areas of Capability Development for Business Incubators

Academics have adopted various theoretical lenses spanning different disciplines to study the complexities of the business incubation process and to understand the mechanisms that make a business incubator more effective [9–13]. Those frameworks used in incubator capability assessment can be found to be divergent. For instance, Mian [14] proposed the assessment framework of the University Technology Business Incubator (UTBI) and determined four features that combined the goal approach, the system resource approach, the stakeholder approach, and the internal process approach. On the other hand, Irshad [15] in the "Incubator Support Programme Evaluation Report 2008" by the Ministry of Economic Development of New Zealand utilized three key phases of incubator lifecycles as the framework (the startup phase, the growth phase, and the maturity phase).

According to the literature review, the capability of business incubators can be grouped into seven areas as hereafter described.

3.1. Strategy and Organizational Structure

Strategy and organizational structure are key components of the survival and sustainability of incubators. Incubators need to create their own differentiation strategy, position themselves as specialists, and focus on particular domains [16–21]. Eccles, Perkins, and Serafeim [22] also highlight the three fundamental elements of organization culture as innovation, trust, and capacity for transformational change. In other words, the dimension of strategy and organizational structure should be created with a focus on specialized areas, the continuity of process, and the adaptation to dynamic environmental changes.

3.2. Finance

The conceptual resource-based views of Barney [23] and Gassmann and Becker [24] are applied. Both contributions lead to the connection between the incubation process and resource allocation. In general, incubators utilize two types of tangible and intangible resources. Tangible resources are used through the flows of finance, infrastructure, and explicit knowledge [25–27], while intangible resources are managed through the flows of implicit knowledge and branding [28,29]. These authors' works point to the importance of financial resources for the incubation process. The incubators need efficient financial management, which involves investment and subsidy as well as salary and wages. Moreover, based upon a relevant literature review and preliminary interviews with the sampled incubators, these authors found that incubators earn revenue from five sources, namely (1) subsidy, (2) activity-based revenue, (3) asset-based revenue, (4) fundraising, and (5) revenue from investments.

3.3. Knowledge Body

Since the degree of incubators' service excellence and specialization depends on their proprietary knowledge body [29], the incubator's capability assessment requires the determination of the management of the knowledge body [25,30,31]. This dimension employs Nonaka's A Dynamic Theory of Organizational Knowledge Creation [32] as the conceptual framework to understand the management of the knowledge body within incubators. Nonaka's work found that there is more explicit knowledge than tacit knowledge at the ratio of 80:20. The tacit nature and explicitness of knowledge can be shifted over time depending on the emergence of new knowledge from influential situations. This continuous shifting of knowledge forces firms to adopt the process of knowledge management through the cyclic of a continuous knowledge management process, which is composed of socialization, externalization, combination, and internalization (SECI) [33].

3.4. Human Resource Development

Human resource development (HRD) involves the process of improving working approaches, knowledge, skills, and attitudes among employees in order to achieve organizational objectives [34–36]. Human resource development needs techniques, tools, and measures in order to align the goals of individuals and organization as well as to support and solve problems for employees. Tseng [37] investigated the relationship between HRD and incubator management and development and revealed that the effectiveness of the incubation process is influenced by HRD's six roles: As a catalyst, a failure rate reducer, a multiplier effect generator, a pilot demonstration center, an entrepreneurship and innovation promoter, and a productive endeavor inspirer.

3.5. Infrastructure

This dimension adopts Smilor's incubation model [38] to understand the roles of infrastructure for incubators. The model indicates that business incubators need a support system that contains four elements: (1) Administrative (such as documentation and file processing); (2) secretarial (such as service work); (3) facilities (such as space, tools, equipment, and other supporting objects); and (4) business expertise (such as technical knowhow and market knowledge). These four elements support the agility and the continuity of services and other activities provided for incubatees [39–41]. Hence, incubators need to invest in facilities in order to be able to provide services with minimum dependence on other incubators within the network.

3.6. Network

It is essential for business incubators to foster a relationship with other agencies, such as research centers, industrial agencies, government agencies, funding organizations, experts, and the market. Moreover, the management of an incubator needs to engage with the networks of local, national, and international incubators to obtain benefits from the pool of shared resources [41–44]. This dimension attempts to understand the adoption of networks among incubators by focusing on the New Economy Incubator Model developed by Lazaroeich and Wojciechwoski [45]. The model highlights that incubators who have a broad network at the regional, local, and international levels tend to have a high degree of service excellence.

3.7. Services

The core function of business incubators is to support incubatees to survive and thrive in the market through the delivery of service excellence [46–49]. This dimension of services employs the Customer Satisfaction Model developed by Zeithaml et al. [50]. The model highlights that service quality can be divided into three levels, depending upon the distance between customer perceptions and expectations. The base level of meeting basic customer requirements is achieved when organizations reach the customer requirements and prevent customer complaints. The mid-level of satisfying unstated customer needs is accomplished when organizations reach the customer requirements and develop customer confidence. The top level of achieving customer delight is reached when organizations provide services that exceed customer expectations and build customer loyalty.

4. Methodology: Development and Validity of Assessment Model

This study addresses two research questions of how the capabilities of business incubators can be assessed and what dimensions of capabilities and their measuring parameters should be used for assessment. In this study, the operation of business incubators from many countries have been reviewed from the literature and their official websites. Although various operations of business incubators can be found, the characteristics can be categorized into seven dimensions as described in Section 3.

Three rounds of focus group interviews were held to test the content validity of seven dimensions and to obtain insights on how business incubators in Thailand should be developed toward sustainability. Fifty managers and executives from various business incubators across the country were invited to participate in three rounds of the focus group interviews organized and moderated during March–April 2018, lasting for 4–5 h per round. The first round focused on opportunities and challenges of business incubators in Thailand, aiming to shed light on the potential of Thai business incubators toward sustainability. The second round discussed the necessary capabilities required toward the future development of business incubators. The participants were allowed to reveal their perspectives of ideal business incubators. As a result, all participants agreed with the seven dimensions as earlier described. For the third round, participants were asked to discuss about the assessment model including what proper indicators and measuring parameters should be used for assessment. The 5-point scale was introduced, and participants were encouraged to give the description of each of the five levels (from initial, defined, established, systemized to matured) of each dimension.

Findings from the three rounds of focus groups led to the conclusions on seven dimensions along with their descriptions, parameters, and measuring indicators. Then, the assessment model was validated through the workshop organized during December 2018 with business incubator managers and operational teams. The assessment model was tested with a prepared case study. The details of model development and the case demonstration are presented in Sections 5 and 6, respectively.

The scales developed in this study may have to be adjusted in the future as many disruptions arrive and the role of business incubators may have to change overtime.

Consequently, the validity of assessment model might have to be revisited from time to time in order to adapt with changing circumstances.

5. Capability Assessment Model: Lessons Learned from Thailand

A capability assessment model for business incubators that is widely accepted and used internationally is still being debated. However, there are a few related studies. Each study has similarities, but they use different assessment models. For instance, ANPROTEC in Brazil uses the CERNE framework (Centro de Referencia para Apoio a Novos Empreendimentos) by setting four levels of capabilities according to the process and the ability in operation. In New Zealand, The Humaira Irshad (2014) [15] and Incubator Support Programme Evaluation Report (2008) by the Ministry of Economic Development [51] defined three levels of capability according to the lifecycle of incubators from the early state, growth state, and maturity state of operational capability.

The community of business incubator managers is still in search of the ideal incbation strategy and models. The UBI Global benchmark 2015/2016 highlights that it is a much more complex endeavor due to the particularities of each center's business model [2]. The different contexts have different types of problems and diverse cultural underpinnings embedded in their structural systems and social relations. Therefore, the consequences of the same incentives and assessment mechanisms applying/functioning in different individual contexts might not be identical [52]. This is the same case for applying any model for capability assessment of incubators in countries that have specific contextual factors. Mian et al. [9] pointed out that although a model needs to develop a unified theory of incubation, which covers the business incubation mechanisms, the key challenge is how to address varying policy objectives, organizational forms, and contexts.

In this study, the capability assessment model is developed by incorporating multidimensional perspectives into the consideration. Not only the vision, mission, and objective of business incubators, but also the opportunities and possibilities to further develop the incubators are considered.

The capability dimensions and maturity levels required for the operation of an incubator were defined during a brainstorming session with experts and top management from leading incubators in Thailand. The meeting was held to define the capability dimensions and their description. The following seven dimensions were agreed up-on (as shown in Figure 1): (1) Strategy and organizational structure; (2) finance; (3) knowledge body; (4) human resource development; (5) infrastructure; (6) network; and (7) services.

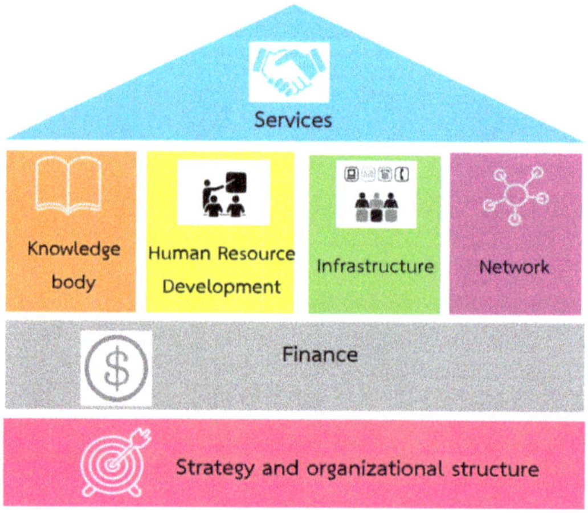

Figure 1. Dimensions used in capability assessment model.

The experts then discussed and agreed upon the capability rating scale by using five maturity levels ranging from initial (the lowest level), defined, established, systemized, and matured (the highest level). The description of an incubator in each level is shown in Table 1.

Table 1. Maturity level used in the capability assessment model for incubators in Thailand.

Level	Description
Initial	An incubator is established with minimum infrastructure to operate. Staff are assigned to cover basic day-to-day operations only. The organization still lacks organizational structure, clear work procedures, etc. The organization requires 100% financial support from the government to operate.
Defined	An incubator has defined the work procedure aligned with strategic goals and targets. However, the strategic implementation plan is still not effectively in place. The achievements are measured based on outputs not outcomes. Currently, an incubator has sufficient infrastructure but faces challenges in coping with the increasing demand requested by incubatees.
Established	An incubator has a well-established organizational structure and is perceived as a stable organization. The strategic implementation plan is in place with clear KPIs. Key risks are identified. An incubator is capable of providing a wide range of services throughout the value chain and stages of incubatees. An incubator begins to focus the outcomes on economic value. An incubator is able to generate incomes from services accounting for around 20% of the annual expenses.
Systemized	An incubator has a well-established organizational structure following the international standards, such as having an advisory board, applying a systematic approach for risk management, etc. An incubator is actively linked with other incubators, domestically and internationally. An incubator is also capable to strategically adapt to changing environments. The issues related to sustainable development of an incubator are always brought up for discussion. An incubator is able to generate income from services accounting for around 21–50% of its annual expenses.
Matured	An incubator has been perceived as a sustainable organization with many achievements contributing to economic value creation. An incubator is able to effectively adapt its strategies to cope with changing environments. An incubator is able to generate income from services of more than 50% of its annual expenses. An incubator takes an active role in many incubator networks and has been internationally recognized for one of the best practice incubators.

6. Description of Capability Assessment Model for Business Incubators in Thailand

This study proposes seven dimensions to be employed in the capability assessment model for business incubators in Thailand. They are strategy and organizational structure, finance, knowledge body, human resource development, network, services, and infrastructure.

6.1. Dimension 1: Strategy and Organizational Structure

The organization can be sustained with a corporate culture consisting of three elements: Innovation, trust, and capacity for transformational change [53,54]. It is crucial to establish a corporate identity to build the corporate culture. The corporate identity features reframing identity, codifying new identity, and leadership commitment. Leadership commitment is essential since any changes or any operations in the organization require an organizational structure that appoints the leader who is distinctly responsible for managing and attending to the specific matters. Furthermore, leaders in the sustainable organization are different from the leaders in the traditional organization. Leaders from the sustainable organization exercise long-term vision in decision making and have tolerance against changes and risks.

The detailed description of each maturity level under Strategy and organizational structure are illustrated in Table 2. At the initial level, business incubators have the operation plan but still lack a distinct goal. At the established level, business incubators have a strategic roadmap, risk assessment, and capability to solve the immediate problems.

At the matured level, they can adapt their strategies due to the changing situations as well as predict the future shortcomings that may affect the business incubators.

Table 2. Maturity levels and their description on "Strategy and organizational structure" dimension.

Level	Strategy and Organizational Structure
Initial	The routine works have been assigned to responsible persons. The day-to-day operations are fine but still lack strategic goals and an organizational structure.
Defined	The work procedure has been defined. The strategic goals and targets are set and are in line with the direction of the governance of the organization. However, the strategic implementation plan is still not in place.
Established	An incubator has a well-established organizational structure and is perceived as a stable organization. The strategic implementation plan is in place with clear KPIs. Key risks are identified. The organization is capable of effectively handling routine problems.
Systemized	An incubator has a well-established organizational structure, which follows the international standards, such as having an advisory board, applying a systematic approach for risk management, etc. The organization is also capable to strategically adapt to changing environments.
Matured	An incubator can integrate change management as a part of day-to-day operations. It is capable to strategically initiate and transform with a forecasting and predictive systems in place in order to cope with changing environments.

6.2. Dimension 2: Finance

The sources of funds supporting business incubators can be divided into five categories: (1) Government-related subsidies; (2) activity-based revenue (e.g., business consultant fee, training fee); (3) asset-based revenues (spaces and equipment rental fees); (4) grant from graduated incubatees or large private corporations; and (5) revenue from other investments.

This study considers the percentage of revenue that business incubators are able to generate by themselves. The detailed description of each maturity level under Finance is illustrated in Table 3. At the initial level, business incubators obtain a subsidy from an outside source of funds. At the established level, business incubators can generate some revenue by themselves, but they still need some subsidization from the external sources of funds. At the matured level, business incubators will create enough revenue to run their operational activities and allocate support to other business incubators.

6.3. Dimension 3: Knowledge Body

The key elements to assess the knowledge body include basic knowledge, value-added knowledge, knowledge for innovation, knowledge management system, knowledge of international standards, and knowledge sharing.

The detailed description of each maturity level under Knowledge body is illustrated in Table 4. At the initial level, business incubators have a knowledge body that is able to solve fundamental problems of incubatees. At the established level, they have a knowledge body that is able to support the incubatees to enter the new market or create innovative products. The knowledge body at the established level also includes proprietary intellectual service. At the matured level is the creation of a new knowledge body and alteration and application the knowledge body to be practical for each incubatee.

Table 3. Maturity levels and their description on "Finance" dimension.

Level	Finance
Initial	An incubator requires 100% financial subsidy from the government to support its operations.
Defined	An incubator mostly requires financial subsidies from the government to support its operations. Some limited revenue is generated from providing services through contracted government projects.
Established	An incubator requires major financial subsidies from the government to support its operations. However, the incubator can generate some revenue by providing services through not only contracted government projects, but also other projects hosted by private organizations, communities, non-profit organizations, etc. The amount of revenue generated is around 20% of required annual expenses.
Systemized	An incubator is capable to generate revenues from its own services accounting to around 20–50% of required annual expenses and relies less on the financial subsidiary from the government. An incubator also allocates the budget to support the future growth of an organization.
Matured	An incubator is capable to generate revenues from its own services accounting more than 50% of annual expenses required and relies less on the financial subsidiary from the government. An incubator also allocates the budget to drive the future growth of an organization as well as support other incubators contributing to the development of incubator networks.

Table 4. Maturity levels and their description on "Knowledge body" dimension.

Level	Knowledge Body
Initial	An incubator has fundamental business knowledge with abilities to provide services to incubatees but still lacks the system for storing and archiving knowledge.
Defined	An incubator has a system to store and archive knowledge, but it still needs an additional system supporting data analysis and synthesis for value creation of knowledge applications.
Established	An incubator has a knowledge management system in place to store, archive, analyze, and synthesize knowledge.
Systemized	An incubator has applied the knowledge management system with case evidence that presents the incubator's abilities to create economic value from knowledge sharing within an organization as well as with other outside incubators.
Matured	An incubator has extensively applied the knowledge management system with many cases that evidence the presentation of the incubator's abilities to create economic value from knowledge sharing within an organization as well as with other outside incubators. Its effective approach in managing knowledge has been internationally recognized as one of the best practice examples.

6.4. Dimension 4: Human Resource Development

Human resource development includes the efficiency of human resource management as well as the development of human resources (skills and career path). The maturity level of human resource development for a business incubator can be clarified in five levels. The detailed description of each maturity level under human resource development is illustrated in Table 5. At the initial level, the support for human resource development is very limited and unplanned. At the established level, an incubator specifies the personnel capability characteristics required for each job position as well as providing the support for staff to complete training and skill development activities. At the matured level, each person is not only aware of his/her role, duty, and responsibility but is also able to set personal working goals in line with the business incubator's goal.

Table 5. Maturity levels and their description on "Human resource development" dimension.

Level	Human Resource Development
Initial	The process supporting human resource development is not well defined. There is no clear plan or program for training or coaching new staff (impromptu).
Defined	The process for human resource development has been defined but the activities are mainly done through on-the-job training.
Established	The process for human resource development has been well-structured in order to assure the alignment between the personnel goal and the organization's goal. The career advancement path is also defined and presented to staff.
Systemized	All staff have clear knowledge about the role of business incubation. Their understanding is in line with international standards. Each staff member is allowed to conduct self-assessment in order to determine his/her level of competencies and identify his/her competency gaps for further improvement.
Matured	All staff understand their roles and responsibilities. They are willing to engage in organization activities in which they strive for success and sustainable development of an organization.

6.5. Dimension 5: Infrastructure

The key elements of infrastructure are comprised of administrations that are related to standard operation procedures (SOPs) for providing services to incubatees as well as facility management (e.g., office, maker space, equipment). The detailed description of each maturity level under Infrastructure is illustrated in Table 6. At the initial level, business incubators have rental space services, essential facilities, and staff. However, it is inadequate for all incubatees. At the established level, there is sufficient infrastructure and ability to appropriately and sufficiently meet the requirements of all incubatees. At the matured level, they can construct or procure the new resources and modify or develop the existing resources to be concurrent with the external changing factors and continuous requirements of the incubatees.

Table 6. Maturity levels and their description on "Infrastructure" dimension.

Level	Infrastructure
Initial	An incubator has some working space, equipment, and infrastructure, but it is not adequate. It still needs to acquire some more resources.
Defined	An incubator has most of its required working space, equipment, and infrastructure. However, these are not enough to support the increasing demands of incubatees.
Established	An incubator has most of its required working space, equipment, and infrastructure, and it can support the increasing demands. Yet an incubator still needs to improve the efficiency and effectiveness of its resource usage.
Systemized	An incubator can effectively manage the working space, equipment, and infrastructure that it has and is able to provide in a form of virtual services.
Matured	An incubator can regularly update current, or acquire new, working space, equipment, and infrastructure to cope with the change requirements of industries and incubatees.

6.6. Dimension 6: Network

It is essential for business incubators to have a relationship with other agencies, such as the knowledge institutes, research centers, industry sectorial agencies, government institutions, fund agencies from both government and private sectors, experts from various areas, and the market. Moreover, it is necessary for the management of incubators to engage in the networks of local, national, and international incubators.

The detailed description of each maturity level under Network is illustrated in Table 7. At the initial level, business incubators have very limited alliances. At the established level, they will be part of the national level alliance network that can make an impact or create national level economic value. At the matured level, business incubators have roles as critical mechanisms or the central nodes of alliance networks.

Table 7. Maturity levels and their description on "Network" dimension.

Level	Network
Initial	An incubator has limited networks of partners that are not sufficient to cover possible services requested by incubatees.
Defined	An incubator has the networks of adequate partners to support the majority of services needed. However, the economic impacts from the collaborations are very limited.
Established	An incubator is a part of networks that can help create economic value from the projects of its incubatees.
Systemized	An incubator is a part of international networks that can exchange knowledge and/or activities that lead to economic value creation.
Matured	An incubator can be a node of international networks that can be a center of economic value creation.

6.7. Dimension 7: Services

The analysis of service capacity is based on balancing customer perceptions with expectations. The acceptance and satisfaction of services is considered to range from meeting basic customer requirements, satisfying unstated customer needs, achieving customer delight that exceeds expectations, and building customer loyalty.

The detailed description of each maturity level under services is illustrated in Table 8. At the initial level, the variety and capacity of services offered by an incubator are still limited. There are the knowledge transfer activities to the locals and the public promotion of the duties of the business incubators. At the established level, an incubator can provide services covering the whole value chain of operations as needed by incubatees. At the matured level, there are unique services. They can see opportunities and offer services that support the dynamics of the business environment.

Table 8. Maturity levels and their description on "Services" dimension.

Level	Services
Initial	The variety and capacity of services offered by an incubator are still limited. The activities are mainly focused on knowledge sharing to create public awareness.
Defined	An incubator has the abilities to identify and solve some basic problems of incubatees. However, the scope of its services still does not cover the whole value chain addressing the different development phases of incubatees.
Established	An incubator is able to offer services that cover the whole value chain, addressing every phase of the lifecycle (from beginning to survival) of incubatees.
Systemized	An incubator offers proactive services to help current incubatees as well as to encourage people to become new incubatees. The system is in place to monitor the operation progress and risks. An incubator also provides the linkage connecting services offered by other incubators within its network.
Matured	An incubator offers a full range of services with some unique or specialized services. An incubator can seize the future opportunities and be able to actively adapt their services to match the changing business environment. In a case in which there is a request, an incubator can also provide services to incubatees who are working with other incubators within its networks.

7. Case Demonstration for Assessing the Maturity Level of a Business Incubator

This case study demonstrates how to operate the proposed model to assess the capability level of a business incubator. This demonstration case will be presented in three steps: (1) Data collection; (2) analysis; and (3) result presentation of the capability assessment level.

7.1. Step 1: Data Collection

To collect the inputs for assessment, the triangular interview approach is applied. The interview sessions are organized into three rounds. The first round is with executives of the incubator and the second round is with employees of the incubator. The third round is with clients of the incubators (see Figure 2). The interviewees are asked questions related to the seven dimensions of the capability assessment model. All interviews are recorded and transcribed.

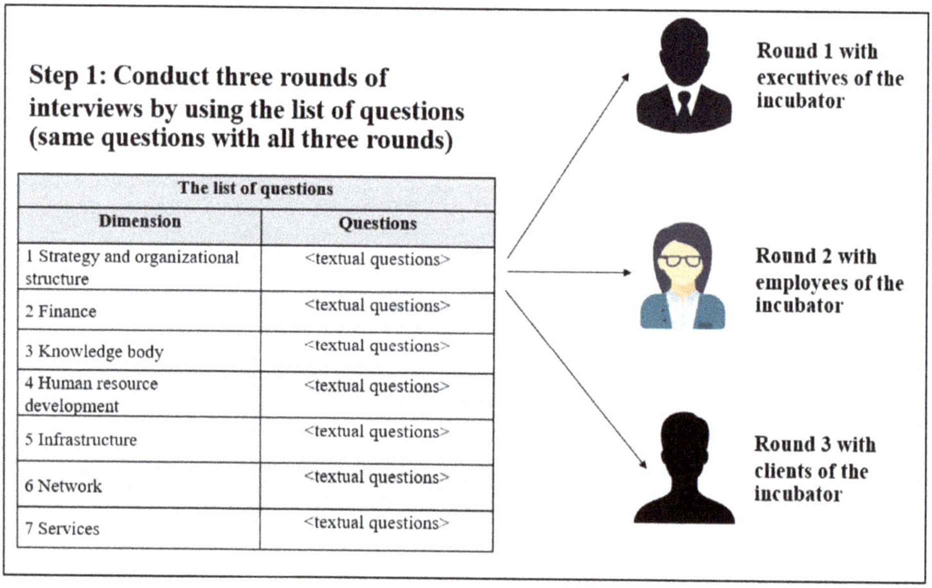

Figure 2. Data collection—triangular interview approach.

The outputs of interviews in step 1 represent the case background and the conditions in which each business incubator operates. One example is shown in Table 9.

7.2. Step 2: Analysis

Interview transcription in step 1 is analyzed along the seven dimensions. Key interview quotations are then extracted and assessed according to the maturity levels specified for each dimension (see Figure 3). For example, in Figure 3, quotations regarding to dimension 1 Strategy and Organizational Structure have been analyzed and can be connected with the definition of level 3 Established.

Table 9. Case description.

XYZ University Business Incubator *
This university business incubator (referred to as XYZ in this case example) has been established for a decade. It is operated under the supervision of the university committee with the mission to promote and support new entrepreneurs through potential commercialization of the university research. With this mission, XYZ plays a role in enhancing the capabilities and competitive advantages for businesses, co-developing innovation projects between academia and practitioners, and forming a network of experts from various fields. XYZ has a flat structure, governed by the science park of the public university. This incubator is composed of five units including 1. Technology Licensing (TLO) 2. Innovation Design Office (IDO) 3. Office of Industrial Liaison (OIL) 4. University Business Incubator and 5. Development Unit for Startup (DUS). The executive meeting for strategic modification is held every three years. The board consists of executives from governing university, government, association, and business sector. This incubator has large service areas; however, the primary services are focused on food products, agricultural products, IoT (Internet of Things), local wisdom, and area-based creativity. Nowadays, XYZ still rents the building space from the governing university. Over the past years, XYZ has prepared sufficient facilities, laboratories, and equipment to serve the entrepreneurs' basic needs along with customized designs and services for individual entrepreneurs. The incubator produces a case study report every six months. However, most reports are still related to local food, agricultural products, and herbs. Internal knowledge-sharing activities among academic researchers, employees, and entrepreneurs are regularly held. Moreover, it has international linkages with countries in Asia, including Taiwan, Indonesia, and Vietnam, in activities of site visits, business matching, and cooperation. The incubator used to experience financial obstacles, but it overcame them by seeking a variety of revenue sources and cutting unnecessary expenses. During the first three years, the incubator received 100% total funding support from the government. Nowadays, the incubator can generate revenue by itself and needs less support from the government. Furthermore, this incubator plans to operate with self-reliance in the long run. As for internal management, this incubator provides financial rewards and honors for researchers and employees. Although the incubator accepts that the financial incentive might not be high, it attempts to use other non-financial incentives such as freedom, open and flexible working conditions, and training. Currently, the proportion between the number of employees and the number of incubation projects is 1:12. The rate of terminated projects (when incubatees do not keep in touch for longer than three months) is 12%. This incubator does not clearly limit the period of incubation service, but it recruits applicants in 3–4 rounds a year. For each round, the interviews are conducted by professionals to screen applicants into 20 incubatees. However, this incubator still provides services by itself without any linkages with other incubators or with its networks.

* Based on the actual organization but the name has not been disclosed.

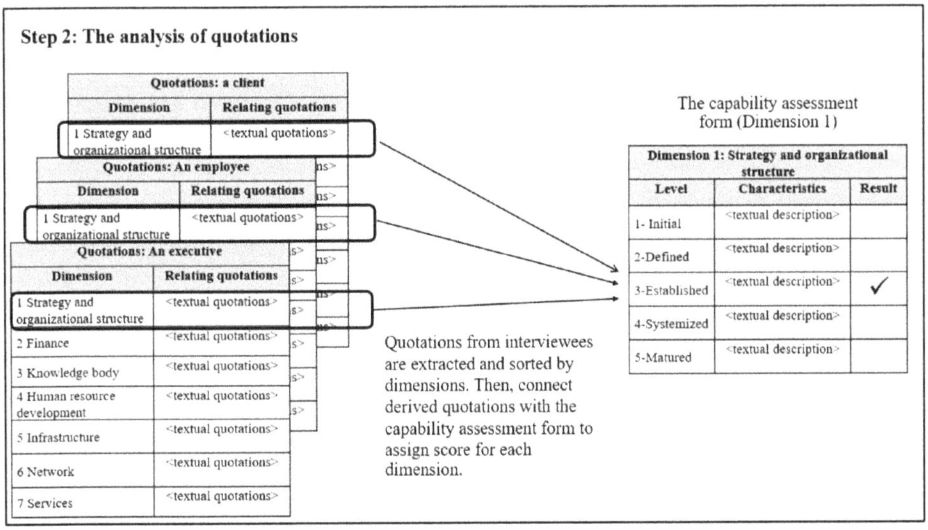

Figure 3. Case analysis.

As the result of step 2, Table 10 shows the analysis linking the interview quotations to the maturity assessment level for each dimension.

Table 10. Linking the extracted quotations to the assessment.

Dimension	*Quotations*	Assessed Level-Characteristics
Strategy and organizational structure	*"Nowadays, the direction of our strategies is the same as the governing university."*	3-Established level The organizational structure has been set and it has been perceived as a stable organization. The strategic implementation plan is in place with clear KPIs. Key risks are identified. The organization is capable of effectively handling routine problems.
Finance	*"Today, the proportion of revenue between from government and from itself today is 50:50. We have sufficient budget for operations and we also have some savings."*	4-Systemized level An incubator is capable of generating revenues from its own services accounting for around 20–50% of total annual expenses. An incubator also allocates the budget to support the future growth of an organization.
Knowledge body	*"Once the incubation has been accomplished, the Development Unit for Startups(DUS) will collect the information, decode into explicit knowledge and profile in both digital and paper formats for future knowledge exchange activities."* *"For intellectual management process, we follow the university policy and the mutual agreement between the incubator and entrepreneurs."*	2-Defined level An Incubator has a system to store and archive knowledge, but it still needs the additional system supporting data analysis and synthesis for value creation of knowledge applications.
Human Resource Development	*"However, we accept that some personnel feel insecure to work for here due to unclear career path."*	2-Defined level The process for human resource development has been defined but the activities are mainly done through on-the-job training.
Infrastructure	*"Most services provided us are regarded as in wall services."* *"When we do not have tools as the clients request, we attempt to acquire them from governing university."*	2-Defined level An incubator has most of its required working space, equipment, and infrastructure. However, these are not enough to support the increasing demands requested by incubatees.
Network	*"We have domestic networks with experts, academic researchers in other universities, other public science parks, trade councils, and even the ministry of culture."* *"However, our international linkage activities have been taken occasionally."*	3-Established level An incubator becomes a part of networks that can help create economic value from its incubatee projects.
Services	*"We offer three types of services including research and development (R&D), preparing for market entry, and creating customers' perception."*	3-Established level An incubator is able to offer services that cover the whole value chain, addressing every phase of the lifecycle (from beginning to survival) of incubatees.

7.3. Step 3: The Presentation of Capability Assessment Results

The numeric results of capability assessment in Step 2 as shown in the right column of Table 10 are presented in a radar chart as shown in Figure 4. For example, the right column in Table 10 reveals that XYZ University Business Incubator performs seven dimensions at levels of 3, 4, 2, 2, 2, 3, and 3, respectively. These numeric levels of seven dimensions are visualized in the form of radar chart format (see Figure 4 below).

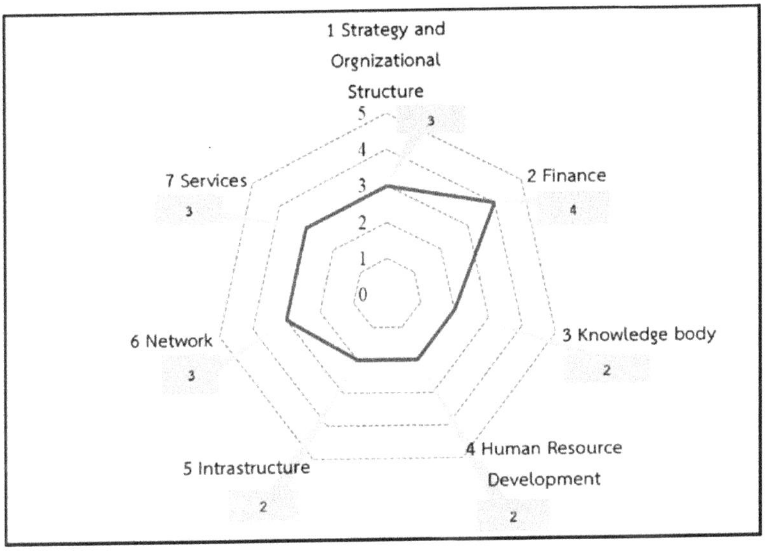

Figure 4. A radar chart representing the capability level of XYZ University business incubator.

8. Discussions and Managerial Implications

This section addresses the managerial implications of capability assessment from three aspects: (1) The development of a proper strategy and strategic roadmap toward becoming an effective business incubator; (2) the cluster development among business incubators according to their capabilities and not just by size or geographical location; and (3) the development of a knowledge-based community among incubators. The details of each aspect are hereafter described.

First, the radar chart (as shown in Figure 4) reveals the current capability level of business incubators in each dimension. In a case in which the capability level is below expectations, managers need to focus on how to close the gap. The wider gap the between the assessed level and the expectation, the more seriously managers need to pay attention. In other words, the results on a radar chart analysis can lead to the priority for closing the gaps. The extended approach of technology and strategic roadmapping can be applied [55–57]. Managers can begin to draft a strategic roadmap by using a radar chart as the reference to identify what gaps they need to bridge and when to do so (see Figure 5 below). The extended details of integrating capability assessment into road mapping can be found in the study by Chutivongse and Gerdsri [58].

Second, the capability assessment results can be analyzed together with operational performance (such as number of incubatees, number of graduate incubates, the survival rate, etc.). The consideration of both capability and performance can be visualized in the form of a performance–capability matrix (see Figure 6). This matrix reveals the positions of business incubators indicating how high/low are their performances and capabilities. This matrix leads to four clusters. Business incubators in different clusters require different strategies to drive their development. Clustering can help policymakers or business incubator promotion agencies at the national level to customize their decisions on effective budgeting and resource allocation to strategically serve the needs of business incubators in each cluster rather than focusing on their size or geographical location. Figure 6 also shows the possible pathways to drive business incubator positioning in Q3 to eventually become the high-capability/high-performance incubator (Q1).

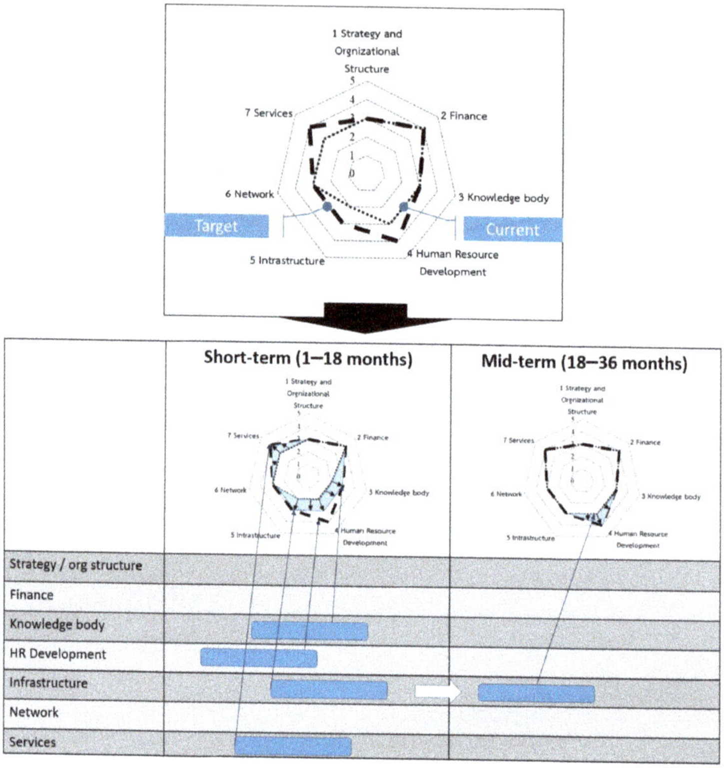

Figure 5. A strategic roadmap guiding development activities for the specified incubator.

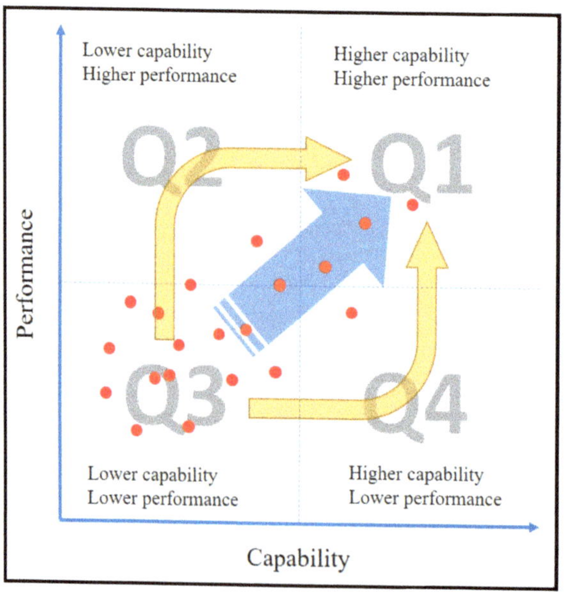

Figure 6. The performance–capability matrix.

Third, the capability assessment results can be used to develop the knowledge-based community among incubators by focusing on knowledge exchange [59] and the development of knowledge cluster [60]. The incubator with the highest level of capability in each dimension is considered as an incubator champion that is expected to act as a coach or a mentor sharing experiences on its developmental journey with other business incubators. Furthermore, the incubator champion can actively engage in community development by leveraging its capabilities and resources to work with other incubators to develop their capabilities. Engaging activities include holding regular meetings and seminars to transfer knowledge, setting up a talent mobility program, or collaborating in some projects with less capable incubators. These approaches have been practiced into develop the sectoral innovation system [61,62]. Figure 7 reveals that business incubator U performs better than business incubators C and N in the three dimensions of finance, services, and network. Business incubator U is expected to act as the incubator champion who shares experiences and its journey of development in light of how to manage finance effectively, improve service quality, and coordinate with partners of business incubators C and N.

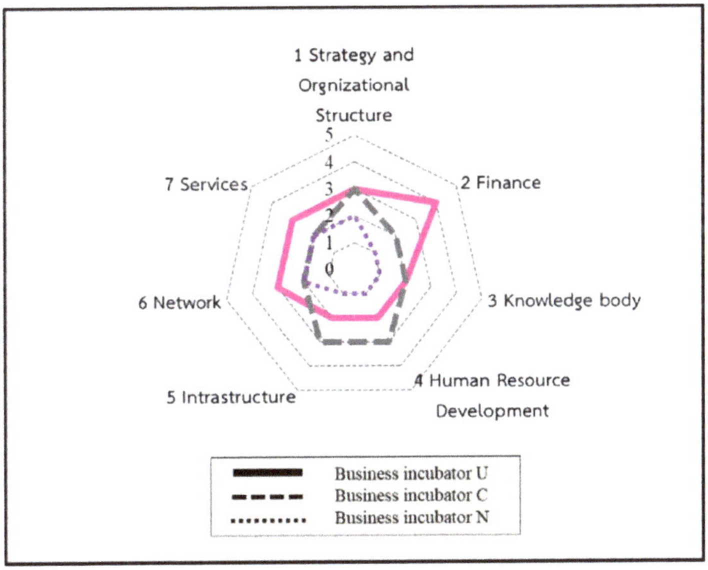

Figure 7. The radar charts of Business incubators U, C, and N.

9. Conclusions

The assessment of business incubators is significant for the country's incubation development. In this paper, the proposed model for capability assessment of business incubators is developed and applied to business incubators. The demonstration of the model is contextualized with the case of a business incubators in Thailand since these business incubators still rely on governmental supports through various forms (e.g., funding, creating business networks and communities, developing specialties in particular areas, etc.). For the long-term development, these incubators have to find the ways to become self-reliance in order to sustain their operation. It is very important for the management and executives of any business incubator to understand their current capabilities and limitations so that they can properly plan for their future development.

The capability assessment model consists of seven dimensions: (1) Strategy and organizational structure; (2) finance; (3) knowledge body; (4) human resource development; (5) infrastructure; (6) network; and (7) services. Each dimension is divided into a capability rating scale by deploying five maturity levels ranging from initial, defined, established, systemized to matured levels. The assessment result in the form of radar chart reports the current status of incubators' capabilities. Managers and executives of any incubator can use it as the reference to determine the areas for development and the degrees to which it needs to be developed. The proposed model can be utilized as the assessment platform for both individual units and national levels. Due to the dynamic of business environment, monitoring progress and re-assessing the capabilities are periodically recommended.

Author Contributions: N.G. coordinated the project and drafted this paper. B.I., K.R., N.M. and J.P. undertook case study and analysis. W.W. coordinated with informants and led the interview study. All authors have read and agreed to the published version of the manuscript.

Funding: This research received no external funding.

Institutional Review Board Statement: Not applicable.

Informed Consent Statement: Not applicable.

Data Availability Statement: Not applicable.

Conflicts of Interest: The authors declare no conflict of interest.

References

1. InfoDev. *Business Incubation Definitions and Principles*; The World Bank Group: Washington, DC, USA, 2009.
2. Bhatli, D.; Marin, M.; Singh, A.; Sala, G. *Global Benchmark 15/16 Report. Top. University Business Incubator*; UBI Global: Stockholm, Sweden, 2015.
3. Chandra, A.; Fealey, T. Business incubation in the United States, China and Brazil: A comparison of role of government, incubator funding and financial services. *Int. J. Entrep.* **2009**, *13*, 67.
4. OECD. *OECD Skills Strategy Diagnostic Report*; Organisation for Economic Co-operation and Development: Seoul, Korea, 2015.
5. KOBIA. Business Guide; Korea Business Incubator Association, Korea. 2020. Available online: http://kobia.or.kr/kobia/koability.do (accessed on 22 March 2020).
6. Chinese Business Incubation Association. C.B.I.A. 2019 Website. Available online: http://www.cbia.org.tw/english.html (accessed on 18 March 2020).
7. Munkongsujarit, S. Business incubation model for startup company and SME in developing economy: A case of Thailand. In Proceedings of the 2016 Portland International Conference on Management of Engineering and Technology, Honolulu, HI, USA, 4–8 September 2016.
8. UNCTAD. *Science, Technology & Innovation Policy Review: Thailand*; United Nations Conference on Trade and Development (UNCTAD): Geneva, Switzerland, 2015.
9. Mian, S.; Lamine, W.; Fayolle, A. Technology Business Incubation: An overview of the state of knowledge. *Technovation* **2016**, *50–51*, 1–12.
10. Bennett, D.; Yábar, D.P.-B.; Saura, J.R. University incubators may be socially valuable, but how effective are they? A case study on business incubators at universities. In *Entrepreneurial Universities*; Springer: Cham, Switzerland, 2017; pp. 165–177.
11. Dee, N.; Gill, D.; Lacher, R.; Livesey, F.; Minshall, T. A review of research on the role and effectiveness of business incubation for high-growth start-ups. In *CTM Working Paper Series*; Cambridge University Press: Cambridge, UK, 2019.
12. Sareen, S.; Acharya, S.R.; Dhochak, M. Assessing the effectiveness of business incubators. *Int. J. Innov. Learn.* **2019**, *26*, 177. [CrossRef]
13. Galiyeva, N.; Fuschi, D.L.; Nikulshin, B. Initial results in measuring the effectiveness of the activities of business incubators. *Int. J. Open Inf. Technol.* **2019**, *7*, 78–89.
14. Mian, S.A. Assessing and managing the university technology business incubator: An integrative framework. *J. Bus. Ventur.* **1997**, *12*, 251–285. [CrossRef]
15. Irshad, H. *Business Incubation in Canada Literature Review & List of Business Incubators in Alberta and Canada*; Rural Development Division, Alberta Agriculture and Rural Development: Edmonton, AB, Canada, 2014.
16. Vanderstraeten, J.; Matthyssens, P. Service-based differentiation strategies for business incubators: Exploring external and internal alignment. *Technovation* **2012**, *32*, 656–670. [CrossRef]
17. Prasetyawan, Y.; Agustiani, E.; Jumayla, S. Aligning business strategy of incubator center and tenants. In Proceedings of the AIP Conference 2017, Sozopol, Bulgaria, 8–13 June 2017.
18. Ahmad, A.J.; Thornberry, C. On the structure of business incubators: De-coupling issues and the mis-alignment of managerial incentives. *J. Technol. Transf.* **2018**, *43*, 1190–1212. [CrossRef]
19. Hausberg, J.P.; Korreck, S. Business incubators and accelerators: A co-citation analysis-based, systematic literature review. *J. Technol. Transf.* **2020**, *45*, 151–176. [CrossRef]
20. De Bem Machado, A.; Catapan, A.H.; Sousa, M.J. Incubators management models. In *Sustainable Business: Concepts, Methodologies, Tools, and Applications*; IGI Global: Hershley, PA, USA, 2020; pp. 1646–1656.

21. Stahl, B.C.; Obach, M.; Yaghmaei, E.; Ikonen, V.; Chatfield, K.; Brem, A. The responsible research and innovation (RRI) maturity model: Linking theory and practice. *Sustainability* **2017**, *9*, 1036. [CrossRef]
22. Eccles, R.G.; Perkins, K.M.; Serafeim, G. How to become a sustainable company. *MIT Sloan Manag. Rev.* **2012**, *53*, 43.
23. Barney, J. Firm Resources and Sustained Competitive Advantage. *J. Manag.* **1991**, *17*, 99–120. [CrossRef]
24. Gassmann, O.; Becker, B. Towards a resource-based view of corporate incubators. *Int. J. Innov. Manag.* **2006**, *10*, 19–45. [CrossRef]
25. Binsawad, M.; Sohaib, O.; Hawryszkiewycz, I. Knowledge-Sharing in Technology Business Incubator. In Proceedings of the 26th International Conference on Information Systems Development, Larnaca, Cyprus, 6–8 September 2017.
26. Zattar, I.C.; Lima, G.P.; Rasoto, V.I.; de Oliveira, N.; Silva, V.F. Classification of R&D infrastructure models in basic business incubators technology in the state of Paraná. *Braz. J. Oper. Prod. Manag.* **2017**, *14*, 239–248.
27. Samaeemofrad, N.; Herik, J.V.D. The effectiveness of finance mobilization by business incubators on the performance of NTBFs. In Proceedings of the 2018 IEEE International Conference on Engineering, Technology and Innovation, Stuttgart, Germany, 17–20 June 2018.
28. Santarino, L.B. What is the Difference Between Business Incubation and Business Acceleration Programs? Master's Thesis, Universidade de Coimbra, Coimbra, Brazil, 2017.
29. Shakhlova, N. Knowledge Management System for a Business Incubator. Master's Thesis, St. Petersburg University, Saint Peterburg, Russia, 2018.
30. Wann, J.-W.; Lu, T.-J.; Lozada, I.; Allain, G.C. University-based incubators' performance evaluation: A benchmarking approach. *Benchmarking Int. J.* **2017**, *24*, 34–49. [CrossRef]
31. Binsawad, M.; Sohaib, O.; Hawryszkiewycz, I. Factors impacting technology business incubator performance. *Int. J. Innov. Manag.* **2019**, *23*, 1950007. [CrossRef]
32. Nonaka, I. A Dynamic Theory of Organizational Knowledge Creation. *Organ. Sci.* **1994**, *5*, 14–37. [CrossRef]
33. Nonaka, I. Managing globalization as a self-renewing process: Experiences of Japanese MNCs. In *Managing the Global Firm*; Routledge: London, UK, 1990; pp. 69–94.
34. Grimaldi, R.; Grandi, A. Business incubators and new venture creation: An assessment of incubating models. *Technovation* **2005**, *25*, 111–121. [CrossRef]
35. Ascigil, S.F.; Magner, N.R. Business incubators: Leveraging skill utilization through social capital. *J. Small Bus. Strategy* **2009**, *20*, 19–34.
36. Bakkali, C.; Messeghem, K.; Sammut, S. Toward a typology of incubators based on HRM. *J. Innov. Entrep.* **2014**, *3*, 1–10. [CrossRef]
37. Tseng, C. Connecting business incubator development with human resource development. *J. Multidiscip. Res.* **2011**, *3*, 29–42.
38. Smilor, R.W. Managing the incubator system: Critical success factors to accelerate new company development. *IEEE Trans. Eng. Manag.* **1987**, *EM-34*, 146–155. [CrossRef]
39. Vedovello, C.; Godinho, M. Business incubators as a technological infrastructure for supporting small innovative firms' activities. *Int. J. Entrep. Innov. Manag.* **2003**, *3*, 4–21. [CrossRef]
40. Robinson, S.; Stubberud, H.A. Business incubators: What services do business owners really use? *Int. J. Entrep.* **2014**, *18*, 29.
41. Dahms, S.; Kingkaew, S. University business incubators: An institutional demand side perspective on value adding features. *Entrep. Bus. Econ. Rev.* **2016**, *4*, 41–56. [CrossRef]
42. Bøllingtoft, A. The bottom-up business incubator: Leverage to networking and cooperation practices in a self-generated, entrepreneurial-enabled environment. *Technovation* **2012**, *32*, 304–315. [CrossRef]
43. Cooper, C.E.; Hamel, S.A.; Connaughton, S.L. Motivations and obstacles to networking in a university business incubator. *J. Technol. Transf.* **2012**, *37*, 433–453. [CrossRef]
44. Soetanto, D.P.; Jack, S.L. Business incubators and the networks of technology-based firms. *J. Technol. Transf.* **2013**, *38*, 432–453. [CrossRef]
45. Lazarowich, M.; Wojciechowski, M.J. *Russian Business Incubator Program, Phase One: Prospect. Development and Strategic Plan*; University of Waterloo: Waterloo, ON, Canada, 2002; Available online: http://www.aucc.ca/_pdf/english/programs/cepra/Rus (accessed on 20 March 2020).
46. Kumar, A. Empirical investigation of business incubation service components in indian technology business incubators (TBIs). In Proceedings of the International Conference on Research and Business Sustainability 2017, Greater Noida, India, 16–17 December 2017.
47. Godeiro, D.P.d.O.; Dantas, M.L.R.; Silva, D.C.D.; Celestino, M.D.S. Application of importance and performance matrix to assess the quality of services provided by business incubators. *Iberoam. J. Entrep. Small Bus.* **2018**, *7*, 1–30. [CrossRef]
48. Ribeiro, H.; Meneses, R.; Sousa, J. A study of the benefits for startups coming from incubators. In Proceedings of the 12th Annual Conference of the EuroMed Academy of Business, Thessaloniki, Greece, 18–20 September 2019.
49. Solberg Hjorth, S.; Brem, A.M. How to assess market readiness for an innovative solution: The case of heat recovery technologies for SMEs. *Sustainability* **2016**, *8*, 1152. [CrossRef]
50. Zeithaml, V.A.; Parasuraman, A.; Berry, L.L. *Delivering Quality Service: Balancing Customer Perceptions and Expectations*; Simon and Schuster: New York, NY, USA, 1990.
51. Ministry of Economic Development. *Incubator Support. Programme Evaluation Report*; Ministry of Economic Development: Wellington, New Zealand, 2008. Available online: https://www.mbie.govt.nz/dmsdocument/2263-incubator-support-programme-evaluation-report-pdf (accessed on 22 March 2020).

52. Rungfamai, K. Governance of national research university in Southeast Asia: The case of Chiang Mai University in Thailand. *Stud. High. Educ.* **2018**, *43*, 1268–1278. [CrossRef]
53. Jahanshahi, A.A.; Brem, A. Sustainability in SMEs: Top management teams behavioral integration as source of innovativeness. *Sustainability* **2017**, *9*, 1899. [CrossRef]
54. Jahanshahi, A.A.; Brem, A.; Bhattacharjee, A. Who takes more sustainability-oriented entrepreneurial actions? The role of entrepreneurs' values, beliefs and orientations. *Sustainability* **2017**, *9*, 1636. [CrossRef]
55. Park, H.; Phaal, R.; Ho, J.Y.; O'Sullivan, E. Twenty years of technology and strategic roadmapping research: A school of thought perspective. *Technol. Forecast. Soc. Chang.* **2020**, *154*, 119965–119979. [CrossRef]
56. Kockan, I.; Daim, T.U.; Gerdsri, N. Roadmapping future powertrain technologies: A case study of Ford Otosan. *Int. J. Technol. Policy Manag.* **2010**, *10*, 157–184. [CrossRef]
57. Vishnevskiy, K.; Karasev, O.; Meissner, D. Integrated roadmaps for strategic management and planning. *Technol. Forecast. Soc. Chang.* **2016**, *110*, 153–166. [CrossRef]
58. Chutivongse, N.; Gerdsri, N. Creating an innovative organization: Analytical approach to develop a strategic roadmap guiding organizational development. *J. Model. Manag.* **2019**, *15*, 50–88. [CrossRef]
59. Intarakumnerd, P.; Gerdsri, N.; Teekasap, P. The Roles of External Knowledge Sources in Thailand's Automotive Industry. *Asian J. Technol. Innov.* **2012**, *20*, 85–97. [CrossRef]
60. Gerdsri, N.; Kongthon, A.; Puengrusme, S. Profiling the Research Landscape in Emerging Areas Using Bibliometrics and Text Mining: A Case Study of Biomedical Engineering (BME) in Thailand. *Int. J. Innov. Technol. Manag.* **2017**, *14*, 1740011–1740037. [CrossRef]
61. Intarakumnerd, P.; Gerdsri, N. Implications of Technology Management and Policy on the Development of a Sectoral Innovation System: Lessons Learned Through the Evolution of Thai Automotive Sector. *Int. J. Innov. Technol. Manag.* **2014**, *11*, 1440009–1440028. [CrossRef]
62. Lazaro-Mojica, J.; Fernandez, R. Review paper on the future of the food sector through education, capacity building, knowledge translation and open innovation. *Curr. Opin. Food Sci.* **2020**, *38*, 162–167. [CrossRef]

sustainability

MDPI

Article

Factors Facilitating the Implementation of the Sustainable Development Goals in Regional and Local Planning—Experiences from Norway

Kjersti Granås Bardal *, Mathias Brynildsen Reinar, Aase Kristine Lundberg and Maiken Bjørkan

Nordland Research Institute, N-8049 Bodø, Norway; mbr@nforsk.no (M.B.R.); akl@nforsk.no (A.K.L.); mbj@nforsk.no (M.B.)
* Correspondence: kgb@nforsk.no

Abstract: Successful implementation of the Sustainable Development Goals (SDGs) depends on regional and local authorities' ability to implement the goals in their respective contexts. Through a survey and interviews with informants in Norwegian municipalities and county councils, this paper explores and offers new empirical insight into (1) which factors can be identified as facilitating the implementation of the SDGs in Norwegian local and regional planning; (2) how the facilitating factors are conditioned by the different local and regional institutional contexts; and (3) how these factors from the Norwegian context correspond or differ from those in the international literature. We find that the existing Planning and Building Act is considered a suitable framework for the implementation of the SDGs in the Norwegian context, and that the SDGs are high on the national and regional governmental agendas. However, work remains in integrating the SDGs into underlying governmental activities. They must be incorporated into action plans and planning tools, which will require involvement, collaboration and development work across sectors and authority levels, and the development of guidelines for how this can be done. Allocating enough resources for this work will be crucial, and smaller municipalities may need other types and degrees of support than larger ones.

Keywords: sustainable development goals; facilitating factors; implementation; regional and local planning; Norway

Citation: Bardal, K.G.; Reinar, M.B.; Lundberg, A.K.; Bjørkan, M. Factors Facilitating the Implementation of the Sustainable Development Goals in Regional and Local Planning—Experiences from Norway. *Sustainability* **2021**, *13*, 4282. https://doi.org/10.3390/su13084282

Academic Editor: Jacques Teller

Received: 25 March 2021
Accepted: 10 April 2021
Published: 12 April 2021

1. Introduction

Sustainable development has been a guiding norm and political objective ever since the Brundtland Commission published the report Our Common Future in 1987. The most frequently quoted definition of sustainable development is from this report [1] (p. 43): "Sustainable development is development that meets the needs of the present without compromising the ability of future generations to meet their own needs". The concept of sustainable development aims to maintain economic advancement and progress while protecting the long-term value of the environment [2]. In a common understanding of the concept, both economic, social and environmental aspects need to be integrated in decision making and balance each other."

In 2015, the United Nation's General Assembly further followed up the Brundtland report and adopted the 2030 Agenda and Sustainable Development Goals (SDGs). The agenda, with its 17 SDGs and 169 targets, is a universal call to action to end poverty, protect the planet and ensure that all people enjoy peace and prosperity by 2030 [3]. The 17 SDGs are integrated and recognize that development must balance social, economic and environmental sustainability. In order to reach the ambitious agenda within this decade, all parts of society, in all countries and regions, must contribute. Although the 2030 Agenda and the SDGs have relatively newly been adopted, an extensive literature that concerns various topics related to them has already emerged—see for instance Alibašić [4], Monkelbaan [5] and Nhamo et al. [6].

As a holistic framework, the SDGs challenge actors across different levels and sectors not only to understand how they influence the prosperity of people and planet, but to act to progress towards a more sustainable and just world. Even though the SDGs have been adopted at the supranational level in the UN, implementation has to be bottom-up. It has been estimated that as much as 65 percent of the targets cannot fully be achieved without the involvement of local actors [7]. Through the concept of localizing the SDGs, the central role of local authorities, civil society organizations and other local stakeholders has been recognized [8]. Thus, successful implementation of the SDGs depends on national, regional and local authorities' ability to translate the goals and targets into their respective contexts, and their ability to implement measures that ensure a holistic approach to the SDGs [9].

Countries and regions differ when it comes to geographic, demographic and economic situations, as well as their legal, democratic and governing systems. This again impacts what national, regional and local authorities experience as challenges when putting the SDGs into action [10,11]. Advantages, conflicts and tensions occurring in the localization process will be formed by the various economical, institutional, social and cultural territories in which the localization process is embedded [12]. Scholars have argued for a greater emphasis on the territorial embeddedness and multi-scalar nature of sustainability transitions, since this can enable a richer understanding of the different ways spatial contexts shape transition processes and the multiplicity and heterogeneity of transition pathways [12]. Furthermore Kulonen et al. [13] emphasize the scientific and political motivations for spatial considerations for sustainable development and that spatial dimensions need to be accounted for in the SDGs framework.

The adoption in 2015 of Agenda 2030 and the 17 UN SDGs has, therefore, led to a growing literature describing factors that either facilitate or challenge the implementation of the SDGs in various contexts.

In 2020, Norway was ranked as the sixth country in the world on overall SDG score, measuring countries' total progress toward archiving all 17 SDGs [14]. Internationally, Norway was an advocate for adopting Agenda 2030, while the Norwegian prime minister headed the UN-appointed SDG advocate group. In the national expectations for local and regional planning, the government stated in 2019 that the SDGs should be the main political framework to address the greatest challenges of our time, such as poverty, climate change and inequality. Further, they underlined that the regional and local authorities' efforts are crucial for Norway's contribution to meeting Agenda 2030, since they are closest to people, local businesses and organizations. The government also emphasized that local and regional authorities are responsible for much of the social and physical infrastructure impacting peoples living conditions and opportunities for development. Therefore, the government expects the SDGs to be implemented and become a foundational part of regional and local planning [15] (p. 3).

In Norway, municipalities (total of 356) have the principal authority to make decisions about land use. However, both regional and national authorities have a say in these processes and seek to influence local planning through national expectations, guidelines and planning provision. At the regional level, county councils are responsible for ensuring holistic and coordinated planning across the municipalities but have no formal authority to dictate local planning. Rather, the counties seek to influence local planning through contributing knowledge, advice and guidance, but also facilitating networks and arenas for local planners, which is particularly important in small municipalities.

That both social and spatial planning should play a central role in delivering sustainable development is not new in the Norwegian context. The purpose of the Norwegian Planning and Building Act of 2008 (PBA) is to "promote sustainable development in the best interests of individuals, society and future generations." However, evaluations of how the PBA works in practice show that it is difficult to balance the three dimensions of sustainability and it has proven difficult to integrate the inter- and intra-generational perspective to sustainability in specific planning decisions [16]. At the same time, previous studies have shown that many small and rural municipalities in Norway have challenges related to resources and capacity when it comes to planning [17].

Despite the underlining of the importance of commitment and effort by county councils and municipalities for Norway's contribution to meeting Agenda 2030, the Norwegian Auditor General [18] recently criticized the government for not coordinating the national implementation of the SDGs sufficiently. The lack of a comprehensive national plan for implementation has resulted in a fragmented approach, in stark contrast to the holistic and cross-sectoral approach needed to realize Agenda 2030. In turn, this has also affected the pace of the national implementation, which the Auditor General describes as being behind other Nordic countries. With this as a backdrop, Norway is an interesting case to examine experiences at the local and regional level with the implementation of the SDGs.

With this article, we aim to expand the existing literature on factors facilitating the implementation of the SDGs in local and regional planning, by providing new empirical evidence from Norwegian municipal and county administrations. Local and regional planning covers both rural and urban municipalities and regions of various size. We focus on the institutional context in which the implementation process takes place. In Norway the various municipal administrations and county councils differ in how far they have come in implementing the sustainable development goals in planning [9]. Thus, our hypotheses are that there are various factors that facilitates or hinders the implementation process, and these factors are affected by institutional contexts among local and regional authorities. The research questions explored are, therefore, as follows:

1. What factors can be identified as facilitating the implementation of the SDGs in Norwegian local and regional planning?
2. How are the facilitating factors conditioned by the different local and regional institutional contexts?
3. How do these factors from the Norwegian context correspond to or differ from those in the international literature?

We understand facilitating factors broadly as key factors for succeeding with the implementation of policies such as the SDGs in local and regional planning. They are success factors dealing with potential barriers related to the implementation. The factors may be seen as products of the local context, the type of policy in question, and the policy process [19].

Data have been collected through both an electronic survey among professionals working with planning and/or environmental issues in Norwegian municipalities and through semi-structured interviews with key informants from six municipalities and five county councils in Norway.

By contributing with new empirical knowledge, this study provides valuable learning for local, regional and national authorities, politicians and the academic field working with the implementation of the SDGs. Knowledge about facilitating factors can contribute to reinforce and strengthen the capacity of local and regional governments to deliver on Agenda 2030. Furthermore, understanding the facilitating factors of local SDG implementation is vital for stepping up the pace in the coming decade.

The article is structured as follows. In Section 2 we present our analytical framework for studying factors facilitating the implementation of SDGs. The data and methods are described in Section 3. In Section 4 we present the results and discuss them in relation to the three research questions. The article ends with some concluding remarks in Section 5.

2. Analytical Framework: Factors Facilitating the Implementation of the SDGs

Policy implementation is what develops between the establishment of an apparent intention on the part of government to do something, or to stop doing something, and the ultimate impact in the world of action [20]. Policy implementation reflects a complex change process where government decisions are transformed into programs, procedures, regulations, or practices with various aims [21]. Implementation of the SDGs in local and regional planning will involve such complex change processes. Tools and guidance on how to localize the SDGs and implement them in planning have been developed by, for example, the Global Taskforce of Local and Regional Governments [22]. Alibašić [4] describes how local governments can design and implement sustainability policies, initiatives and programs by offering guidance, strategies, and methods in applying sustainability and resilience planning, while Nhamo et al. [6] show how to draw national level baselines for the localization of the SDGs, aiming to provide a clear roadmap toward achieving Agenda 2030. The authors are cognizant of various institutions' common but differentiated responsibilities and capabilities within their socio-political, environmental, and economic conditions.

Implementation of the SDGs can be both hindered and facilitated by various factors. DeGroff and Cargo [21] identify three factors affecting policy implementation processes that they argue are of particular importance: networked governance, socio-political context, and democratic factors. Other literature also emphasizes how contextual and cultural influences affect policy implementation in significant ways, and that the effectiveness of any policy is also shaped and molded by context and culture [23].

Factors influencing policy implementation may be categorized in various ways. We have been inspired by the framework of Åkerman et al. [24] on barriers and success factors, and have categorized the factors facilitating the implementation of the SDGs in planning into the seven categories: cultural, political, legal, organizational, knowledge-related, and financial and technology-related factors. We find the framework useful for structuring the discussion, although the factors are sometimes partly overlapping and are not mutually exclusive.

Financial factors relate to the funding for SDG implementation activities. This includes ensuring that enough resources and capacity are available for planning, data collection and data analysis. Technological factors are related to having available the necessary technological solutions and tools for, for example, data collection and data analysis. Knowledge-related factors are related to knowledge about how to operationalize the SDGs, measure sustainability, collect and analyze data, etc. Political factors are related to having the support of democratic institutions at the national, regional or local governmental levels, or from organized interest groups. Cultural factors ensure that policies implemented do not conflict with norms and values, and therefore lack public and/or stakeholder acceptance. Planning cultures might also act as a powerful barrier to new ways of thinking and therefore need to be adjusted [25]. Organizational factors concern issues related to the collaboration within and between institutions. Legal factors are factors that help integrate the policies within existing laws and regulations, ensuring that they do not counteract each other.

We find that a considerable amount of literature already exists on factors facilitating the implementation of the SDGs. We have summed up some of the findings in the literature in Table 1. The publications both include peer-reviewed articles and reports. As the table shows, there are examples of facilitating factors within all the seven categories defined above. However, despite the large amount of literature on factors facilitating the implementation of the SDGs, scientific literature from the Nordic and Norwegian contexts is scarce.

In Table 1 we have also indicated which methodological approaches the literature draws on. Some of the literature provides new empirical knowledge collected through interviews, surveys, and document studies. However, many of the articles and reports are theoretical, some in the sense that they discuss empirical knowledge collected by others. Although the list of literature is not extensive, it indicates an overweight of theoretical or discussion papers and reports. Empirical literature from the Nordic countries include Gassen et al. [26] and SWECO [27].

Our article builds on the existing literature and provides extended and updated empirical knowledge on facilitating factors for implementation of the SDGs in the Norwegian context.

Table 1. Identified key factors for successful implementation of the sustainable development goals (SDGs) in local and regional planning.

Category	Key Factors Facilitating the Implementation of the SDGs	Examples of Literature and Methodological Approaches
Financial	Provide sufficient resources for planning, data collection and data analysis Finance strategic activities such as workshops, campaigns and education	Gassen et al. [26]—Interviews Satterthwaite [10]—Theoretical Smoke [28]—Theoretical SWECO [27]—Interviews and survey UCLG [29]—Theoretical UN Department of Economic and Social Affairs [30]—Document studies Wymann et al. [31]—Interviews
Technological	Make available relevant, disaggregated, high-quality data that permits comparisons with other local and/or regional authorities Avoid using too many indicators Make available tools for data collection and analysis of data Encourage technology and innovation that positively contribute to the implementation of the SDGs	Lucci [32]—Theoretical Mischen et al. [33]—Literature review Nordtveit [34]—Document studies Patel et al. [35]—Document studies SWECO [27]—Interviews and survey UN Department of Economic and Social Affairs [30]—Document studies
Knowledge and plan processes	Increase competence on data collection and analysis Increase knowledge about how to work with the SDGs Develop plan processes that deal with conflicting considerations and ensure broad participation Use bottom-up approaches that ensure anchoring in local realities Ensure open and inclusive processes Use methods that engage stakeholders Give room for experimentation, trials and failures Ensure mechanisms that hold societal actors responsible for decisions, investments and actions Perform a cost analysis of the implementation of the SDGs Share good examples and solutions to inspire others	Bowen et al. [36]—Theoretical Gassen et al. [26]—Interviews Hofstad & Vedeld [37]—Survey, document studies and interviews Leal-Arcas [38]—Theoretical Moallemi et al. [39]—Theoretical Satterthwaite [10]—Theoretical Slack [40]—Theoretical SWECO [27]—Interviews and survey UCLG [29]—Theoretical UN Department of Economic and Social Affairs [30]—Document studies
Political	Inclusive and representative decision-making at all levels Trust-building between inhabitants and authorities through dialogue Clear communication of national priorities and activities in Agenda 2030 Political support for the work with the SDGs	Awan [41]—Literature review Gassen et al. [26]—Interviews Oosterhof [42]—Theoretical UCLG [29]—Theoretical
Cultural	Awareness about the SDGs among stakeholders Promote culture as driver for development Relate the SDGs to local activities	Fleming et al. [43]—Interviews Gassen et al. [26]—Interviews Tjandradewi & Srinivas [44]—Theoretical UCLG [29]—Theoretical UN Department of Economic and Social Affairs [30]—Document studies
Organizational/institutional	Involve all local authority departments Integrate the SDGs into key steering documents, plans and processes Involve the local population and encourage young people to participate Support sustainable businesses and organizations Form strong partnerships between different local authorities, inhabitants, businesses and voluntary organizations Engage existing partners in long-term commitments for the SDGs Promote collaboration between sectors at all levels Better integration and coordination of management systems between various levels of authority Integrate the SDGs in the institutions' mandates Efficient, transparent and responsible institutions	Bhattacharya et al. [45]—Literature review, document studies, interviews Fenton & Gustafsson [46]—Literature review United Nations [47]—Empirical, documents studies Garcia-Alaniz et al. [48]—Theoretical Gassen et al. [26]—Interviews Hofstad & Vedeld [37]—Survey, document studies and interviews Klopp & Petretta [49]—Theoretical Lucci [32]—Theoretical Oosterhof [42]—Theoretical Shulla et al. [50]—Survey Slack [40]—Theoretical UCLG [29]—Theoretical Valencia et al. [51]—Pilot study Veldhuizen et al. [52]—Theoretical
Legal—laws and regulations	Formalizing commitments Adopting buyer requirements Establishment of new procurements deals Legislation securing the implementation of the SDGs (e.g., legal equality, equal right to education) and not counteract it	Awan et al. [41]—Literature review Biermann et al. [53]—Theoretical Gassen et al. [26]—Interviews Gladun [54]—Context analysis, interviews Mokoena & Jegede [55]—Theoretical

3. Materials and Methods

This article is based on data collected for a project financed by the Ministry of Local Government and Modernization [9]. We have used a mixture of qualitative and quantitative data and methods as further described below.

3.1. Survey

An electronic survey was sent by e-mail to professionals working with planning and/or environmental issues in all Norwegian municipalities (356). In the small municipalities, only one person received the invitation to participate, while in the larger municipalities, two or more persons were invited. In total, 715 persons received the e-mail, whereas 132 persons completed the survey, which gave a response rate of 18.5 percent. Although this is not so high at the individual level, in total, 30 percent of the municipalities were represented among the respondents. The represented municipalities showed good coverage in geographical location and size.

The survey included 37 questions with predefined alternative answers, and various themes related to the implementation of the SDGs in municipal planning. In two of the questions, the respondents were directly asked about potential barriers related to the implementation of the SDGs.

The first question was as follows:

1. To what degree do the following represent a barrier for using the SDGs as a planning tool in your municipality?
    a. Lack of knowledge about the SDGs in the municipality
    b. Different understanding of sustainability in various parts of the municipal administration
    c. Lack of relevance of the SDGs for local planning
    d. Lack of time/resources
    e. Lack of methods and tools for using the SDGs
    f. Lack of guidance from the county council
    g. Lack om guidance in Norwegian
    h. Existing guidance is too comprehensive and complicated
    i. Lack of good indicators for monitoring status and progress.
    j. Lack of coordination and dialogue across sectors in the municipality
    k. Lack of political anchoring
    l. Lack of engagement in the municipal administration
    m. Lack of engagement among inhabitants in the local community.

The respondents were asked to rank the barriers on a scale from 1 to 5 where 1 represented "to a very small degree" and 5 represented "to a very large degree". The respondents were also asked to comment and give their thoughts about how each of the barriers could be overcome by writing in open text boxes.

The second question was:

2. Describe other potential barriers and their importance.

This question was followed by an open text box for the respondents to describe barriers and their importance in their own words.

In the open text boxes, many respondents provided rich descriptions of challenges and success factors related to the implementation of the SDGs, which have been valuable in the data analysis.

3.2. Interviews

Interviews were performed with key informants from six municipalities (Ålesund, Narvik, Gloppen, Lunner, Asker, and Arendal) and five county councils (Viken, Nordland, Møre og Romsdal, Vestland, and Agder). Common to all is that they have started to implement the SDGs in their planning, but they differ when it comes to size, geographic location, population, degree of urbanity and rurality, and whether they had recently merged with other municipalities/counties. The five county councils cover the six municipalities, which allows a multilevel analysis. The municipality of Ålesund is located in the county of Møre og Romsdal, Narvik in Nordland, Gloppen in Vestland, Lunner and Asker in Viken, and Arendal in Agder. Figure 1 shows a map of the location of the case counties.

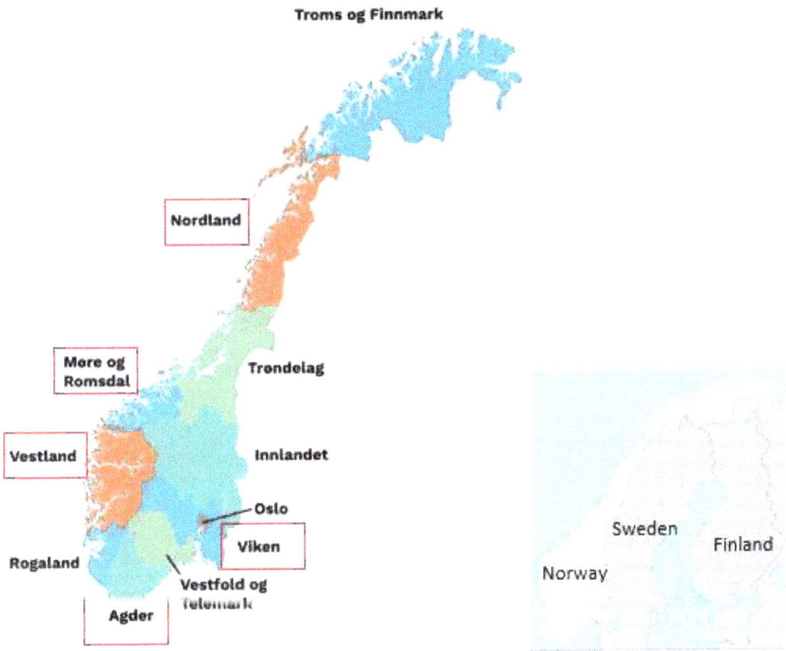

Figure 1. Map of Norway showing the location of the case counties (highlighted in the red boxes).

In total, 16 key informants were interviewed. In four of the interviews, two informants participated. The informants all had central roles in implementing the SDGs in planning in their respective municipalities and counties. Several of the informants had management roles in planning, such as head of the planning department, and they had been involved in both overall societal planning and spatial planning.

The interviews were conducted as semi-structured interviews, guided by an interview guide, but at the same time giving the informants room to add relevant information and comments. Eight of the interviews were performed on Skype, one by telephone, and two face-to-face with the informants. The interviews lasted from 50 to 90 min. All interviews, except for the telephone interview, were recorded, and notes were taken both during the interviews and afterwards on the basis of the recordings. In the presentation of findings, the informants are anonymized.

3.3. Data Analysis

The notes and transcripts from the interviews were first analyzed by thematic analysis [56]. The data were coded for themes, and a theme was identified when the coder noticed something in the data reflecting the research questions and themes of interest in a patterned way. The informants were allowed to read and comment on the written reports from the analysis of each interview. This was done to ensure that our interpretation of the interviews was in line with what the informants had meant and to allow for corrections of misunderstandings. Several of the informants added information which further enlightened the research questions. Next, both the data from the interviews and surveys were analyzed by the theoretical framework on facilitating factors for implementation of policies described in Section 2. This includes categorizing the facilitating factors into cultural, political, legal, organizational, knowledge-related, financial and technology-related factors.

The analysis was carried out in four steps. First, we analyzed data as part of the project financed by the Ministry of Local Government and Modernization [9]. Second, from the data, we subtracted issues with particular relevance for facilitating factors. Third, the semi-structured interviews, survey data, and theoretical literature allowed us to analyze our findings from different angles and hence triangulate data [57]. Finally, we analyzed the data by organizing them into the predefined categories that our theoretical framework provides.

4. Results and Discussion

A common experience among the respondents and key informants in the survey is that working with the SDGs has created enthusiasm and has been useful, important, and exciting. Several informants also mentioned that they had learned a lot in the process. However, working with the SDGs could be challenging in various ways. In this section, we present and discuss the respondents' and key informants' thoughts and experiences about factors hindering the implementation of the SDGs in local and regional planning, and what they think may be key factors in overcoming these and facilitating the implementation process.

The discussion is organized in accordance with the three research questions. In Section 4.1, we present and discuss identified factors facilitating the implementation of the SDGs in Norwegian local and regional planning, while Section 4.2 discusses how the factors are conditioned by the different Norwegian local and regional contexts and how they differ from those in the international literature.

4.1. *Identified Factors Facilitating the Implementation of the SDGs*

In Table 2 the identified facilitating factors have been categorized by type. The categories of factors are discussed in their respective sections below the table.

4.1.1. Financial—Capacity and Resources for Development Work

Lack of capacity and resources was mentioned as one of the most important barriers for implementing the SDGs in local and regional planning by the respondents to the survey. This comment from a respondent illustrates this well: "Daily operations 'eat up' time and capacity for development work". Particularly the smaller municipalities pointed at the need to rely on the work of others, since they experience it as challenging to manage to do development work themselves. This often led to copying from larger municipalities with the danger of not being able to adequately consider their specific context. Moreover, they pointed at the fact that it is a huge task for small municipalities to make good plans with the SDGs as a framework when they often only have one or two employees working with planning.

Further, our findings show that there is also a need to have the capacity to become acquainted with the literature on the SDGs and particularly the implementation process. In the interviews and survey, the informants mostly concentrated on reading the guidance material from the national authorities. While about 60 percent of the respondents answered that they were familiar with two specific Norwegian guidelines (see [58,59]), about 70–80 percent did not know about relevant English guidelines such as those by Gassen et al. [26], SWECO [27], and The Global Taskforce of Local and Regional Governments [22,60–62]. A large share (39 percent) of the respondents in the survey, considered a lack of methods and tools for implementing the SDGs as a large or very large barrier. However, this may not only be because the methods and tools are lacking, but rather also a question of making existing tools accessible and having the time to explore them.

The results clearly indicate that successful implementation of the SDGs in local and regional planning requires that both the municipalities and counties have the capacity and resources needed to work with SDG implementation. In order to prioritize time for working with SDG implementation, respondents mentioned the significance of administrative leaders expressing that this was an important task with high priority. Inter-municipal collaboration was also mentioned as a potential strategy to overcome the capacity barrier. However, as one of the respondents commented, the work with the SDGs needs to be integrated into existing activities, not become something separated from the service provision activities for which the municipalities and county councils are responsible.

Table 2. Facilitating factors identified in the survey and the interviews with key informants.

Category	Key Factors Facilitating the Implementation of the SDGs in Norwegian Local and Regional Planning
Financial—esources and capacity	Allocate time and capacity for development work Allocate time for getting familiar with literature and guidance material
Technological	Access to adequate methods and tools for implementing the SDGs Access to guides written in Norwegian Access to indicators for measuring status and progress on sustainability—particularly how to consider qualitative issues not measurable to ensure that they are not left out Access to guides on how to operationalize the goals locally Access to guidance materials and information relevant for small urban municipalities
Knowledge	Necessary to have access to comprehensible and relevant knowledge about the SDGs and which role they should play—need for a systematic representation of the overwhelming literature Access to knowledge about relevant networks that exist Access to good examples on implementation of the SDGs in local and regional planningHave systems for sharing knowledge within and between organizations
Political	Need for clear messages from the national government of what they want and expect Need for engagement in the municipality and county council organizations and the local communities Necessary with good anchoring and support from regional and local politicians The SDGs must be incorporated in the financial plans Avoid "green washing" of existing activities instead of change Avoid budgets being tied up with statutory tasks and earlier priorities
Cultural	Need for a common understanding of what sustainable development means and what the work with the SDGs means and how to interpret the SDGs locally and regionally
Organizational/institutional	Need for internal coordination and dialog across sectors in the municipality/county council Need for common methodology for working with the SDGs across levels Need for a cross-sectorial understanding of the goals at the national level Need for consistency in the state authorities' principles, directions and guidelines at underlying levels, particularly within spatial and transport planning Avoid work with the SDGs being a top-down process not ensuring involvement by those who are going to implement the goals. Need for a good connection between community plan and financial plan Need for county councils to give guidance and help to municipalities Need for coordination of activities across local, regional and national levels Need for arenas for collaboration with actors such as businesses and academia
Legal—laws and regulations	Use the opportunities lying within the Planning and Building Act

It has to be mentioned that the United Nations Institute for Training and Research (UNITAR) has developed activities, courses, webinars and conferences, aiming to help national, regional and local authorities to build the capacity to implement SDGs and monitor progress. In addition, the SDG Accelerator and Bottleneck Assessment tool is aimed at supporting countries in identifying policies and measures that can help to solve bottlenecks and accelerate the implementation of the SDGs [63]. Nevertheless, our findings show that these guidelines are not widely known or used among municipal planners in Norway at the moment.

4.1.2. Technological—The Need for Accessible Methodologies, Tools and Indicators

Many of the informants experienced a lack of (knowledge of) suitable methods, tools and indicators for implementing the SDGs and difficulty in relating the tools and indicators to their local context. One-fourth of the respondents in the survey replied that guides that are too complex and extensive were a barrier to a large or very large degree, and approximately the same amount answered that the lack of guides in Norwegian was a barrier. They also expressed that the international literature was not always relevant for the Norwegian context. As one respondent in the survey commented: "Guides in English are experienced as far away from our daily lives. Norwegian guides focus on Norwegian contexts and that feels more relevant".

Another point that was mentioned was that for smaller municipalities, the existing literature and guides on SDG implementation were seen as not being that relevant since the examples often came from bigger municipalities with specifically urban challenges. A respondent in the survey made the point clear:

> The large themes concerning environmental issues in planning are more targeting the larger cities and their problems. In our case it is the scattered settlements, small business areas and issues related to this, for which finding solutions is urgent. Safeguarding green spaces, handling surface water and densification, are minor problems for us. This means that the existing guides are not so relevant for us.

A similar comment was made from another respondent in the survey, supporting the impression that the smaller municipalities do not experience that existing guidelines fit their situation to the same degree as the larger ones:

> I receive e-mails from the Ministry and others and have noticed that reports and seminars (often international) exist, however often of the type 'sustainable cities'. We have not considered these as relevant for a small rural municipality as ours, which barely has small villages.

Some respondents to the survey argued that it ought to be a task for the Ministry to find "best practices" that the municipalities could use: "The municipalities do not need to be unique in everything, and the politicians in different parts of the country do not disagree on everything", as one respondent in the survey commented.

One municipality had good experiences with using a "significance analysis" for making the goals relevant in their local context. Through the "significance analysis", goals and targets were systematized on the basis of which goals the municipality could influence directly and the areas that the municipality ought to take direct responsibility for in the work of developing the local community and services for the inhabitants.

On the one hand, there seems to be a wish for clear methods, indicators, and systems for the measuring and reporting of effects in such a way that it is possible to make comparisons with others. On the other, there is also a recognition that not everything can be measured. Several informants pointed at the need to know about status, in order to measure progress and in some way "put a number" on the efforts made; however, it is not possible to capture in the existing indicators. Several areas were mentioned which were difficult to describe due to the lack of data. In addition, there are themes that the SDGs and their targets themselves do not capture very well. As examples of this, the preservation of cultural heritage and the conservation of nature were mentioned. The management of marine resources and agriculture were also mentioned as not being handled properly in the SDGs. This shows that although the SDGs seem to embrace broadly, there are still themes that are not sufficiently covered.

If too much focus is on those issues that can be measured, there is also a danger of leaving out important areas. The informants therefore considered it essential to make subjective judgments in addition to quantitative measurements, and perhaps it was not necessary to produce more indicators but rather to develop methods and tools for considering issues that cannot be measured. As stated by one informant in one county council:

> Not all we want ... can be measured. You can measure depopulation and the occurrence of muscle and skeleton diseases, and you can ask people if they are depressed. However, youths leaving high school is a complex issue... There is a reason behind the leaving, which is complex. So, in my opinion, it is important that we measure what we are able to measure, but then it is extremely important to obtain a subjective consideration of what you really want, where you can add another dimension.

One informant from a county council participating in the U4SSC network [64] underlined the need for them to consider key performance indicators more relevant from a regional perspective, and that not all the indicators developed within the network was necessarily suitable for them. To help with this, it was necessary to develop a digital tool. On the one hand, it is necessary to operationalize the goals locally, and some of the informants expressed that this was a demanding task. On the other, several informants argued that it is important to make use of what methods and indicators already exist, and that one has to make sure that not too many resources are spent on establishing unique reports on the SDGs. One municipality used external expertise to get started with implementing the SDGs but then took over the control of the process and experienced that this increased the municipality's ownership of the matters in question.

In several of the interviews, the relationship between a holistic approach and a more selective approach to the SDGs was problematized. While there is a need for a holistic approach in order to capture the whole picture and the dynamics between goals, some emphasized the need to narrow down the work and prioritize between the goals; otherwise, it would be too overwhelming and difficult to get started. These two approaches are very different.

4.1.3. Knowledge—Internal and External Knowledge about the SDGs

As already mentioned, a lack of knowledge about the SDGs and existing methodologies, guides and indicator tools represents an obstacle to the implementation of the SDGs. Attending courses, seminars and workshops were mentioned as strategies that could increase knowledge of the SDGs and existing materials. "We at least need to talk about it," as one respondent in the survey commented. However, several recognize that increasing knowledge and competence takes time and requires continuous focus in many arenas.

While all municipalities, independent of size, reported using the Norwegian Association of Local and Regional Authorities (KS) as a source, the larger municipalities responded to using the United Nations Association of Norway and Norwegian Smart Cities to a greater degree compared to the smaller ones. The informants mentioned participation in networks as being useful for obtaining knowledge about SDGs, tools and indicators, and sharing experiences from implementing the SDGs in planning. However, it was mentioned that it is challenging to gain an overview of all the networks that exist.

In order to make the existing literature on the SDGs and their implementation more accessible, there is both a need to systematize the literature and to translate the findings to the local (here Norwegian) language and contexts, as mentioned in the previous sections. As one informant commented:

> There is no doubt that a lot of information exists, however you drown in all the information and do not know how to use it ... There is a need for a clear guide that thematically approaches how to use the information, both in a simple way and more advanced. It also has to give guidance on where the information can be used most effectively—in the municipality plan and its regulations, or in zoning plans and associated regulations, or both.

Another respondent to the survey commented:

> As planners in small municipalities, we have many roles and tasks ... It is difficult to have enough time to familiarize oneself with new knowledge.

This illustrates that there is need to establish routines and systems for sharing information and competence within the municipalities and county councils. It is also seen as crucial that regular employees get involved in the implementation process to ensure that it becomes part of the daily work throughout the organizations and not only something going on in specific parts of the administration.

4.1.4. Political—Commitment among Politicians, Stakeholders and Inhabitants

All informants agreed that anchoring and support from the politicians is crucial for successful implementation of the SDGs locally and regionally. As one informant from a municipality commented:

> I wish sometimes that they were more concrete and visionary, because it is much more difficult for the administration to be visionary. The latter implies going beyond your mandates, promoting a case no one has asked for. It is much easier the other way around.

Although the informants reported that many politicians put the SDGs high on the agenda, work still needs to be done to achieve a broad commitment across parties and to reduce the polarization of the debates regarding the SDGs. One respondent from a county council argued that it was important not to depoliticize the SDGs and make such rigid systems that have all priorities set out in advance. There had to be room for prioritizing, the informant argued, but it was important to establish a good decision base so politicians were able to see the consequences of their decisions. However, several emphasized that the SDGs must be something more than just a checklist of which goals, various measures and decisions contribute to achieving. One barrier mentioned was that statutory tasks and earlier priorities often reduced the room for action in budgets, as pointed out by one respondent in the survey:

> One important barrier lies in the budget being to a large degree tied up with statutory tasks and earlier priorities. You need time to turn this around.

It was mentioned as important to provide arenas for participation and knowledge-sharing with politicians (e.g., meetings, seminars, and workshops) to increase politicians' knowledge about the SDGs. This was also suggested as a good strategy to increase awareness and commitment to the SDG work internally in the municipalities and county councils. In addition, it was mentioned as important to be able to move beyond just talking about the SDGs. It is important to be able to show some results from the work, either internally, or good examples from "first movers". Use of public meetings and increased communication of knowledge about the SDGs, and results from working with them, were mentioned as factors which could increase awareness and engagement among inhabitants. However, it was recognized that differentiated measures needed to be implemented in order to reach various groups of the population.

Twenty-eight percent of the respondents also considered lack of engagement and involvement throughout the organization in the work with the SDGs as representing a barrier to a large or very large degree. They call for better internal communication, collaboration and awareness-building about the SDGs and sustainable development throughout the municipal administration. It was considered a challenge to be able to involve all politicians, and it was recognized that it was important to remember that not all politicians have the same knowledge about the SDGs. Thus, extensive information activities are necessary to enlighten both politicians and employees in the municipality and county council administrations. One informant emphasized the importance of associating what is done in the municipal organization with the rest of the local community. It should be noted that several of the municipalities in the study had performed various, partly innovative, co-creation activities in the work with the municipal plan in order to ensure support and commitment from all stakeholders.

Several informants were concerned about making politicians aware of how various political decisions impacted the SDGs. One of the county council interviewees highlighted striving for economic growth as "the elephant in the room". It was questioned to what degree the SDGs contributed anything new to the discussion about societal development, in line with the objectives of the Planning and Building Act (PBA) that planning should contribute to sustainable development. Several were worried that the SDGs would only lead to "green washing" of existing politics. As one informant from a municipality commented: "I am a little bit afraid that the SDGs are associated with everything you do and do not lead to any changes". The informant highlighted it as important to make sure that the SDGs impact priorities in practice, and not only become "a new wrapping".

A key issue that informants emphasized in the interviews was that although the work with the SDGs is long term, it is also important to show that things are happening now, and that the involvement of inhabitants also commits the municipalities to take the issues raised by inhabitants seriously and act upon them, although they may not relate to the specific plan in question.

4.1.5. Cultural—A Common Understanding of Sustainability

The survey revealed that 33 percent of the respondents consider different perceptions and interpretations of sustainability in the municipal administrations as representing a barrier for SDG implementation to a large or very large degree, and that it was challenging to secure social, environmental, and economic sustainability simultaneously.

Most informants agreed that participation and involvement of a broad spectrum of stakeholders is a key factor for successful implementation of the SDGs. However, during the interviews, challenges related to dialog and collaboration with stakeholders were thematized, as people interpret the SDGs differently and want different things from them. As one informant from a county council commented:

> Everyone is working with the SDGs; however, they have implicit objectives that may be diverging … If you don't uncover these implicit objectives, then you are not talking about the same things when you meet up.

These challenges assert themselves across administrative levels and between different actors, illustrating the need for a common understanding of what the work with the SDGs means.

In order to develop a common understanding of sustainability among stakeholders internally in the municipalities and county councils, the respondents suggested several potential measures such as increased collaboration and communication across sectors, arranging of shared meetings and projects, and shared competence-building programs across sectors. Anchoring and commitment in the top management of the municipalities and county councils were seen as crucial for this to happen. Manager development programs incorporating the SDGs were also mentioned as a measure which could help develop a shared understanding of what sustainability means and how it impacts the work of the municipalities and county councils.

4.1.6. Organizational—The Importance of Cross-Level and Cross-Sectorial Work and a Coherent Goal Structure

The different national authorities also have a key role in facilitating the implementation of the SDGs locally and regionally by coordinating their activities and following their own principles, directions and guidelines at all levels and sectors, and in all their meeting points with the local and regional level.

Since the SDGs are cross-sectorial by nature, one municipality pointed to the importance of developing cross-sectorial priority areas where different service areas are working with the same goals. The same municipality also stated the importance of building a "red thread" from the overarching goals in the municipal plan, to the more concrete strategies and measures in the subordinated plans. In particular, the connection to the financial plan was pointed out as important, because then you had to make specific priorities. A direct connection between the community plan and the financial plan will enable reporting against the financial plan which can be directly associated with the goal structure at the base of the community plan. The SDGs then become a part of the financial plan, and the community part of the municipal plan will become more useful, a respondent argued.

Several mentioned the involvement of and co-creation with stakeholders as important when developing plans, by contributing to anchoring the plan among stakeholders and developing the plans based on what stakeholders find important and relevant, as well as making it easier to achieve commitment to the plans across sectors. "Sustainability breakfasts" had successfully been used by one county council to make the SDGs better known internally in the organization and create engagement in sectors other than the planning sector.

Twenty-eight percent of the respondents from the municipalities thought that lack of guidance from the county councils represented a barrier for implementation to a large or very large degree. Some commented that they were waiting to become part of the county council's development work and that guidance from the county councils was particularly important for smaller municipalities with few employees dedicated to working with SDG implementation. However, the interviews with the representatives from the county councils suggested that several found it challenging to provide guidance to the municipalities because they were themselves trying to figure out what to do. It was recognized that small and large municipalities may have different needs regarding guidance from the county councils. Although the situation may be more transparent in smaller municipalities, the competence and capacity to work with SDG implementation may be scarcer here, compared to larger municipalities.

The interviews revealed that there is a need for good examples illustrating the variety of approaches and methodologies of working with the SDGs that have been tested and of which experience has been gained.

It was pointed out as important to get started with implementing the SDGs and that much could be learned along the way by trial and error. Several of the interviewees also emphasized the importance of work done by enthusiasts and inspiration gained from networks and other actors in order to get started. A good strategy may be to give room for enthusiasts, participate in networks, and make allowances for trial and error. One small municipality specifically mentioned the regular meetings in a planning forum as important. It helped them to shift their focus from daily operations to important community issues.

4.1.7. Laws and Regulations—The Planning and Building Act as Tool for SDG Implementation

A perception among most of the informants was that the Norwegian Planning and Building Act (PBA) (https://lovdata.no/dokument/NL/lov/2008-06-27-71 accessed on 15 January 2021) is well suited to serving as the framework for a cross-sectorial and holistic approach to the SDGs. The informants also thought that the Planning and Building Act would contribute to a coordinated approach to the SDGs across the local, regional and national levels. In planning, working with holistic societal development and coordinating various stakeholders and interests is nothing new. However, the informants from the municipalities stated that the SDGs had not made this work any easier and that a more exciting dialog had emerged between the municipalities and other stakeholders in the community. The SDGs, in a way, serve as a common language across subjects and sectors. The county councils reported that the SDGs had contributed to making the county councils' broad spectrum of responsibilities visible in society, and subjects that had not been prioritized earlier were now on the agenda. Some of the informants also thought that the SDGs could contribute to highlighting and finding solutions to conflicts between goals, and to shedding light on political dilemmas.

The implementation of the SDGs in planning concerns many actors in and around the municipalities and county councils, and therefore requires systematic efforts from many actors over a long period of time. One respondent argued for it to be a prerequisite that the PBA be used as an active management tool in the process.

4.2. The Impact of the Norwegian Local and Regional Context

In this section we discuss the identified factors presented in the previous section in relation to the two research questions: (1) how are the facilitating factors conditioned by the different local and regional institutional contexts, and (2) how do these factors from the Norwegian context correspond to or differ from those in the international literature?

Although countries and regions may be geographically, culturally, and economically different, they seem to struggle with many of the same issues when addressing the SDGs in their respective contexts. The size of the municipality administration seems to be more important than geography in explaining the differences observed. This is not surprising giving that small and rural municipalities often have limited planning capacity and smaller professional environments [17]. Our findings indicate that this affects how the municipalities can relate to the national guidelines and implement the SDGs in their planning. However, realizing Agenda 2030 depends on the ability of both large and small municipal administrations in urban as well as rural areas to initiate sustainable transition processes. Thus, the literature needs to further develop the understanding of the factors enabling implementation of the SDGs in different institutional contexts, and how cross-sectorial synergies can be achieved in these different institutional settings. Small municipalities have challenges different to the vast literature on "sustainable cities," and thus, need tools to handle their specific challenges.

When we compare the results from the literature (Table 1) and our findings from the Norwegian cases (Table 2), we find many similarities. In general, our findings from the Norwegian context seem to align with experiences elsewhere. A common challenge across different institutional contexts relates to the provision of sufficient resources for planning activities, data collection and analyses. This also seems to be more prominent in smaller municipalities. In our study, 58 percent of respondents in the municipalities with less than 10,000 inhabitants reported that the lack of capacity and resources represented a barrier to a large or very large degree, while only 39 percent of respondents in the larger municipalities answered the same. The same trend was also seen regarding knowledge about the SDGs in the municipalities. Here, 42 percent of the respondents in the smaller municipalities said that this represented a barrier to a large or very large degree, while only 26 percent said so in the larger municipalities. An interesting finding is that, for reasons unknown, the smallest municipalities with less than 1500 inhabitants departed from this trend both with regard to capacity, resources, and lack of knowledge being barriers. A reason might be that often fewer stakeholders are involved in smaller municipalities. The municipal administrations are also smaller and the variation in activity in the municipality is often limited.

Both in Norway and other countries, there seems to be a strong need for relevant tools and indicators for data collection and analyses, monitoring of sustainability, and guidelines for implementing the SDGs in the planning processes. However, the lack of such methods expressed by the informants in the Norwegian context seems to be just as related to making existing tools and guidelines available for Norwegian-speaking planners in a more comprehensive manner. The extensive literature, partly consisting of large-scale reports, need to be systematized, translated, and made relevant probably not only for Norwegian contexts, but for other regional and local contexts as well. The extensive work by, for example, the UN is in practice difficult to use for local and regional governments, and even more so for smaller municipalities than larger ones. The interviews indicate that the smaller municipalities may need more guidance from the regional authorities and could particularly benefit from making the existing methodologies, guides and indicator tools more widely available. In addition, they may also benefit from the development of "best practice" methodologies, so that each and every little municipality does not have to do all the development work by themselves.

Both in the Norwegian context and in the international literature, there is a need to ensure the anchoring of the SDGs at all levels and with stakeholders within the community. The informants in our study emphasized the need to develop a common understanding among stakeholders of what the work with the SDGs actually means. They also highlighted the polarizing debate between those skeptical to climate change and their opposites as challenging for the implementation of the SDGs. The need to involve the local population was also highlighted in the literature, in addition to the participation of young people being seen as important.

Both the literature and this study point to the importance of institutional factors, such as better coordination of plans, activities and processes across different sectors, actors and government levels. Clear communication of national priorities was seen as important. The Norwegian government has yet to take a stand on any national priorities, leaving this in the hands of regional and local governments. On the basis of the literature review and our findings, the formulation of a national agenda seems crucial to successfully implementing the SDGs. In the Norwegian context, the role of the national authorities was highlighted as important, especially the need for the state to follow up its own principles, directions and guidelines at underlying levels. On the one hand, the national authorities call for the SDGs to frame all planning activities, while on the other hand, they use other measurement tools, which are not based on the SDGs, to prioritize the financing of activities such as developing transport infrastructure. In order for the SDGs to have a real impact, the national authorities need to integrate the SDGs in the work performed by the various underlying levels as well.

One interesting finding in the Norwegian context, which we did not find in the literature, was the usefulness of existing legal instruments, most specifically the Norwegian Planning and Building Act (PBA), in implementing the SDGs. While the literature pointed to a need for legislation to secure implementation, our findings indicate that the formal legal framework in Norway is, in general, considered appropriate for implementation. There is broad agreement among the informants in our study that the PBA may serve as a good framework for implementing the SDGs. This is a tool which has already integrated the focus on sustainable development where the SDGs may be seen as an operationalization of this. Using an already institutionalized framework for the implementation of the SDGs may prove easier than developing entirely new ones.

There were also fewer mentions of issues such as a need for efficient, transparent and responsible institutions in the Norwegian context compared to the international contexts. The same goes for a need to ensure mechanisms that hold societal actors responsible for decisions, investments, and actions. This seems to indicate a high degree of trust in institutions amongst our informants and respondents. Lundberg et al. [9] found that SDG 16, which relates to strong institutions, is prioritized to a lesser degree than many of the other goals amongst Norwegian municipalities. This might also be interpreted as a sign of relatively well-functioning institutions. Considering the strong institutions and the appropriate legal framework, as noted above, we might question how far the Norwegian effort to achieve the SDGs has come. At the same time, sustainable development has been a topic in Norwegian planning for several years, and even though measures might not be linked directly with the SDGs, they might, in practice, cover these issues. This is perhaps in opposition to other places where the SDGs are seen to bring something substantially new to local planning and policymaking.

5. Conclusions

In this study, we have examined the barriers that challenge the implementation of SDGs in planning at the local and regional level in Norway, and how municipal and regional planners perceive them in their practical work. Relating these to the framework of facilitating factors, we contribute empirical insights into how these challenges can be handled. The SDGs are implemented in very different institutional contexts and, as we have shown, this affects how barriers are perceived and what facilitating factors are needed to enable a successful implementation. Knowledge about these factors, and measures that can be taken to improve them to enable capacity-building at the local and regional level is vital. In practice, financial, technological, knowledge, political, cultural, institutional, and legal aspects will affect planning and planners' ability to implement the SDGs.

The expectations on local and regional planning in addressing the SDGs in Norway are high, and municipalities and counties play a key role in the Norwegian efforts to realize Agenda 2030. However, as mentioned in the introduction, the Auditor General [18] has criticized the lack of national coordination in the implementation of the SDGs. Our study shows that there is a need for a clearer voice from the national authorities about what this means in practice for local and regional authorities: what the challenges and objectives are. There is a demand for better knowledge and competence in how the SDGs can be implemented locally and regionally. This can be responded to by systematizing the existing literature on tools and guidelines, by making them accessible to planners, and by encouraging participation in existing networks and establishing new ones to enhance the sharing of experiences and learning.

There are reasons to believe that for the SDGs, existing planning tools may be both a promising route and a barrier. The findings in this study indicate that the Norwegian Planning and Building Act is an example of the former. A suggestion for further studies would be to do comparative studies of similar planning systems in different countries and assess their potential as tools for implementing the SDGs. This could provide valuable knowledge transfer, improving the implementation of the SDGs in planning at local and regional levels.

If the SDGs are to make a difference through concrete actions, they have to be incorporated into action plans and existing planning tools. This will require collaboration and development work across different sectors and authority levels, as well as the development of guidelines on how this can be done. Further, we would like to emphasize the need to develop a broader understanding and approach to SDG 11, concerning sustainable cities and local communities, to make it relevant for smaller municipalities, often in rural areas. This will likely apply to other European rural municipalities as well. In line with Kulonen et al. [13], who draw on research form remote mountain regions in Europe, we argue for the need to develop approaches and framework that can account for spatial and institutional considerations at the sub-national levels. Smaller municipalities face different challenges to larger ones, and our study indicates that they may need other types and degrees of support when implementing the SDGs. Expanding on their role and opportunities to plan and develop more sustainable communities may be vital for the implementation of the SDGs outside the large urban areas too.

Author Contributions: Conceptualization, K.G.B., M.B.R., A.K.L. and M.B.; methodology, K.G.B., M.B.R., A.K.L. and M.B., validation, K.G.B., M.B.R., A.K.L. and M.B.; data analysis K.G.B., M.B.R., A.K.L. and M.B.; investigation, K.G.B., M.B.R., A.K.L. and M.B.; writing—original draft preparation and review and editing, K.G.B., M.B.R., A.K.L. and M.B. All authors have read and agreed to the published version of the manuscript.

Funding: This research received no external funding.

Institutional Review Board Statement: The study was conducted according to the guidelines of the Declaration of Helsinki, and approved by the Institutional Review Board of NSD (Norwegian Data Protection Service, January 2020).

Informed Consent Statement: Informed consent was obtained from all subjects involved in the study.

Data Availability Statement: The data presented in this study are available on request from the corresponding author. The data are not publicly available due to privacy restrictions.

Conflicts of Interest: The authors declare no conflict of interest.

References

1. United Nations General Assembly. *Report of the World Commission on Environment and Development: Our Common Future*; United Nations General Assembly, Development and International Co-Operation: Oslo, Norway, 1987.
2. Emas, R. *The Concept of Sustainable Development: Definition and Defining Principles*; Brief for GSDR; GSDR: Hereford, UK, 2015.
3. UNDP. Sustainable Development Goals. 2020. Available online: https://www.undp.org/content/undp/en/home/sustainable-development-goals.html (accessed on 6 July 2020).
4. Alibašić, H. *Sustainability and Resilience Planning for Local Governments. The Quadruple Bottom Line Strategy*; Sustainable Development Goals Series; Springer: Berlin/Heidelberg, Germany, 2018.
5. Monkelbaan, J. *Governance for the Sustainable Development Goals. Exploring an Integrative Framework of Theories, Tools, and Compentencies*; Sustainable Development Goals Series; Springer: Berlin/Heidelberg, Germany, 2019.
6. Nhamo, G.; Odularu, G.O.A.; Mjimba, V. *Scaling up SDGs Implementation. Emerging Cases from State, Development and Private Sectors*; Sustainable Development Goals Series; Springer: Berlin/Heidelberg, Germany, 2020.
7. Kanuri, C.; Revi, A.; Espey, J.; Kuhle, H. *Getting Started with the SDGs in Cities: A Guide for Stakeholders*; Sustainable Development Solutions Network: New York, NY, USA, 2016.

8. Wright, C. *Sustainable Development through Local Action. Sustainable Development Goals and Local Government Associations*; Report Prepared for the United Cities and Local Government Capacity and Institution Building Working Group; UCLG CIB Working Group: Den Haag, The Netherlands, 2018.
9. Lundberg, A.K.; Vangelsten, B.V.; Bardal, K.G.; Reinar, M.B.; Bjørkan, M.; Richardson, T.K. *Strekk i Laget: En Kartlegging av Hvordan FNs Bærekraftsmål Implementeres i Regional og Kommunal Planlegging*; NF Report No.: 7/2020; NMBU: Ås, Norway, 2020.
10. Satterthwaite, D. Successful, safe and sustainable cities: Towards a New Urban Agenda. *Commonw. J. Local Gov.* **2016**, 3–18. [CrossRef]
11. Ascher, B.; Halvorsen, A.H.; Johansson, U. *Attractive Nordic Towns-Strategies towards a More Sustainable Future*; SWECO Report; SWECO: Stockholm Sweden, 2019.
12. Coenen, L.; Truffer, B. Places and spaces of sustainability transitions: Geographical contributions to an emerging research and policy field. *Eur. Plan. Stud.* **2012**, *20*, 367–374. [CrossRef]
13. Kulonen, A.; Adler, C.; Bracher, C.; von Dach, S.W. Spatial context matters in monitoring and reporting on Sustainable Development Goals: Reflections based on research in mountain regions. *GAIA-Ecol. Perspect. Sci. Soc.* **2019**, *28*, 90–94. [CrossRef]
14. Sachs, J.; Schmidt-Traub, G.; Kroll, C.; Lafortune, G.; Fuller, G.; Woelm, F. *The Sustainable Development Goals and COVID-19*; Cambridge University Press: Cambridge, UK, 2020.
15. The Ministry of Local Government and Modernistation. *National Expectations for Regional and Municipal Planning 2019–2023*; Government Document; The Ministry of Local Government and Modernistation: Oslo, Norway, 2019.
16. Hofstad, H. Bærekraftig Planlegging for Framtida? In *Plan-og Bygningsloven-en Lov for Vår Tid?*; Hanssen, G.S., Aarsæther, N., Eds.; Universtitetsforlaget: Oslo, Norway, 2018.
17. Bjørkan, M.; Rybråten, S.; Lundberg, A.K. Navigating visions, knowledge and practical challenges in coastal zone planning. *J. Environ. Plan. Manag.* **2020**. [CrossRef]
18. Auditor General (Riksrevisjonen). *Riksrevisjonens Undersøkelse av Styring av og Rapportering på den Nasjonale Oppfølgingen av Bærekraftsmålene*; Document 3:3 (2020–2021); Riksrevisjonen: Oslo, Norway, 2020.
19. Sørensen, C.H.; Isaksson, K.; Macmillen, J.; Åkerman, J.; Kressler, F. Strategies to manage barriers in policy formation and implementation of road pricing packages. *Transp. Res. Part A Policy Pract.* **2014**, *60*, 40–52. [CrossRef]
20. O'Toole, L.J., Jr. Research on policy implementation: Assessment and prospects. *J. Public Adm. Res. Theory* **2000**, *10*, 263–288. [CrossRef]
21. DeGroff, A.; Cargo, M. Policy implementation: Implications for evaluation. *New Dir. Eval.* **2009**, *2009*, 47–60. [CrossRef]
22. The Global Taskforce of Local and Regional Governments. *Roadmap for Localising the SDGs: Implementation and Monitoring at Subnational Level*; UCLG: Barcelona, Spain, 2016.
23. Harris, A.; Jones, M. Why context matters: A comparative perspective on education reform and policy implementation. *Educ. Res. Policy Pract.* **2018**, *17*, 195–207. [CrossRef]
24. Åkerman, J.; Gudmundsson, H.; Hedegaard Sørensen, C.; Isaksson, K.; Olsen, S.; Kessler, F.; Macmillan, J. *How to Manage Barriers to Formation and Implementation of Policy Packages in Transport*; OPTIC Report; Optimal Policies for Transport in Combination: Oslo, Norway, 2011.
25. Booth, C.; Richardson, T. Placing the public in integrated transport planning. *Transp. Policy* **2001**, *8*, 141–149. [CrossRef]
26. Gassen, N.S.; Penje, O.; Slätmo, E. *Global Goals for Local Priorities: The 2030 Agenda at Local Level*; Nordregio Report 2018:2; Nordregio: Stockholm, Sweden, 2018.
27. SWECO. *Suggested Indicators & Toolbox-Attractive and Sustainable Nordic Town and Regions*; SWECO Report; SWECO: Stockholm Sweden, 2018.
28. Smoke, P. Fiscal decentralisation frameworks for Agenda 2030: Understanding key issues and crafting strategic reforms. *Commonw. J. Local Gov.* **2017**, 3–23. [CrossRef]
29. UCLG. *The Sustainable Development Goals: What Local Governments Need to Know*; United Cities and Local Governments: Den Haag, The Netherlands, 2016.
30. United Nations Department of Economic and Social Affairs-Division for Sustainable Development. *Voluntary National Reviews: Synthesis Report*; United Nations: New York, NY, USA, 2017.
31. Wymann von Dach, S.; Bracher, C.; Peralvo, M.; Perez, K.; Adler, C. *Leaving No One in Mountains Behind: Localizing the SDGs for Resilience of Mountain People and Ecosystems*; Issue Brief 2018. Sustainable Mountain Development; Centre for Development and Environment and Mountain Research Initiative, with Bern Open Publishing (BOP): Bern, Switzerland, 2018.
32. Lucci, P. *'Localising'the Post-2015 Agenda: What Does It Mean in Practice?* Note Prepared for UCLG (United Cities and Local Governments) with Support from Development Partners Working Group on Decentralisation and Local Governance (DeLoG); ODI: London, UK, 2015.
33. Mischen, P.A.; Homsy, G.C.; Lipo, C.P.; Holahan, R.; Imbruce, V.; Pape, A.; Zhu, W.; Graney, J.; Zhang, Z.; Holmes, L.M.; et al. A foundation for measuring community sustainability. *Sustainability* **2019**, *11*, 1903. [CrossRef]
34. Nordtveit, I. *Bærekraftsrapport for Buskerudamfunnet*; Asplan Viak Report; Asplan Viak: Drammen, Norway, 2018.
35. Patel, Z.; Greyling, S.; Simon, D.; Arfvidsson, H.; Moodley, N.; Primo, N.; Wright, C. Local responses to global sustainability agendas: Learning from experimenting with the urban sustainable development goal in Cape Town. *Sustain. Sci.* **2017**, *12*, 785–797. [CrossRef]

36. Bowen, K.J.; Cradock-Henry, N.A.; Koch, F.; Patterson, J.; Häyhä, T.; Vogt, J.; Barbi, F. Implementing the "Sustainable Development Goals": Towards addressing three key governance challenges—collective action, trade-offs, and accountability. *Curr. Opin. Environ. Sustain.* **2017**, *26*, 90–96. [CrossRef]
37. Hofstad, H.; Vedeld, T. Planlegging og byutvikling i klimaomstillingens tidsalder. *Plan* **2017**, *49*, 14–19.
38. Leal-Arcas, R. Re-thinking global climate change: A local, bottom-up perspective. *Whitehead J. Dipl. Int. Rel.* **2018**, *20*, 4.
39. Moallemi, E.A.; Malekpour, S.; Hadjikakou, M.; Raven, R.; Szetey, K.; Moghadam, M.M.; Bandari, R.; Lester, R.; A Bryan, B. Local Agenda 2030 for sustainable development. *Lancet Planet. Health* **2019**, 3, e240–e241. [CrossRef]
40. Slack, L. The post-2015 Global Agenda–a role for local government. *Commonw. J. Local Gov.* **2015**, 3–11. [CrossRef]
41. Awan, U.; Kraslawski, A.; Huiskonen, J.; Suleman, N. Exploring the Locus of Social Sustainability Implementation: A South Asian Perspective on Planning for Sustainable Development. In *Universities and Sustainable Communities: Meeting the Goals of the Agenda 2030*; Springer: Berlin/Heidelberg, Germany, 2020.
42. Oosterhof, P.D. *Localizing the Sustainable Development Goals to Accelerate Implementation of the 2030 Agenda for Sustainable Development*; Governance Brief; Asian Development Bank (ADB): Mandaluyong, Philippines, 2018.
43. Fleming, A.; Wise, R.M.; Hansen, H.; Sams, L. The sustainable development goals: A case study. *Mar. Policy* **2017**, *86*, 94–103. [CrossRef]
44. Tjandradewi, B.I.; Srinivas, H. Localization of SDGs: Role of Local Governments. *Yokohama City Univ. Inst. Repos.* **2018**, *70*, 335–351.
45. Bhattacharya, S.; Patro, S.A.; Vaidyanathan, V.; Rathi, S. *Localising the Gender Equality Goal through Urban Planning Tools in South Asia*; Working Paper; Center for Study of Science, Technology and Policy: Bengaluru, India, 2016.
46. Fenton, P.; Gustafsson, S. Moving from high-level words to local action—Governance for urban sustainability in municipalities. *Curr. Opin. Environ. Sustain.* **2017**, *26*, 129–133. [CrossRef]
47. United Nations. *Working Together: Integration, Institutions and the Sustainable Development Goals*; World Public Sector Report 2018; Guidance and systemization of experiences Report from Department of Economic and Social Affairs; United Nations: New York, NY, USA, 2018.
48. Garcia-Alaniz, N.; Equihua, M.; Pérez-Maqueo, O.; Benítez, J.E.; Maeda, P.; Urrutia, F.P.; Martínez, J.J.F.; Gaytán, S.A.V.; Schmidt, M. The Mexican National Biodiversity and Ecosystem Degradation Monitoring System. *Curr. Opin. Environ. Sustain.* **2017**, *26–27*, 62–68. [CrossRef]
49. Klopp, J.M.; Petretta, D.L. The urban sustainable development goal: Indicators, complexity and the politics of measuring cities. *Cities* **2017**, *63*, 92–97. [CrossRef]
50. Shulla, K.; Filho, W.L.; Lardjane, S.; Sommer, J.H.; Salvia, A.L.; Borgemeister, C. The contribution of Regional Centers of Expertise for the implementation of the 2030 Agenda for Sustainable Development. *J. Clean. Prod.* **2019**, *237*, 117809. [CrossRef]
51. Valencia, S.C.; Simon, D.; Croese, S.; Nordqvist, J.; Oloko, M.; Sharma, T.; Buck, N.T.; Versace, I. Adapting the Sustainable Development Goals and the New Urban Agenda to the city level: Initial reflections from a comparative research project. *Int. J. Urban Sustain. Dev.* **2019**, *11*, 4–23. [CrossRef]
52. Veldhuizen, L.J.; Giller, K.E.; Oosterveer, P.; Brouwer, I.D.; Janssen, S.; van Zanten, H.H.; Slingerland, M. The Missing Middle: Connected action on agriculture and nutrition across global, national and local levels to achieve Sustainable Development Goal 2. *Glob. Food Secur.* **2020**, *24*, 100336. [CrossRef]
53. Biermann, F.; Kanie, N.; Kim, R.E. Global governance by goal-setting: The novel approach of the UN Sustainable Development Goals. *Curr. Opin. Environ. Sustain.* **2017**, *26*, 26–31. [CrossRef]
54. Gladun, E. Sustainable Development of the Russian Arctic: Legal Implications. *NISPAcee J. Public Adm. Policy* **2019**, *12*, 29–60. [CrossRef]
55. Mokoena, C.; Jegede, A.O. SDGs as a framework for realizing the right to education of the girl child in South Africa: Challenges and possibilities. *Gend. Behav.* **2017**, *15*, 9873–9889.
56. Braun, V.; Clarke, V. Using thematic analysis in psychology. *Qual. Res. Psychol.* **2006**, *3*, 77–101. [CrossRef]
57. Storbjörk, S.; Hedrén, J. Institutional capacity-building for targeting sea-level rise in the climate adaptation of Swedish coastal zone management. Lessons from Coastby. *Ocean Coast. Manag.* **2011**, *54*, 265–273. [CrossRef]
58. DOGA. Smartbyene, and Nordic Edge. In *Nasjonalt Veikart for Smarte og Bærekraftige Byer og Lokalsamfunn*; En Guide for Kommuner og Fylkeskommuner Utarbeidet av Design og Arkitektur Norge (DOGA), Smartbyene og Nordic Edge; DOGA: Oslo, Norway, 2019.
59. Nørgaard, E.; Rognerud, L.M.; Storrud, A. *Indikatorer til FNs Bærekraftsmål-Kartlegging av Tilgjengelig Statistikk i Norge for Måling av FNs Bærekraftsmål*; Document 2018/01; Statistics Norway (SSB): Oslo, Norway, 2018.
60. The Global Taskforce of Local and Regional Governments. *Towards the Localization of the SDGs*; The Global Taskforce of Local and Regional Governments: Barcelona, Spain, 2019.
61. The Global Taskforce of Local and Regional Governments. UNDP, and UN-Habitat. In *Learning Module 1: Localizing the SDGs/Introduction. The Trainer's Guide*; The Global Taskforce of Local and Regional Governments: Barcelona, Spain, 2017.
62. The Global Taskforce of Local and Regional Governments. *Learning Module 2: Territorial Planning to Achieve the SDGs. The Trainer's Guide*; The Global Taskforce of Local and Regional Governments: Barcelona, Spain, 2019.
63. UNDP. *SDG Accelerator and Bottleneck Assessment*; Tool Developed by United Nations Development Programme (UNDP); UNDP: New York, NY, USA, 2017.
64. U4SSC. *Connecting Cities and Communities with the SDGs*; Report Developed by United for Smart Sustainable Cities (U4SSC) Initiative; Posted by Fabienne Perucca; U4SSC: Geneva, Switzerland, 2017.

sustainability

MDPI

Article

Sustainability-Oriented Project Scheduling Based on Z-Fuzzy Numbers for Public Institutions

Dorota Kuchta , Ewa Marchwicka * and Jan Schneider

Faculty of Computer Science and Management, Wroclaw University of Science and Technology, 50-370 Wroclaw, Poland; dorota.kuchta@pwr.edu.pl (D.K.); jan.schneider@pwr.edu.pl (J.S.)
* Correspondence: ewa.marchwicka@pwr.edu.pl

Abstract: A new approach to sustainable project scheduling for public institutions is proposed. The approach is based on experts' opinions on three aspects of sustainability of project activities (human resources consumption, material consumption and negative influence on local communities), expressed by means of Z-fuzzy numbers. A fuzzy bicriterial optimization model is proposed, whose objective is to obtain a project schedule of an acceptable sustainability degree and of acceptable duration and cost. The model was inspired and is illustrated by a real-world infrastructure project, implemented in 2019 by a public institution in Poland.

Keywords: sustainability; project management; scheduling; public infrastructures; Z-fuzzy number

Citation: Kuchta, D.; Marchwicka, E.; Schneider, J. Sustainability-Oriented Project Scheduling Based on Z-Fuzzy Numbers for Public Institutions. *Sustainability* **2021**, *13*, 2801. https://doi.org/10.3390/su13052801

Academic Editor: Margarita Martinez-Nuñez

Received: 20 January 2021
Accepted: 1 March 2021
Published: 5 March 2021

1. Introduction

In this paper the problem of project scheduling in the context of sustainability is considered. The work concentrates on public administration, where sustainability issues are an immediate consequence of the public administration ethics principles [1]. Ethics pose an even more sensitive issue for government than for corporations or other private sector organizations because government, by definition, must serve all interests in a society [2]. Public administration has to observe the law, organize work in a just and ethical way, be totally transparent, serve the public interest in all its endeavors. Thus, exploiting and harming human beings, damaging human health, damaging the environment, acting against the interests of individuals: all this is irreconcilable with the mission of public administration [1]. Also, public organizations differ from private organizations and have other basic goals than profit-oriented organizations. These goals include public accountability, honesty, openness, responsiveness to policy, fairness, due process, social equality, balanced criteria for the distribution of manufactured goods, and correct moral behavior [3]. This means that public administration has to include sustainability as a factor in all their everyday activities.

This obligation results also from the generally accepted conviction that public administration is generally less advanced in the adoption of modern project management standards (see [4,5] for more details). Public projects are often not quite successful [6,7], which means that public money is not spent in an efficient way or is spent not according to the society's expectations. Thus, all possible steps should be undertaken in order to increase public money spending efficiency and acceptability, and sustainable project management may be one the remedies [8]. Also, developed project stakeholder management, strongly linked to sustainability [9], is necessary for efficient and effective project management. In short, sustainable project management may be a way to increase the success rate of projects realized by public institutions.

Sustainability means taking care of people and the world. But of course, even though public organizations are not profit-oriented, their goals (public accountability, responsiveness to policy, etc.) also obligate them to control expenditures and, as far as possible, minimize them, if more important goals are not compromised through the savings. In public organizations trade-off decisions have to be made every day, and sustainability and cost-related criteria are an integral part of them.

The goal to be addressed in this paper is connected to project schedules in public organizations, while we adopted the following definition of schedule: "A schedule is a timetable showing the forecast start and finish dates for activities or events within a project, program or portfolio" [10]. Scheduling will be understood as "a collection of techniques used to develop and present schedules that show when work will be performed" [10], and the goal will consist in introducing a sustainable project scheduling model for organizations, focusing on public institutions. To this end a model that aims to generate, under certain assumptions, a project schedule that will be sustainable and yet acceptable also from the economic perspective is proposed. A first attempt in this direction was undertaken in [11]. This present paper offers a modification and extension of the proposal put forward there. No other attempt related to sustainable scheduling was identified in the literature.

The aim of this paper is to propose an optimization model for public project scheduling that allows to take sustainability issues into account. Individual project activities will be evaluated by experts as to their suitability from three perspectives (in three dimensions), while expert expertise and available sources of information will be considered, too. This is important because sustainability evaluation can be a sensitive subject and may depend on the subjective perspective of a given expert (some public institution employees are more eco-oriented or emphatic, some less, some are more knowledgeable, some less, also local citizens and organizations may have different expertise and attitudes). In order to model such a complex situation (for the first time in the context of project scheduling) Z-fuzzy numbers are used. The proposed bicriteria model will generate a trade-off project schedule —of an acceptable duration and cost, but also of an acceptable degree of sustainability (or the answer that such a schedule does not exist). The approach will be illustrated by means of a real-world example of a public schedule, which constituted, by the way, an inspiration for this paper.

The structure of the paper is as follows. In Section 2 a literature review is presented, which proves that our approach is a novel one with respect to sustainable project management and the application of Z-fuzzy numbers in this context. In addition to reviewing literature on sustainable project scheduling, approaches to modeling multi-objective project scheduling as adopted in this article are reviewed. Section 3 presents the theoretical model proposed in this article. Basic assumptions and notations related to sustainability, scheduling problem formulation, the bicriterial problem solution adopted and Z-fuzzy numbers are presented. The section closes with the final model proposal. In Section 4 the model is applied to a real-world public project that was implemented by a certain Polish public institution, which showed itself open to new management solutions. This institution also cooperated in the research and helped to establish the applicability of the method proposed in this paper. The obtained results are discussed in Section 5. Finally, the conclusions of the study are presented in Section 6.

2. Literature Review

2.1. Sustainable Project Scheduling

Our literature research included publications from two different databases: Scopus and Google Scholar. The literature was reviewed according to different combinations of the following search phrases: "project," "planning," "scheduling," "sustainable," "sustainability," "green," "public project." For several databases, it was necessary to modify initial test criteria, so that the number of results produced was appropriate. In most cases too many results were obtained and narrowing down the criteria was needed, but there were also cases when the search had to be extended (e.g., S6 from Table 1) because no interesting results were found initially. The search terms and search results are summarized in Tables 1 and 2. Finally, 18 literature outputs were found in total. Because some of the results from different databases overlapped with each other, the final number of articles described was smaller than this number and equaled 15 (four related to project scheduling, four related to project planning and seven related to public projects). It is also worth mentioning that the literature review was restricted to publications from recent years. The year 2016 was added as lower constraint. Three different categories were assigned to the search criteria and literature is discussed in each of the categories separately, for better readability. At the end the results are discussed with special attention to the tools and methods that were used in this article, namely, in relation to the usage of fuzzy numbers. The literature review showed that there exists a gap in the literature in the area of sustainable public project planning and sustainable public project scheduling.

Table 1. Search criteria used in the literature review together with their categories.

ID	Search Term	Category
S1	"project" AND "scheduling" AND ("sustainable" OR "green" OR "sustainability")	scheduling
S2	"project" AND "planning" AND ("sustainable" OR "green" OR "sustainability")	planning
S3	"project planning" AND ("sustainable" OR "green" OR "sustainability")	planning
S4	"public project" AND ("sustainable" OR "green" OR "sustainability")	public projects
S5	"project scheduling" AND ("sustainable" OR "green" OR "sustainability")	scheduling
S6	"sustainable scheduling" AND "project management"	scheduling
S7	"sustainable scheduling" AND "project"	scheduling
S8	"sustainable planning" AND "project"	planning
S9	"sustainable project planning"	planning
S10	"public project" AND "sustainable"	public projects
S11	"sustainable project" AND "public project"	public projects

Table 2. Summary of search results obtained for different databases.

Search Term	Year Filter	Database	Results Found	Results Reviewed
S1	≥2016	Scopus	34	1
S2	≥2016	Scopus	1742	0
S3	≥2016	Scopus	35	3
S4	≥2016	Scopus	15	4
S5	≥2016	Google Scholar	1700	0
S6	≥2016	Google Scholar	10	0
S7	≥2016	Google Scholar	95	3
S8	≥2016	Google Scholar	7680	0
S9	≥2016	Google Scholar	54	3
S10	≥2016	Google Scholar	2800	0
S11	≥2016	Google Scholar	101	4

First, the literature related to sustainable scheduling is summarized. In [12] the authors presented a scheduling method by solving the discrete time/cost trade-off problem (DTCTP) for more than 500 project activities using a genetic algorithm and a heuristic. The sustainability aspect was reduced to the fact that in sustainable project management resources should be used economically, so that the shorter project duration is achieved at lower cost. No relation to the three dimensions of sustainability (i.e., social sustainability, environment sustainability, economic sustainability) was given. Sustainable Job-Shop scheduling approaches were presented in [13,14], but the three dimensions of sustainability were not considered, either. In [13], authors considered energy efficiency, while in [14] carbon emission was considered. Similarly, reference [15] described sustainable production scheduling in the context of minimizing pollution.

As far as sustainability in relation to project planning is considered, even fewer relevant references were found (although the set of reviewed articles was equally represented as in the case of project scheduling). The idea of integrating sustainability into traditional project management methods in the context of construction projects was presented in [16]. Although the authors often used the term "sustainable project planning" and indicated the need for a proper definition of "sustainable project planning," they rather focused on the sustainable aspects of project control, risk response strategies and communication. No concept of sustainable project scheduling was presented. An evolutionary optimization method for sustainable project planning shown through the example of developing a sustainable port was presented in [17], but the sustainability aspect was reduced to carbon emission only. A sustainable project pre-planning phase was discussed in [18], including the observation that green project planning requires more effort than traditional project planning. The authors of [19] tried to list the barriers to integrating sustainability rules into construction projects. Lack of a systematic approach to sustainable project planning is on the list of barriers.

The most relevant set of publications was found for public projects, but here also a lack of in-depth insight into sustainability can be observed. Sustainable public projects were discussed in [20]. The authors presented an evaluation measure of the sustainability of a public project that included carbon emission, resource utilization, renewable energy and impact on the surroundings, when evaluation information is fuzzy. A fuzzy analytical hierarchy process (FAHP) was used for evaluation. This work considered many important dimensions of sustainability in the context of public projects, but it focused on public project management in general. Project scheduling was not considered. Different sustainability objectives for major public projects and their evaluation by different project stakeholders (government, owner, contractor, designer, end user, university, NGOs) were presented in [21]. The evaluation was based on interviews. Five economic, nine social and four environmental objectives were listed. The results came from the Hong Kong region. A sustainability-oriented evaluation method for public government projects was presented in [22]. It was based on a judgment matrix and an analytic hierarchy process. The authors developed a bid evaluation index that can be used for selecting the best bidder according to sustainability criteria. As in the case of [21], the method was evaluated in a Chinese context. The social dimension of public infrastructural projects in an Italian context was presented in [23]. The aspect of social sustainability was discussed in the context of inconveniences that appear when building a new infrastructure, like relocation of the residents. The authors of [24] proposed an interesting classification of infrastructure projects that takes into account a sustainability dimension. Public infrastructure projects in the context of sustainable project controlling were presented in [25]. Sustainability controlling methods were analyzed by the example of a road tunnel construction project. The paper showed that project control mechanisms are used differently for different sustainability dimensions. The authors of [26] focused on a method for estimating the social sustainability of a public infrastructure project, motivated by the observation that this aspect of social sustainability is often mentioned in the literature.

Summing up, there are not many literature positions that refer to sustainable project planning or sustainable project scheduling, in the context of public projects. What is more, only two positions were found [20,22] that present sustainable methods based on the theory of fuzzy numbers, and in none of them Z-fuzzy numbers were used. The method presented in this article tries to fill in this research gap.

2.2. *Multi-Objective Project Scheduling*

Sustainable project scheduling is bound to be a multi-objective scheduling problem, and sustainability can never constitute the sole criterion. Therefore, literature concerning general multi-objective project scheduling problems was also reviewed here. In [27] one can find a review of the state of art—not only with respect to the criteria used (needless to say that sustainability has not been used as a criterion in such models), but also to the parameters and decision variables of the models.

Parameters include precedence relationships between project activities, information on the nature and quantity of needed and available resources, activity duration, cost and project budget. The prevalent decision variables are start times and finish times of activities. Constraints refer to requirements set for the project deadline, the available budget, the available quantity of resources, etc.

Of course, project scheduling can be seen as a more general problem. It may cover problems where it was possible to reduce the duration of activities on the critical path against a certain cost [28] or preempt or interrupt them [29]. In this article, however, the setting was limited to the basic version of the project scheduling problem [30] where the decision variables were the start times of activities, with the other activity features remaining fixed.

As far as computational complexity is concerned, exact mathematical models tend to be complex if an integer solution is required. In the case of resource-constrained project scheduling, oftentimes 0–1 problems occur [30], and the integer nature of the model cannot be avoided. If the exact solution of the problem is too complex computationally, various heuristics are known and may be applied.

3. Theoretical Model

3.1. Sustainability Modeling

It is assumed that each project activity per se may violate sustainability requirements in three different aspects: human resources consumption, materials consumption and influence on the environment. Thus, each activity can be evaluated with respect to its sustainability in each of the three domains.

- In the domain of the human resources consumption, the activity is evaluated on the basis of the answer to the question of whether it causes work overload, tiring or exhaustion of humans, forcing humans to work overtime without their consent, in their free time. The higher the evaluation value, the less the activity is a cause of these negative phenomena.
- In the domain of materials consumption, we evaluate the over-wasting of materials and the usage of materials that by themselves or whose extraction is harmful to the environment. The higher the evaluation value, the less the activity wastes materials, and the less the materials and their extraction are harmful to the environment.
- In the domain of the influence on the environment (understood here above all as the local community where the public institution in question operates), the evaluation of an activity is higher the less negative influence the activity has on both the physical and human environment (e.g., the less noise or dirt its execution generates, the less reduction in green spaces it causes, etc.).

It is also assumed that in the case of some activities, it is possible to increase their sustainability degree in one or more of the aspects, but this will take time (which means delaying their start times with respect to the earliest possible one) and generate an additional cost. The improvement of the sustainability degree, linked to delaying the activity start times, can be achieved in the following ways:

- In the human resources consumption aspect: other human resources than initially planned can be assigned to the activity. They may be less experienced, but it will save the initially assigned employees from overworking and give their colleagues an opportunity to develop their experience and show their potential. The activity costs more money, but its sustainability index will also increase. Another possibility here is to outsource the activity, even for higher costs, so that the workload of the employees diminishes. Both measures—assigning other employees or outsourcing an activity—may mean a delay in the activity start times; the unexperienced employees have to be trained and freed from their normal duties, an external company has to be searched for and a contract with it has to be negotiated. Both measures are also linked to an additional cost: a training cost for the employees and the additional cost for paying the external company. However, in relation to sustainability, the more balanced the human consumption is, the more social benefits it brings, which increases social sustainability.
- In the materials consumption aspect: substitute materials can be used, whose usage is less harmful from the environmental point of view. However, this will require

additional time (searching for possible suppliers, choosing the best offer, contract signing, waiting for delivery, etc.) and additional cost (prices of eco-materials are usually higher, but also the cost of the extra work needed from the purchase department of the public institution will generate additional cost) of bringing the materials to the organization. But better material used increases environmental sustainability.

- In the environmental (local community-related) influence aspect: different measures are possible, depending on the context; for example, in case of too much noise caused by cumulation of various projects in one area, the activity start time can be postponed in order not to coincide with other projects. Such a measure does not only mean waiting longer for the activity to start but may also generate additional cost: if a contract was signed for hiring heavy equipment, some fixed fees for waiting will have to be incurred.

It is assumed that the sustainability degree of an activity in each of the three aspects will depend on the moment when the activity is started and that if this moment is shifted forward in time, the sustainability degree will not diminish (in other words, postponing the start of an activity cannot decrease the sustainability degree, but can increase it). This assumption may be restricting in some cases, but usually the start of the activities is postponed in order to be more sustainable (e.g., time is used to train employees, or to supply environmentally friendly materials, or to develop/learn greener technology, or to shift nighttime tasks to daytime to avoid noise, etc.). Also, it is assumed that postponing the activity start is not desirable from the point of view of staying within deadlines and may be the cause of extra cost. Additionally, it is assumed that the sustainability measures in the three aspects can be aggregated, to give one total sustainability measure of each activity, which is a non-decreasing function of the starting moment of the activity.

Let us denote the measures of the three dimensions of sustainability mentioned above as $S_i^r(s),\ r = 1,2,3,\ i = 1,\ldots,N$, where $r = 1$ refers to human resources consumption, $r = 2$ to materials consumption, $r = 3$ to the influence on the environment and $s \in [0, H)$, where H is the furthest possible acceptable horizon for the project to terminate and stands for the starting time of the respective activity. It is assumed that $S_i^r(s),\ r = 1,2,3,\ i = 1,\ldots,N$ are non-decreasing functions of s. The total sustainability degree of the ith activity is defined as in (1) (the formula used here is only an example, weighted sums or other formulas can be considered):

$$S_i(s) = \frac{1}{3}\sum_{r=1}^{3} S_i^r(s). \quad (1)$$

3.2. The Scheduling Problem Formulation

It is assumed that at the starting decision moment ($s = 0$) the activities have a certain level of sustainability (measured as in (1)), which in some cases may be increased at an additional cost if the activities start later than at $s = 0$. The two criteria chosen as optimization criteria are thus:

- The average level of sustainability of all project activities,
- The total cost of improving the sustainability of project activities with respect to moment $s = 0$.

It is assumed that other parameters of the project (like activity duration or cost not related to sustainability improvement) cannot be changed. The decision variables are the start times of individual activities. The main problem parameters are the planning horizon H, the number of activities N, and the duration of ith activity $T_i,\ i = 1,\ldots,N$.

For the ith activity, $C_i^*(s)$ stands for the cost of achieving a sustainability degree of $S_i(s)$. As stated above, both $C_i^*(s)$ and $S_i(s)$ are non-decreasing functions of s. As can be seen from (1), the sustainability degree $S_i(s)$ of each activity is dependent on its start time, which means that sustainability is connected to the project's schedule. The same observation is true for $C_i^*(s)$. In this article an assumption is made that $C_i^*(s)$ is a linear function of s (an example of such a function can be found in Results section and is described as the total cost per one time unit of waiting). This means that two sustainability-related measures used in this article are connected to the project's schedule.

It is assumed, for the sake of simplicity, that there are no dependencies between the considered activities. This per se unrealistic assumption is true for the majority of activities in the example project referred to in this paper and can be easily lifted in eventual applications. Also, there are no budgeting or resource constraints considered, but this omission can also be easily amended. Such a future extension will not break the assumptions made, and the problem will be solvable in polynomial time, provided that, e.g., heuristics are used to obtain a sufficiently good solution.

The problem is to determine a schedule SH, which will be understood as the set of starting times for each activity $\{s_i\}_{i=1}^{N}, 0 \leq s_i \leq H - T_i, i = 1, \ldots, N$, maximizing objective (2) (the total sustainability of the project) and minimizing objective (3) (the cost of achieving respective sustainability degrees).

$$O1 = \frac{1}{N} \sum_{i=1}^{N} S_i(s_i) \rightarrow max, \quad (2)$$

$$O2 = \sum_{i=1}^{N} C_i^*(s_i) \rightarrow min. \quad (3)$$

The resulting problem is a bicriterial one, with one objective being maximized and the other one minimized. Multi-objective problems, in order to be solved, have to be transformed into single objective ones. There exists a vast spectrum of possible approaches. Here we chose an interactive approach, called max-min compromise solution [31]. According to this method, the decision makers are asked to indicate the lower and upper limits to their satisfaction with the values of both objectives. Let us denote them as: O_L^1, O_U^1 for $O1$ and O_L^2, O_U^2 for $O2$. They mean that the decision makers are totally unsatisfied if the value of the first objective (maximized) is equal to or lower than O_L^1 and completely satisfied if its value is at least O_U^1. In-between satisfaction is assumed to be growing linearly. In case of the other objective, $O2$, which is minimized, the satisfaction is full below O_L^2, 0 over O_L^2 and diminishes linearly in-between. The objective function to be maximized in the final model is the minimum of the two satisfaction degrees. Such an approach will generate a max-min compromise solution (Figure 1), in which the decision makers are satisfied to some extent with both objectives.

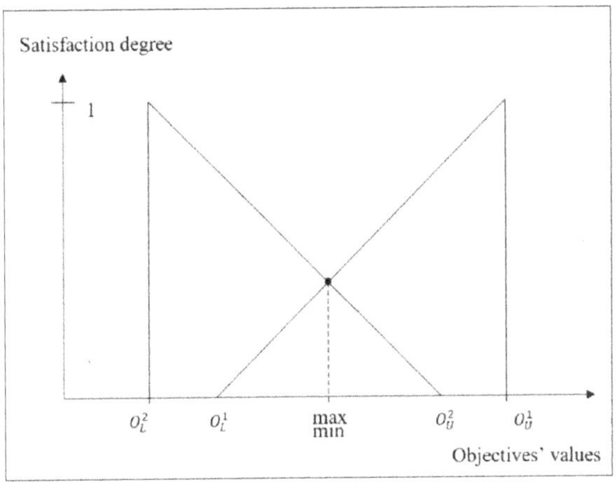

Figure 1. Illustration of the max-min compromise solution of a bicriteria problem.

3.3. The Use of Z-Fuzzy Numbers

Of course, an important problem is the determination of $S_i^r(s)$, $r = 1,2,3$, $i = 1,\ldots,N$. The sustainability degree in all the three aspects is not an easily measurable feature, its evaluation is subjective and has to be based on expert opinion. What is more, various experts may have different views on the subject. Project stakeholders representing the local community, the employees, ecologists, etc., are examples of expert groups who should be asked about the sustainability degree of individual activities, but they are bound to have conflicting views. What is more, these experts will also differ with respect to their credibility. There will be more experienced and less experienced employees, more selfish and less selfish representatives of the local community, more honest and less honest ecologists, etc. Of course, it is desirable to have the possibility to consult only experienced and cooperative experts, but the reality of public institutions is such that organizational and hierarchical connections play an important role. As research on public institutions shows, executives in the public sector are much less willing to delegate power, even to more experienced employees. Also, financial restrictions make it often impossible to have access to highly qualified external experts and the necessary expertise is not always available [32]. That is why one aim considered in this paper is to include in the model both the subjectivity and the undetermined nature of the sustainability degree evaluation and the problem of experts' credibility. In order to achieve this goal, $S_i^r(s)$, $r = 1,2,3$, $i = 1,\ldots,N$ is modeled by means of so-called Z-fuzzy numbers [33]. The ultimate bicriterial optimization model is then a fuzzy model with Z-fuzzy numbers as model parameters. This is the main difference with respect to the only known approach to sustainability-oriented scheduling in public institutions [11]—where the problem of credibility is not taken into account and classical fuzzy numbers are applied.

The notion of Z-fuzzy numbers was proposed in [33] and discussed in numerous other papers. A Z-fuzzy number is an ordered pair $\left(\widetilde{A},\widetilde{Z}\right)$, where $\widetilde{A}$ and $\widetilde{Z}$ are fuzzy numbers and the support of $\widetilde{Z}$ is included in the interval [0,1]. It is assumed here that both $\widetilde{A}$ and $\widetilde{Z}$ are triangular fuzzy numbers, which is of sufficient generality from the point of view of potential applications in public institutions: more complicated fuzzy numbers may occur in future implementations, for now, triangular fuzzy numbers are more than sufficient.

$\widetilde{A}$ represents a magnitude whose exact value is not known at the moment. $\widetilde{A}$ is represented by three crisp numbers $\underline{a}, \hat{a}, \overline{a}$, such that $\underline{a} \leq \hat{a} \leq \overline{a}$. Its so-called membership function μ_A, defined as in (4) and based on expert opinions, represents for each x the possibility that, according to the experts, x will actually be of the unknown magnitude.

$$\mu_A = \begin{cases} 0 \text{ for } x \leq \underline{a} \\ \frac{x-\underline{a}}{\hat{a}-\underline{a}} \text{ for } \underline{a} < x \leq \hat{a} \\ \frac{\overline{a}-x}{\overline{a}-\hat{a}} \text{ for } \hat{a} < x \leq \overline{a} \\ 1 \text{ for } x > \overline{a} \end{cases}. \tag{4}$$

$\widetilde{Z}$ is a triangular fuzzy number defined analogously by three crisp numbers $\underline{z}, z, \overline{z}$, such that $\underline{z} \leq \hat{z} \leq \overline{z}$, and an analogous formula for μ_Z as in (4), with the additional condition $0 \leq \underline{z} \leq \hat{z} \leq \overline{z} \leq 1$. $\widetilde{Z}$ represents the credibility of the expert opinion $\widetilde{A}$. The closer to 1 the numbers $\underline{z}, z, \overline{z}$ are, the higher the credibility of $\widetilde{A}$. The higher the difference $\overline{z} - \underline{z}$, the higher the non-determinacy of the credibility of $\widetilde{A}$. In the literature it has been proposed to choose $\widetilde{Z}$'s from a list of triangular fuzzy numbers, each of which corresponds to a linguistic expression, like (credibility) high, medium, low, etc. An example of the "dictionary" for the values of $\widetilde{Z}$ can be found in Figure 2.

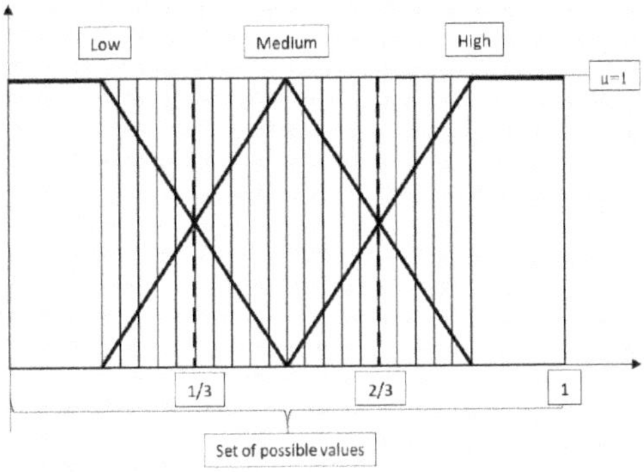

Figure 2. An example of the "dictionary" for the values of $\widetilde{Z}$.

In the literature there have been several proposals of arithmetic operations on Z-fuzzy numbers, e.g., [34–37], which differed in the procedure to encapsulate the information given by the ordered pair $\left(\widetilde{A},\widetilde{Z}\right)$ in a simplified form (as a classical fuzzy number or in a defuzzified form, as a crisp number), in order to make possible various operations and comparisons among Z-fuzzy numbers.

Defuzzification is a method often used in practice in order to summarize the information conveyed by a fuzzy number of any type. Of course, defuzzification always entails a loss of information, but as our approach was focused on public institutions, where the introduction of mathematically complicated tools was bound to encounter considerable resistance, we took the decision to apply defuzzification at his stage.

Let us thus denote by $D\left(\widetilde{A},\widetilde{Z}\right)$ a crisp number being a defuzzification of the Z-number $\left(\widetilde{A},\widetilde{Z}\right)$. Here a modification of the proposal described in [36] is used. The basis of this modification is the following defuzzification Formula (5) for triangular fuzzy numbers [38]:

$$d\left(\widetilde{A}\right) = 0.25(\underline{a} + 2\hat{a} + \overline{a}). \tag{5}$$

This choice is not bounding, any other method might be selected.

Thus, it is proposed to use the following procedure (6) for the calculation of $D\left(\widetilde{A},\widetilde{Z}\right)$:

$$D\left(\widetilde{A},\widetilde{Z}\right) = d\left(\sqrt{d\left(\widetilde{Z}\right)}\underline{a}, \sqrt{d\left(\widetilde{Z}\right)}\hat{a}, \sqrt{d\left(\widetilde{Z}\right)}\overline{a}\right). \tag{6}$$

Equation (6) gives the defuzzification of a Z-fuzzy number as was defined in the literature in [36] (Ref. [36] is the only existing proposal so far). This approach was adopted here, but attention has to be paid to the implications of this concrete defuzzification method. The method as it was adopted here (after [36]) reflects a natural and practical approach: If the credibility $\widetilde{Z}$ is low, the evaluation $\widetilde{A}$ is simply shifted to the left, assuming it to be lower than the original one. Such an approach may be acceptable in many cases, but not always. This problem is discussed further in the conclusions section.

We propose to define $S_i^r(s)$, $r = 1,2,3$, $i = 1,\ldots,N$ in the form of Z-fuzzy numbers $\left(\widetilde{A}_i^r(s),\widetilde{Z}_i^r(s)\right)$, where

$$\widetilde{A}_i^r(s) = \left(\underline{l}_i^r + \underline{k}_i^r s, \hat{l}_i^r + \hat{k}_i^r s, \overline{l}_i^r + \overline{k}_i^r s\right), \tag{7}$$

$$\widetilde{Z}_i^r(s) = \left(\underline{b}_i^r + \underline{d}_i^r s, \hat{b}_i^r + \hat{d}_i^r s, \overline{b}_i^r + \overline{d}_i^r s\right), \tag{8}$$

where all the parameters are non-negative, and $\overline{b}_i^r + \overline{d}_i^r s \leq 1$ for $s < H$.

The fuzzy valued functions (7) and (8) are non-decreasing in the sense that each of the three component classical functions determining them are non-decreasing.

Then Formula (1) for the aggregated sustainability degree of each activity takes the following form:

$$S_i^*(s) = \frac{1}{3}\sum_{r=1}^{3} D\Big(\Big(\widetilde{A}_i^r(s), \widetilde{Z}_i^r(s)\Big)\Big). \tag{9}$$

3.4. Final Model

The objectives (2) and (3) are thus reformulated as follows:

$$O1 = \frac{1}{N}\sum_{i=1}^{N} S_i^*(s_i) \to max, \tag{10}$$

$$O2 = \sum_{i=1}^{N} C_i^*(s_i) \to min \tag{11}$$

.

The bicriterial optimization is accomplished according to the max-min compromise approach described in Section 3.2. The minimum satisfaction of the decision maker with each of the objectives (10) and (11) is maximized. Let λ_1 stand for the satisfaction of the decision maker with objective (10) and λ_2 for the satisfaction of the decision maker with the second objective. The following pair of objectives are then considered:

$$\lambda_1 \to max, \ \lambda_2 \to max. \tag{12}$$

Objective (12) is defined using the two pairs of numbers given by the decision makers:

- O_L^1, O_U^1 for $O1$, such that $\lambda_1 = 0$ if $O1 < O_L^1$, $\lambda_1 = 1$ if $O1 > O_U^1$ and $\lambda_1 = \frac{O1 - O_L^1}{O_U^1 - O_L^1}$ for $O1 \in \left[O_L^1, O_U^1\right]$ (objective (10) is maximized);
- O_L^2, O_U^2 for $O2$, such that $\lambda_2 = 1$ if $O2 < O_L^2$, $\lambda_2 = 0$ if $O2 > O_U^2$ and $\lambda_2 = \frac{O_U^2 - O2}{O_U^2 - O_L^2}$ for $O2 \in \left[O_L^2, O_U^2\right]$ (objective (11) is minimized).

It is proposed here to maximize the minimum of the two satisfaction objective (12), denoted as λ. The optimization model to be solved is thus finally as follows (λ stands for the minimum of the two satisfactions and is maximized):

$$\begin{gathered}
\lambda \to max \\
\lambda_1, \lambda_2 \geq \lambda \\
\frac{1}{N}\sum_{i=1}^{N} S_i^*(s_i) \geq O_L^1 + \lambda_1\left(O_U^1 - O_L^1\right) \\
\sum_{i=1}^{N} C_i^*(s_i) \leq O_U^2 - \lambda_2\left(O_U^2 - O_L^2\right) \\
0 \leq s_i \leq H - T_i, \ i = 1, \ldots, N.
\end{gathered} \tag{13}$$

The result of the solution of (13) is a schedule SH, understood as a set $\{s_i\}_{i=1}^N$ of starting points of individual activities. The decision variables are λ, λ_1, λ_2 and $s_i, i = 1, \ldots, N$ with (5), (6) i (9) and if $C_i^*(s_i)$ is a linear function of s_i (with the coefficient C_i), a quadratic problem is considered. Although in fact the solution should be an integer (it is unrealistic to assume fractional activity start times), it is proposed—in order to avoid problems with computational complexity—to consider in the first place the relaxation with the integer constraints lifted. In practical applications, where the durations of project tasks are usually easily adjustable within small ranges (like hours), the relaxed problem should give an acceptable starting point, from which the final schedule may be constructed manually, in a fairly easy way. The relaxed problem would cause no complexity-related problems. It should be noted that both original objectives (10) and (11) are functions of decision variables s_i and all constraints (presented in (13)) are functions of decision variables λ, λ_1, λ_2 and s_i, $i = 1, \ldots, N$.

Table 3 presents some basic assumptions and the properties of the model proposed in this paper (described above in Section 3). These assumptions can be used as a reference for assessing the applicability of the model. They should also help in understanding the future research and possible extensions of the method.

Table 3. Summary of the assumptions used in the model.

Scheduling Assumptions	
activity durations	fixed
resource constraints	no
predecessor–successor constraints	no
cost constraints	no
Sustainability Assumptions	
dimensions	resource consumption, material consumption, local environment
assessment	experts' assessments
assessment representation	Z-fuzzy numbers
expert selection method	not specified
initial selection of experts	office employees (public projects)
sustainability gain	e.g., postponing activities, new materials
cost of sustainability	e.g., teaching staff, materials transportation, delay costs
Model Assumptions	
input	activities (with durations), experts' assessments of sustainability and of preferences as to the two objectives' values, the planning time horizon H
output	sustainable schedule (defined by activity start times)
model used	bicriterial optimization (max-min compromise approach) + Z-fuzzy numbers
computational complexity	polynomial
optimization criteria	sustainability (max), cost of increasing sustainability (min), reduced through the compromise max-min approach to one criterion: minimum of the satisfaction with the value of each of the objectives (maximized)
decision variables	activity start times, satisfaction with the value of each objective, minimum of the satisfaction with the two objectives
novelty (originality)	sustainability considered in project schedule
Possible Extensions	
predecessor–successor constraints	planned research
cost constraints	planned research
resource constraints	possible research (heuristics needed to maintain complexity)
more sustainability dimensions	planned research
more public projects tested	planned research

4. Results (Based on a Real-World Project)

Model (13) (implemented in Maple) was applied to an example inspired by an infrastructure project implemented by a Polish municipality in Lower Silesia. The area of the municipality covers almost 15,000 hectares and the numbers of inhabitants was almost 5000 in 2020. The management and the employees were open to new project management methods, even the more sophisticated ones. In the past they took part in an experiment regarding the implementation of a fuzzy project risk management method [39] and both the management and the employees were rather positive in their opinions with respect to the endeavor.

The project to which the model described in this paper was applied was the reconstruction of a municipal road. The project budget was about EUR 400,000, and it was realized between June and November of 2019. The following task groups were to be executed: preparatory work, surface milling, channeling, surface construction, surface elevation, widening of surface, roadsides, descents, pavement extension, clearing of culverts, cleaning of ditches, repair of retaining wall, strengthening of slopes. Most of the tasks involved pressure on the workers because of the deadline, usage of a mixture of various resources whose exploitation might damage the environment (some could have been replaced with other materials, more eco-friendly ones, e.g., the asphalt composition for the road surface [40]) and produce noise during the project realization. Thus, all the three sustainability dimensions (i.e., social sustainability, environment sustainability, economic sustainability) were at stake but were not taken into account during project realization. The project team members and the executives were aware of this and they were considering the possibility of introducing a systematic sustainability-related element into the scheduling procedure, being aware of the challenges related to sustainability that public institutions are facing (see Introduction). The authors of the paper were asked to analyze the case from the point of view of such a possibility. Thus, the proposed approach was post factum applied to the project.

The example's objective is to illustrate how the sustainability degree of a project can be influenced within the negotiated deadline and budget, with the aim of increasing the project sustainability degree—thus enhancing the human well-being without disregarding time and cost constraints.

In the example the parameters assumed in (13) were $O_L^1 = 6$ and $O_U^1 = 10$ and $O_L^2 = 80$ and $O_U^2 = 20$ (expressed in units suitable for each case). It was also assumed that the sustainability degree of the tasks could be increased with time, at some cost (thanks to assigning other team members, choosing alternative materials, etc.). At the same time, the suitability evaluation was seen as potentially not fully credible, for example, because of the lack of experience of the experts in the public institution, where sustainable project management had not been fully implemented yet.

Two cases were considered: one where the credibility of experts was fixed and could not be increased (by a higher quality of the experts or because the judgment could not be "bought in" in any form, even if the activity was postponed) and another one, where in case of the postponement of the evaluation of the sustainability degree, higher quality opinions could be gained (where it was possible to search for other experts or other sources of information, but this had to take time). Table 4 presents the data for the first case (where the second element of the Z-fuzzy numbers does not depend on time).

Table 4. Data on four project tasks without dependencies among them, where the sustainability can be increased but the credibility of its evaluation cannot.

i, T_i, C_i	$\tilde{A}_i^1(s_i), \tilde{Z}_i^1(s_i)$	$\tilde{A}_i^2(s_i), \tilde{Z}_i^2(s_i)$	$\tilde{A}_i^3(s_i), \tilde{Z}_i^3(s_i)$
1, 3, 10	$(1+0.6s_i, 2+0.6s_i, 3+0.6s_i)$, $(0.1, 0.2, 0.3)$	$(0.5+2s_i, 1+2s_i, 1.5+3.5s_i)$, $(0.2, 0.3, 0.4)$	$(5+s_i, 10+s_i, 11+s_i)$, $(0.8, 0.9, 1)$
2, 4, 20	$(0.5+2s_i, 1+2s_i, 1.5+2s_i)$, $(0.2, 0.3, 0.4)$	$(1+0.6s_i, 2+0.6s_i, 3+0.6s_i)$, $(0.1, 0.2, 0.3)$	$(1+s_i, 2+s_i, 3+s_i)$, $(0.5, 0.6, 0.7)$
3, 2, 5	$(1+s_i, 2+s_i, 3+s_i)$, $(0.5, 0.6, 0.7)$	$(5+s_i, 10+s_i, 11+s_i)$, $(0.8, 0.9, 1)$	$(0.5+2s_i, 1+2s_i, 1.5+2s_i)$, $(0.2, 0.3, 0.4)$
4, 1, 4	$(5+s_i, 10+s_i, 11+s_i)$, $(0.8, 0.9, 1)$	$(1+s_i, 2+s_i, 3+s_i)$, $(0.5, 0.6, 0.7)$	$(1+0.6s_i, 2+0.6s_i, 3+0.6s_i)$, $(0.1, 0.2, 0.3)$

For example, the first project task had duration $T_1 = 3$ weeks, and its sustainability degrees in the three aspects, expressed by means of $\widetilde{A}_1^r(s_1),\ r = 1,\ 2,\ 3$, depended on the starting time s_1. If the task was started later, certain measures would be taken to increase its sustainability, and the total cost per one-time unit of waiting was equal to $C_1 = 10$. The credibility degrees of the sustainability evaluations, respectively for each sustainability aspect, were as follows $(0.1,\ 0.2,\ 0.3)$, $(0.2,\ 0.3,\ 0.4)$, $(0.8,\ 0.9,\ 1)$, thus the credibility of the experts' opinion was very low for the human resources aspect and very high for the environmental aspect. Here it was assumed that the credibility could not be changed with time, thus the experts could neither be replaced nor their expertise enhanced.

We assumed various time horizons H and for each we solved another problem (13). The following schedules SH were generated:

- for H = 10 (weeks): $s_1 = 0.33,\ s_2 = 0.1,\ s_3 = 8,\ s_4 = 8$, with total satisfaction degree $\lambda = 0.04$;
- for H = 9 (weeks): $s_1 = 1.5,\ s_2 = 0,1,\ s_3 = 7, s_4 = 7$, with total satisfaction degree $\lambda = 0.04$;
- for smaller H there was no schedule with a non-negative total satisfaction degree.

It can be seen that for shorter time horizons H not even the minimal satisfaction with the sustainability of the schedule could be achieved. The two schedules that could be determined had a very small satisfaction degree, close to zero. In this example (Table 4) the credibility of the sustainability evaluation was fixed, it could not be improved, which may have influenced this result: in case of the defuzzification method selected (6), low credibility lowered the overall sustainability evaluation.

In the next example (Table 5) of entry data both the sustainability and the credibility could be increased with time, of course at some cost. This cost here covered both certain measures to increase the sustainability and steps to enhance the experts' knowledge.

Table 5. Data on four project tasks without dependencies among them, where the sustainability can be increased as well as the credibility of its evaluation.

$i,\ T_i,\ C_i$	$\widetilde{A}_i^1(s_i), \widetilde{Z}_i^1(s_i)$	$\widetilde{A}_i^2(s_i), \widetilde{Z}_i^2(s_i)$	$\widetilde{A}_i^3(s_i), \widetilde{Z}_i^3(s_i)$
1, 3, 10	$(1 + 0.6s_i, 2 + 0.8s_i,\ 3 + 1s_i)$, $(0.1 + 0.02s_i,\ 0.2 + 0.02s_i,\ 0.3 + 0.02s_i)$	$(0.5 + 2s_i, 1 + 3s_i,\ 1.5 + 3.5s_i)$, $(0.2 + 0.02s_i,\ 0.3 + 0.03s_i,\ 0.4 + 0.05s_i)$	$(1 + s_i, 1 + 1.5s_i,\ 1 + 2s_i)$, $(0.5 + 0.01s_i,\ 0.6 + 0.02s_i,\ 0.7 + 0.03s_i)$
2, 4, 20	$(0.5 + 2s_i, 1 + 3s_i,\ 1.5 + 3.5s_i)$, $(0.2 + 0.02s_i,\ 0.3 + 0.03s_i,\ 0.4 + 0.05s_i)$	$(1 + 0.6s_i, 2 + 0.8s_i,\ 3 + 1s_i)$, $(0.1 + 0.02s_i,\ 0.2 + 0.02s_i,\ 0.3 + 0.02s_i)$	$(5 + s_i, 10 + 1.5s_i,\ 11 + 2s_i)$, $(0.8 + 0.01s_i,\ 0.9 + 0.01s_i,\ 1)$
3, 2, 5	$(1 + s_i, 1 + 1.5s_i,\ 1 + 2s_i)$, $(0.5\ + 0.01s_i,\ 0.6 + 0.02s_i,\ 0.7 + 0.03s_i)$	$(5 + s_i, 10 + 1.5s_i,\ 11 + 2s_i)$, $(0.8 + 0.01s_i,\ 0.9 + 0.01s_i,\ 1)$	$(0.5 + 2s_i, 1 + 3s_i,\ 1.5 + 3.5s_i)$, $(0.2 + 0.02s_i,\ 0.3 + 0.03s_i,\ 0.4 + 0.05s_i)$
4, 1, 4	$(5 + s_i, 10 + 1.5s_i,\ 11 + 2s_i)$, $(0.8 + 0.01s_i,\ 0.9 + 0.01s_i,\ 1)$	$(1 + s_i, 1 + 1.5s_i,\ 1 + 2s_i)$, $(0.5\ + 0.01s_i,\ 0.6 + 0.02s_i,\ 0.7 + 0.03s_i)$	$(1 + 0.6s_i, 2 + 0.8s_i,\ 3 + 1s_i)$, $(0.1 + 0.02s_i,\ 0.2 + 0.02s_i,\ 0.3 + 0.02s_i)$

The following schedules SH were generated:

- H = 10 (weeks): $s_1 = 0.1,\ s_2 = 0.1,\ s_3 = 8,\ \ s_4 = 2$, satisfaction degree $\lambda = 0.48$;
- for H = 9 (weeks): $s_1 = 1.5,\ s_2 = 0.1,\ s_3 = 7, s_4 = 4$, satisfaction degree $\lambda = 0.43$;
- for H = 8 (weeks): $s_1 = 0.1,\ s_2 = 0.1,\ s_3 = 6,\ s_4 = 5.8$, satisfaction degree $\lambda = 0.39$;
- for H = 7 (weeks): $s_1 = 1,\ s_2 = 0.1,\ s_3 = 5,\ s_4 = 6$, satisfaction degree $\lambda = 0.3$.

If the credibility degree could be increased with time, one can see that even for smaller time horizons, schedules with much higher overall satisfaction degrees could be obtained.

It can be observed that in the first case, where the credibility was fixed and could not be improved with time, even at some cost, the prudent attitude (expressed by adopting (5) and (6)) led to a situation where actually no acceptable schedule could be determined. The possibility of increasing the expertise of experts asked for judgment or of using other sources of information made it possible to obtain schedules with a shorter duration, and all these schedules had a substantially higher satisfaction degree than in the previous case.

Table 6 summarizes the experiment described above that was used to test the method.

Table 6. Summary of the experiment used to test the method.

Item	Description
tested institutions	1 public institution
tested projects	1 public project
expert selection	office employees
testing method	post factum experiment
planned extensions	more institutions researched, activity relationships included
practical aspects	real-world schedule that includes sustainability

5. Discussion

The results indicate a positive answer to the question formulated by the staff of the public institution in question: yes, it is potentially possible to incorporate dimensions of sustainability into the scheduling procedure. For the project in question, it is shown that there exists a mathematical scheduling model that takes sustainability into account, turning it into one of the objectives of a multicriterial model.

In the concrete model that was proposed, the solution also indicates an important feature: the overall satisfaction with the schedule. This indicator shows clearly that in the first version of the problem, where the expert credibility was fixed (one can have experts with a certain expertise or competences and cannot gain access to any other ones, even against payment), it is not possible to generate a schedule with an acceptable degree of satisfaction. This means that the whole project has to be rethought and redesigned.

It can be observed that in the solution the activity start times are not always integers. Of course, the fractional starting times would have to be adjusted to specific hours, days or half-days. This step would have to be performed manually (for smaller projects, like the one in question) or by means of an algorithm for bigger projects, which might introduce complexity-related problems. In such a case the use of heuristics should be considered.

The example shows a potential advantage of using models similar to (10) and (11) in the scheduling of public projects: the multicriteria approach makes it possible to find a trade-off between the cost of increasing sustainability and sustainability itself. It should be emphasized that (10) incorporates several (here, three) criteria related to suitability in its numerous dimensions as well as the credibility of sustainability evaluation. As the example in question shows, taking this credibility into account and elaborating methods and ways of increasing it may add value to the performance of public project scheduling. It is important to take into account all the sustainability dimensions and the credibility of experts evaluating them when scheduling projects that are paid with public money and realized for the public good.

6. Conclusions

In this article a general model that can be helpful in scheduling projects for which sustainability is an important issue is proposed. This is the case for many projects today, but in a special way it is true for public institutions, which have to be sustainable—in terms of taking care of people and environment—often to a higher degree than profit-oriented organizations.

The model is based on the following assumptions:

- Sustainability has to be measured in several dimensions (here, three dimensions were chosen, but this number is not a limitation).
- Being sustainable gives rise to cost and may take time—it is usually cheaper and quicker to take less care of people and the environment.
- The evaluation of sustainability is subjective and depends on the experts and information available. It may be a higher or lower quality.

These assumptions can be considered as general: they will apply to most public projects. However, in each case the approach will have to be concretized. In the paper one possible concretization is proposed.

Thus, three dimensions of sustainability are suggested to be considered: human resources consumption, material consumption and negative influence on the environment and local communities. In other cases, other measures of sustainability can be selected. The method proposed takes advantage of expert opinions and takes into account the uncertainty (credibility) level of individual expert decisions. In order to express such a complex construct in a formal way, Z-fuzzy numbers are used—to the authors' knowledge for the first time in the context of sustainability and project scheduling. Other approaches to modeling of subjective opinions and personal features can be used, too.

A model for the case where project activity sustainability can be changed is proposed, at least in some cases, for example, by paying for other activity resources or changing the time or place of activity execution. The problem of experts being asked to evaluate activity sustainability and their lacking knowledge or experience is also taken into account. Also, their opinion quality can be improved, of course at some cost. In the model a compromise solution is determined, where a trade-off between a high sustainability evaluation based on highly credible expert opinions is balanced against the cost of achieving such sustainability and of having access to high quality experts. As a result, a schedule is generated that will be acceptable in terms of "classical" project success criteria (time and cost), but at the same time will be less destructive for humans and the planet.

In real-life models many more constraints would have to be considered (e.g., the typical constraints [30] of project scheduling). What is more, for the sake of simplicity, we assume that the starting times of other tasks cannot affect the sustainability of a given activity, although in real conditions such an assumption may be untrue (e.g., 10 renovations performed at the same time increase the total noise level to an unsupportable value and thus affect the sustainability of each single renovation). Also, more research is needed on the question of how to conduct the questioning of experts and who should evaluate their credibility. One can suppose, taking into account the specificity of public institutions [32], that this process would have to be conducted intuitively, respecting the organizational and cultural context of public institutions [41], and it is clear that this will not be an easy process.

This approach is proposed, but of course not exclusively, for public institutions, whose mission includes being sustainable in all aforementioned dimensions. Its idea was inspired by a Polish municipality, which, like any other similar organization, implements a lot of large infrastructure projects where humans are often overworked and stressed, harmful materials are used, and local inhabitants are disturbed or annoyed. Moreover, the employees of this municipality are open to new project management methods, so there is a chance that further common research can be conducted. In any case, the next research steps will have to consist of case studies conducted in public institutions or other organizations ready and open for new solutions.

Of course, the acceptability of the organization employees is a sine qua non condition for the introduction of sustainable project scheduling, which will not always be easy to achieve. But it is necessary to analyze a full case study in order to validate the proposed approach. Another limitation of the research is the fact that no dependencies between project tasks were considered in the model and, of course, the difficulty to obtain the Z-fuzzy evaluations. Also, it will be necessary to introduce and investigate alternative Z-fuzzy decomposition and aggregation methods (it was mentioned that the method chosen assumes a prudent attitude, which is not always adequate, but the problem of defuzzifying Z-fuzzy numbers in accordance with real-world contexts is still an open one). The model can be also modified with respect to the aggregation method of both objectives. On the whole, the model presented here constitutes an initial proposal of a certain holistic attitude toward the scheduling of public projects: the classical Gantt charts have to be fed with pieces of information other than technically estimated durations of activities and the most technically suited resource requirements. Numerous questions have to be asked about the consequences of the choice of specific activity features, about possible scenarios and other perspectives of defining activities and project objectives in public institutions.

To sum up, hopefully the method will attract some attention to the topic of sustainable project scheduling—a subject matter that clearly is hardly present in the literature. The matter of sustainable project scheduling is especially important for public projects, because public institutions simply have to take care of humans and their planet, and public money should not be spent without taking into account human well-being and the condition of the planet. Obviously, extensive further research on a holistic approach to project defining, scheduling and using expert opinions in this process is still ahead of us.

Author Contributions: Conceptualization, D.K., E.M. and J.S.; methodology, D.K., E.M. and J.S.; software, D.K., E.M. and J.S.; validation, D.K., E.M. and J.S.; formal analysis, D.K., E.M. and J.S.; investigation, D.K., E.M. and J.S.; resources, D.K., E.M. and J.S.; data curation, D.K., E.M. and J.S.; writing—original draft preparation, D.K., E.M. and J.S.; writing—review and editing, D.K., E.M. and J.S.; visualization, D.K., E.M. and J.S.; supervision, D.K., E.M. and J.S.; project administration, D.K., E.M. and J.S.; funding acquisition, D.K., E.M. and J.S. All authors have read and agreed to the published version of the manuscript.

Funding: This research was funded by the National Science Centre (Poland), under Grant 394311, 2017/27/B/HS4/01881: "Selected methods supporting project management, taking into consideration various stakeholder groups and using type-2 fuzzy numbers".

Informed Consent Statement: Not applicable.

Data Availability Statement: The data presented in this study are available on request from the corresponding author. The data are not publicly available due to privacy.

Conflicts of Interest: The authors declare no conflict of interest.

References

1. Hall, S. What Are Public Administration Ethics? Available online: https://bizfluent.com/info-8014104-public-administrationethics.html (accessed on 30 September 2020).
2. Milakovich, M.E.; Gordon, G.J. *Public Administration in America*; Bedford, St. Martin's: New York, NY, USA, 2001.
3. Gasik, S. A Conceptual Model of National Public Projects Implementation Systems. *PM World J.* **2016**, *5*, 1–13.
4. Bartošíková, R.; Pitrova, K.; Taraba, P. Application of Project Management in Public Administration. *Econ. Manag.* **2013**, *7*, 15–19.
5. Obradović, V. Contemporary Trends in the Public Sector Project Management. *Eur. Proj. Manag. J.* **2018**, *8*, 52–56. [CrossRef]
6. Volden, G.H. Public Project Success as Seen in a Broad Perspective: Lessons from a Meta-Evaluation of 20 Infrastructure Projects in Norway. *Eval. Program Plan.* **2018**, *69*, 109–117. [CrossRef] [PubMed]
7. Senouci, A.; Alaa, I.; Neil, E. Time Delay and Cost Overrun in Qatari Public Construction Projects. *Procedia Eng.* **2016**, *164*, 368–375. [CrossRef]
8. Carvalho, M.; Rabechini, R. Can Project Sustainability Management Impact Project Success? An Empirical Study Applying a Contingent Approach. *Int. J. Proj. Manag.* **2017**, *35*, 1120–1132. [CrossRef]
9. Stiglbauer, M. Strategic Stakeholder Management by Corporate Social Responsibility: Some Conceptual Thoughts. *Risk Gov. Control: Financ. Mark. Inst.* **2011**, *1*, 45–55. [CrossRef]
10. APM. *APM Body of Knowledge Definitions*; Association for Project Management: Princes Risborough, UK, 2012.
11. Kuchta, D.; Marchwicka, E. Sustainability Oriented Scheduling Procedure for Public Projects, in Intelligent and Fuzzy Techniques: Smart and Innovative Solutions. In Proceedings of the INFUS 2020 Conference, Istanbul, Turkey, 21–23 July 2020; pp. 1548–1554.
12. Li, H.; Xu, Z.; Wei, W. Bi-Objective Scheduling Optimization for Discrete Time/Cost Trade-Off in Projects. *Sustainability* **2018**, *10*, 2802. [CrossRef]
13. Jiang, T.; Zhang, C.; Zhu, H.; Deng, G. Energy-Efficient Scheduling for a Job Shop Using Grey Wolf Optimization Algorithm with Double-Searching Mode. *Math. Probl. Eng.* **2018**, *3*, 1–12. [CrossRef]
14. Lao, W.; Wang, T. A Novel Collaborative Optimization Model for Job Shop Production–Delivery Considering Time Window and Carbon Emission. *Sustainability* **2019**, *11*, 2781. [CrossRef]
15. Zhang, R. Sustainable Scheduling of Cloth Production Processes by Multi-Objective Genetic Algorithm with Tabu-Enhanced Local Search. *Sustainability* **2017**, *9*, 1754. [CrossRef]
16. Yu, M.; Zhu, F.; Yang, X.; Wang, L.; Sun, X. Integrating Sustainability into Construction Engineering Projects: Perspective of Sustainable Project Planning. *Sustainability* **2018**, *10*, 784. [CrossRef]
17. Wang, W.; Chen, J.; Liu, Q.; Guo, Z. Green Project Planning with Realistic Multi-Objective Consideration in Developing Sustainable Port. *Sustainability* **2018**, *10*, 2385. [CrossRef]
18. Khun-anod, K.; Limsawasd, C. Pre-project Planning Process Study of Green Building Construction Projects in Thailand. *Eng. J.* **2019**, *23*, 68–81. [CrossRef]

19. Fathalizadeh, A.; Ghoddousi, P.; Javid, A.; Hosseini, M. Integrating sustainability into construction project management: Barriers in developing countries. In Proceedings of the 13th International Conference on Modern Building Materials, Structures and Techniques, Vilnus, Lithuania, 16–17 May 2019.
20. Li, L.; Fan, F.; Ma, L.; Tang, Z. Energy Utilization Evaluation of Carbon Performance in Public Projects by FAHP and Cloud Model. *Sustainability* **2016**, *8*, 630. [CrossRef]
21. Lin, H.; Jin, R.; Ning, X.; Skitmore, M.; Zhang, T. Prioritizing the Sustainability Objectives of Major Public Projects in the Guangdong–Hong Kong–Macao Greater Bay Area. *Sustainability* **2018**, *10*, 4110. [CrossRef]
22. Zhang, Y. Construction of Bid Evaluation Index System in Government Public Project Green Procurement in China Based on D-S Evidence Theory. *Sustainability* **2020**, *12*, 651. [CrossRef]
23. Rosasco, P.; Sdino, L. Infrastructures and Sustainability: An Estimation Model for a New Highway Near Genoa. *Sustainability* **2020**, *12*, 5051. [CrossRef]
24. Butkovic, D.; Sjekavica, M. Infrastructure projects classification—Sustainable development perspective. In Proceedings of the 5th IPMA SENET Project Management Conference, Belgrade, Serbia, 19–21 May 2019.
25. Sierra, L.A.; Yepes, V.; Pellicer, E. Assessing the social sustainability contribution of an infrastructure project under conditions of uncertainty. *Environ. Impact Assess. Rev.* **2017**, *67*, 61–72. [CrossRef]
26. Kivilä, J.; Martinsuo, M.; Vuorinen, L. Sustainable project management through project control in infrastructure projects. *Int. J. Proj. Manag.* **2017**, *35*, 1167–1183. [CrossRef]
27. Yeganeh, T.Y.; Zegordi, S.H. A multi-objective optimization approach to project scheduling with resiliency criteria under uncertain activity duration. *Ann. Oper. Res.* **2020**, *285*, 161–196. [CrossRef]
28. Deckro, R.; Herbert, J. Resource constrained project crashing. *Omega* **1989**, *17*, 69–79. [CrossRef]
29. Ballestin, F.; Valls, V.; Quintanilla, S. Scheduling projects with limited number of preemptions. *Comput. Oper. Res.* **2009**, *36*, 2913–2925. [CrossRef]
30. Herroelen, W.; De Reyck, B.; Demeulemeester, E. Resource-constrained project scheduling: A survey of recent developments. *Comput. Oper. Res.* **1998**, *25*, 279–302. [CrossRef]
31. Gasik, S. Are public projects different than projects in other sectors? Preliminary results of empirical research. *Procedia Comput. Sci.* **2016**, *100*, 399–406. [CrossRef]
32. Sakawa, M.; Nishizaki, I. Max-min solutions for fuzzy multiobjective matrix games. *Fuzzy Sets Syst.* **1994**, *67*, 53–69. [CrossRef]
33. Zadeh, L.A. A Note on Z-Numbers. *Inf. Sci.* **2011**, *181*, 2923–2932. [CrossRef]
34. Aliev, R.A.; Alizadeh, A.V.; Huseynov, O.H. An Introduction to the Arithmetic of Z-Numbers by Using Horizontal Membership Functions. *Procedia Comput. Sci.* **2017**, *120*, 349–356. [CrossRef]
35. Gündoğdu, F.K.; Kahraman, C. Spherical Fuzzy Sets and Spherical Fuzzy TOPSIS Method. *J. Intell. Fuzzy Syst.* **2019**, *36*, 337–352. [CrossRef]
36. Kang, B.; Wei, D.; Li, Y.; Deng, Y. A Method of Converting Z-Number to Classical Fuzzy Number. *J. Inf. Comput. Sci.* **2012**, *9*, 703–709. [CrossRef]
37. Mohamad, D.; Shaharani, S.A.; Kamis, N.H. A Z-Number-Based Decision Making Procedure with Ranking Fuzzy Numbers Method. In Proceedings of the 3rd International Conference on Quantitative Sciences and Its Applications, Langkawi Kedah, Malaysia, 12–14 August 2014. [CrossRef]
38. Kaufmann, A.; Gupta, M.M. *Introduction to Fuzzy Arithmetic: Theory and Applications*; Van Nostrand Reinhold Company: New York, NY, USA, 1985.
39. Marchwicka, E. Project Risk Management Method. Ph.D. Dissertation, Wroclaw University of Science and Management, Wroclaw, Poland, 2012.
40. Kowalski, K.; Król, J.; Radziszewski, P.; Barrasa, R.; Blanco, V.; Pérez, D.; Viñas, V.; Brijsse, Y.; Frosch, M.; Le, D.M. Eco-Friendly Materials for a New Concept of Asphalt Pavement. *Transp. Res. Procedia* **2016**, *14*, 3582–3591. [CrossRef]
41. Rus, M.; Rusu, D.O. The Organizational Culture in Public and Private Institutions. *Procedia Soc. Behav. Sci.* **2015**, *187*, 565–569. [CrossRef]

sustainability MDPI

Article

A Systemic Approach for Sustainability Implementation Planning at the Local Level by SDG Target Prioritization: The Case of Quebec City

David Tremblay [1,*], Sabine Gowsy [2], Olivier Riffon [1], Jean-François Boucher [1], Samuel Dubé [2] and Claude Villeneuve [1]

1 Département des Sciences Fondamentales, Université du Québec à Chicoutimi, Chicoutimi, QC G7H 2B1, Canada; olivier_riffon@uqac.ca (O.R.); jean-francois_boucher@uqac.ca (J.-F.B.); claude_villeneuve@uqac.ca (C.V.)
2 Ville de Québec, Québec, QC G1R 4A2, Canada; Sabine.Gowsy@ville.quebec.qc.ca (S.G.); Samuel.Dube@ville.quebec.qc.ca (S.D.)
* Correspondence: david1_tremblay@uqac.ca

Abstract: The success of the 2030 Agenda hinges on mobilization at the local level. The localization of sustainable development goals (SDGs) and their targets involves adapting them to local contexts. This case study of Quebec City, Canada, illustrates how the use of a systemic sustainability analysis tool can help integrate SDGs in the building of a sustainable development strategy at the local level. Our approach focuses on the use of an SDG target prioritization grid (SDGT-PG) and begins with the mobilization and training of a group of officers representing various city services. We first used an original text-mining framework to evaluate SDG integration within existing strategic documents published by the city. The result provides a portrait of existing contributions to SDG targets and identifies potential synergies and trade-offs between services and existing policies. A citywide prioritization workshop was held to assess the relative importance of SDG targets for the city. Priorities were then identified by combining the importance of the targets as viewed by stakeholders, the current level of achievement of SDG targets as determined by the analysis of existing documents, and the jurisdiction and responsibilities given to Quebec City in regard to federal and provincial legislation. We identified the main focus areas and related SDG targets. Furthermore, we observed whether actions needed to be consolidated or new actions needed to be implemented. The identification of synergies and trade-offs within the city service actions provides information on the links to be made between the different municipal services and calls for partnerships with other organizations. The use of the SDGT-PG allows the vertical and horizontal integration of the SDG targets and demonstrates how participation and inclusion facilitate stakeholders' appropriation of the applied sustainable development strategy.

Keywords: 2030 Agenda; sustainable development goals (SDGs); systemic sustainability analysis; SDG targets prioritization

Citation: Tremblay, D.; Gowsy, S.; Riffon, O.; Boucher, J.-F.; Dubé, S.; Villeneuve, C. A Systemic Approach for Sustainability Implementation Planning at the Local Level by SDG Target Prioritization: The Case of Quebec City. *Sustainability* **2021**, *13*, 2520. https://doi.org/10.3390/su13052520

Academic Editors: Margarita Martinez-Nuñez and Mª Pilar Latorre-Martínez

Received: 23 January 2021
Accepted: 22 February 2021
Published: 26 February 2021

1. Introduction

In 2015, members of the United Nations unanimously adopted the 2030 Agenda for Sustainable Development [1]. The 17 Sustainable Development Goals (SDG) and 169 targets represent a global framework to guide the implementation of sustainable development (SD) by 2030 [2,3]. While the SDGs and targets were first designed for the global level [4], the 2030 Agenda is a universal program that applies to all governments and actors, regardless of their level of intervention [1,5]. Because cities represent the level of government closest to the population [6], they have the capacity to intervene quickly and concretely, according to the powers assigned to them, [7–9], and they are considered essential actors for sustainability [4]. Furthermore, urbanization is accelerating globally, and 68 percent of the world's population is expected to live in an urban area by 2050 [10]. Moreover, all human activities at the city level affect the economy, the environment, people, culture, governance, etc. [4,8,9,11–13].

Implementing the 2030 Agenda and achieving the SDGs require an integrated and systemic approach [1,14,15]. The principle of integration applies (i) horizontally between different policy areas, (ii) vertically from global to national to local levels, and (iii) territorially between local governments [6]. SDG implementation at the local level is termed "localization". In the context of implementing the 2030 Agenda, SDG localization refers to "the process of defining, implementing, and monitoring strategies at the local level for achieving global, national, and subnational SDGs and targets" [16]. Although there are numerous approaches and tools at the national level [11], the localization of SDGs requires documented [3] approaches adapted to applying sustainable development at the local level [17]. Scientific research on cities and the SDGs has increased [7,18,19]; nonetheless, there remains a knowledge gap in regard to how best to implement SDGs at the local level [3,11].

Approaches and tools dedicated to SDG localization involve some critical elements. Although the 17 SDGs and their 169 targets are set as "universal and indivisible," they must be applied in line with the realities, capacities, levels of development, and priorities specific to national or local contexts [1,20]. Cities face various challenges when implementing SDGs. These challenges include contextualizing their approaches to the specific environmental, economic, social, political, and cultural conditions [7,21,22]. This contextualization involves adapting SDG content and their targets to make them locally relevant [3,6] while maintaining integrated and systemic thinking to keep a holistic perspective of the system as a whole [1,14,15,23]. The systemic approach implies horizontal, vertical, and territorial integration [24]. Horizontally, integration aims to maximize synergies and diminish trade-offs [4,24–27]. Vertically, at the local level, integration involves the principle of subsidiarity [7,11,28] and a clear understanding of the division of powers between the various levels of government [11,14,28]. In addition to implementing SDGs, invoking these elements will ensure policy coherence and integrated multilevel governance [6,7,28].

The issue of prioritizing SDG targets is critical for local authorities, as the needs are multiple and urgent; yet, there are often limited capacities and resources. Prioritization is a complex exercise that combines assessing the importance of a target and determining its level of achievement at a specific time, for a given territory [2]. The integrated approach of the SDGs induces a paradigm shift in the elaboration of development plans and strategies at all levels. Today, it is a matter of contributing to the achievement of a global and shared vision. Thus, SDGs and their targets may constitute a normative framework. Given the inability and, most likely, the irrelevancy for cities to implement all the targets, it is essential to use approaches and tools to determine priorities [4,29].

A growing number of studies are looking at a systemic approach for prioritizing and implementing the SDGs [12,27,30,31]; however, very few tools for SDG and target prioritization are documented in the scientific literature. Among papers presenting tools, three include a set of criteria for establishing priorities. First, Allen et al. [30] proposed a multicriteria analysis for prioritizing SDG targets through the use of three main criteria:

1. Level of urgency: identifying historical trends and comparing current baseline values against global benchmarks;
2. Systemic impact: identifying interlinkages between SDG targets, evaluated against a semi-quantitative cross-impact matrix assessment and network analysis;
3. Policy gap: assessing how SDGs align with existing strategies.

The Stakeholder Forum [20] submitted a method of analysis addressed to developed countries to assist them in identifying the goals and targets representing the biggest transformational challenges. They proposed three criteria:

1. Applicability: evaluating the relevance of the goal/target;
2. Implementability: assessing the reality of attaining the goal/target within the time frame;

3. Transformationalism: determining whether the achievement of the goal/target requires new and additional policies beyond those currently in place.

Finally, the Sustainable Development Solution Network (SDSN) [6] proposed broad guidelines to define local SDG targets:

1. Targets should be relevant and achievable;
2. Targets should correspond to local government mandates;
3. Priorities should be established on the basis of development gaps.

This paper presents the case study of Quebec City (Quebec, Canada) using an original and adapted systemic sustainability analysis tool. Quebec City is located in the province of Quebec, Canada. Canada is a federal state where responsibilities are shared between federal and provincial jurisdictions. Local governments, such as cities, are a provincial responsibility; however, as issues related to sustainable development touch multiple jurisdictions, our analysis includes the federal, provincial, and local (city) levels. This approach aims to bolster the implementation of the SDGs at the local level and integrates the key elements of contextualization, adaptation, systemic thinking, subsidiarity, and policy coherence.

The method focuses mainly on the use of the SDG target prioritization grid (SDGT-PG), a participatory prioritization tool for SDG targets applicable at local, national, and regional scales. The SDGT-PG methodology is inspired by the sustainable development analytical grid (SDAG) used at local and national levels since 1988 [32]. The SDAG methodology, which emphasizes participatory processes and scientific robustness, was developed in a partnership between academics (Université du Québec à Chicoutimi, Canada), an international organization (Organisation internationale de la Francophonie), and an international consulting firm (GlobalShift Institute Ltd., Quebec City, QC, Canada). This approach was tested in both developed and developing countries at national and local levels. Burkina Faso, Benin, Niger, and Togo refer to the use of the SDGT-PG as a prioritization tool in their Voluntary National Reviews presented at the United Nations High-level Political Forum on Sustainable Development. In this study, we test our central hypothesis that the use of the SDGT-PG allows the vertical and horizontal integration of the SDG targets. We also demonstrate how participation and inclusion permit stakeholder appropriation.

2. Materials and Methods

The applied approach is an iterative process (Figure 1) inspired by two analytical tools: the sustainable development analytical grid (SDAG) [32] and the rapid integrated assessment (RIA) [33]. Our approach also makes use of best-known practices and guidelines [6,20,30]. The different advancements of this action–research process, in co-construction with Quebec City officials, aimed to generate data and information related to the three main SDGT-PG criteria being evaluated:

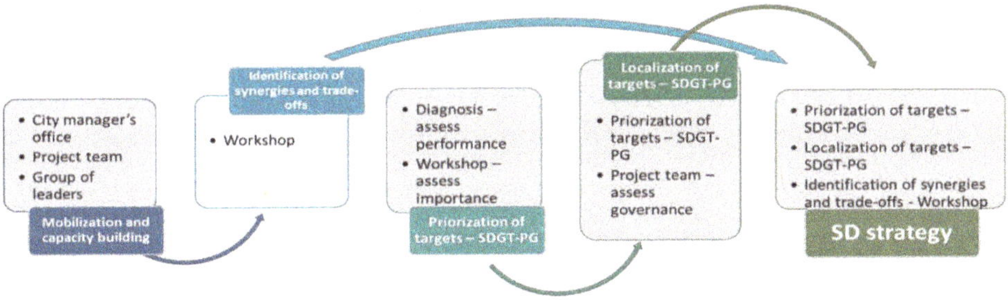

Figure 1. Stages of the applied approach.

1. Performance: Relying on SDG target indicators, what is the current level of achievement of the targets?

2. Importance: Given the specific context of the city, what is the significance level of the targets?
3. Governance: Knowing the constitutional division of powers, what level of governance (from national to local) holds the power and responsibilities associated with the targets?

2.1. Mobilization and Capacity Building

To eliminate silos and apply a systemic approach, we took the initial step of forming a group of leaders. We mobilized 27 leaders from 19 Quebec City's administrative units. To be selected, a leader needed to represent one of the main administrative units and embody the concepts of sustainability through their values, interests, personal, and/or professional activities. The selected leaders were readily available and also had the authorization of their respective managers. The 27 leaders included 17 advisers, 6 managers, 2 engineers, 1 analyst, and 1 police officer. The leaders comprised 15 men and 12 women. The city manager's office coordinated the project. The project team, established by the office, worked in partnership with our research team to organize and structure the complete process.

A series of workshops was co-managed by the partners. The main objective of these workshops was to raise awareness about the SDGs [6]. The specific objectives were to:

1. Understand the concepts of sustainable development;
2. Become familiar with the 2030 Agenda and the SDGs;
3. Understand systemic sustainability analysis;
4. Share practices between city services;
5. Identify potential synergies and trade-offs;
6. Prepare the prioritization activity;
7. Produce and validate proposals for the sustainable development strategy.

2.2. Diagnosis–Performance

The internalization of the SDGs requires identifying those already-existing actions that can be linked to the SDGs [6,14]. To carry out this diagnosis, we analyzed 89 strategic documents produced by Quebec City (A list of the analyzed documents is available in the Supplementary Materials). We aimed to:

1. Identify already-implemented SD initiatives;
2. Align the identified SD initiatives with SDG targets;
3. Assess the potential achievement of the SDG targets (document the performance criteria of the SDGT-PG).

We identified the initiatives using WordStat, a content-analysis and text-mining software within ProSuite (Provalis Research, Montreal, QC, Canada), a collection of integrated text-analysis tools [34]. For our analyses, we developed a specific dictionary linked to the content of the 2030 Agenda. Our dictionary included 1602 expressions found in the labels of the SDGs, SDG targets, and SDG indicators (The full dictionary is available in the Supplementary Materials).

First, we prepared each document for analysis by removing figures, hyphens, brackets, and braces. We imported the strategic documents into QDA Miner, a qualitative data analysis component of ProSuite. We conducted content analysis using WordStat for each document separately. Each SDG target represented a category in the dictionary. To transform textual data into keywords or content categories, we used a lemmatization substitution process. Lemmatization is a "process by which various forms of words are reduced to a more limited number of canonical forms like conversion of plurals to singulars and past tense verbs to present tense verbs" [35].

For each occurrence identified by the software, an expert in charge of the processing validated the result to retain only relevant occurrences. These retained occurrences were then classified within a matrix where they were associated with corresponding targets. The matrix is based on the rapid integrated assessment developed by the United Nations Development Programme [33] (Table 1).

Table 1. Excerpts of the matrix of links between the SDG targets (6.1 to 6.6) and the analyzed Quebec City strategic documents. The X mark indicates an occurrence between a strategic document and an SDG target (The full matrix is available in the Supplementary Materials). The title of each strategic document has been translated by the authors from the original French title.

City Service	Strategic Document	SDG 6—Clean Water and Sanitation						
		6.1	6.2	6.3	6.4	6.5	6.6	Total
Water management/Provide a sustainable and healthy environment/Stimulate the development of the city	Vision for the development of agricultural and food-processing activities							0
Water management	Summary of actions to protect Lake St. Charles and water intake	x		x	x		x	4
Water management	Environmental and water regulations of the Quebec metropolitan community	x					x	2
Water management	2018 annual report on drinking water quality	x		x				2
Water management	Drinking water regulations for the Quebec City agglomeration	x		x	x			3

We processed the data from the matrix to obtain a portrait of the SDG target coverage by the existing strategic documents and to group those documents influencing the same targets. Through identifying occurrences between the strategic documents and the SDG targets, we could assess, in light of the identified actions, the degree to which SDG targets had been achieved (performance). We assessed performance on a four-level scale:

1. The target was not at all achieved;
2. The target was partially achieved: there is much room for improvement, although some results are visible;
3. The target is in the process of being achieved: improvements remain possible;
4. The target has been achieved.

We automatically assigned a performance score of 1 to targets having no occurrences. We awarded a performance score of 2 or 3 when occurrences existed between the city's strategic documents and a target. Our assessment of performance varied in accordance with the number of strategic documents associated with a target and the quality of actions mentioned in those documents. We never assigned a maximum score of 4, as we found no indicator, with verified metrics, for which we could attribute this performance.

2.3. Identification of Synergies and Trade-Offs

To apply systems thinking, we organized a workshop with the aim of identifying the potential interactions between the applied activities within the various city services. The research team identified themes for the 107 operational targets on the basis of target content and their indicators. SDG targets fall into two categories: "operational" and "means of implementation" (MoI). Operational targets relate to those to be achieved, whereas MoI refers to conditions that help attain targets [36]. MoI includes the mobilization of financial resources, technology development and transfer, capacity building, inclusive and equitable trade, regional integration, and the creation of a national environment conducive to the implementation of the sustainable development agenda [1]. MoI targets apply to national competences, which deviate from those at the local level. We therefore discarded MoI targets and SDG 17 [31,37].

As an initial step, the 27 leaders identified areas of activity undertaken both inside and outside of their administrative units and associated these areas with the themes of the SDG targets. Then, they identified potential interactions with other SDG targets. All interactions are directional; hence, for each interaction, there is a source target and an impacted target. In the case of bidirectional interactions, we used two-directional interactions, reversing sources and impact. City leaders characterized interactions as synergies or trade-offs. A member of the research team validated each of the interactions and completed the exercise by adding interactions and adjusting some interactions associated erroneously with targets. Once the validation was completed, we analyzed the synergies and trade-offs by SDG and target. We used a cross-impact matrix [27,30,31,38] where the weight given to an interaction corresponds to the number of activities associated with the interaction.

2.4. Importance and Prioritization

Localizing the SDGs requires adapting the global reference framework to ensure that it is relevant to the local context [3,6]. This contextualization of the 2030 Agenda will promote ownership and mobilization of stakeholders [14,39]. To increase the understanding of the SDGs and their targets for the stakeholders in our study, we adapted the wording of the targets without changing the original meaning. The target labels were adjusted to change references from a national scale to a local scale. For example, Target 1.2: "By 2030, reduce at least by half the proportion of men, women and children of all ages living in poverty in all its dimensions according to **national** definitions" becomes "By 2030, reduce at least by half the proportion of men, women and children of all ages living in poverty in all its dimensions according to the **city** definitions."

Targets adapted to the local context were prioritized during a workshop that brought together 182 city employees. The employees occupied positions at various levels throughout the city services. We sampled city employees to ensure the representativeness of employment sectors, age group, gender, and workplace location. The selected employees did not need any particular skills or knowledge to participate in the workshop. Prioritization is an essential stage in identifying the relevant actions to be implemented at the local level [4].

Twenty-four tables of seven to eight employees, separated across two three-hour sessions, discussed the level of importance of SDG targets in the context of Quebec City. The table composition was predetermined to maximize diversity. Each table weighted the targets of three to four SDGs, for a total of 21 to 22 targets per table. An animator facilitated the discussion, while a second person recorded notes on a prepared canvas. Each participant was provided access to a set of four cards, representing the four levels of importance, to help them judge the importance of a target:

N/A: Not applicable;

1: Unimportant: Not important and not a priority;

2: Important: Priority in the medium to long term;

3: Essential: Priority in the short term.

For each target, (1) the animator announced and explained the appropriate target; (2) the employees expressed their views on the level of importance to be given to the target; (3) the animator initiated a dialogue in regard to the employees' justifications for this importance; and (4) the employees then expressed their final scoring for the importance of the target after these discussions.

We recorded both employee assessments of importance (before and after discussions), and we noted the employees' justifications. To define the final level of importance of the targets to be entered in the SDGT-PG, we averaged (rounded to the nearest unit) the importance score of the final results.

Using the SDGT-PG, we produced a priority index for each target. The more participants that deemed a target to be significant and the poorer the target's performance, the greater the priority given to the target in question. The priority level corresponds to the table shown in Figure 2.

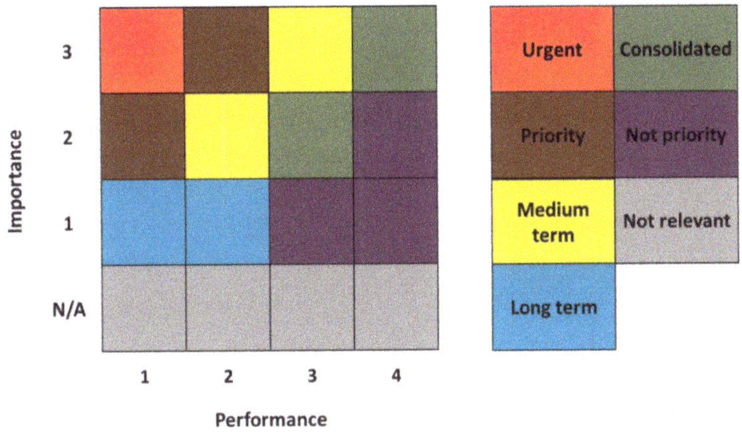

Figure 2. Prioritization index grid in which an urgent target requires immediate intervention: a priority target should be addressed within a three-year horizon, a medium-term target should be addressed within seven years, a long-term target should be addressed within a 10-to-15-year period, and a target to be consolidated requires interventions that make it possible to maintain the current level of performance. The other priority levels do not require specific actions.

2.5. Governance

This stage aims to determine, for each target, the level of governance, from local to national, legally responsible for implementing actions required to achieve the target. In our case, the national level includes both the provincial and federal governments. The project team evaluated the target with reference to legislation at the Quebec (provincial) and Canada (federal) levels. To identify the governance level, we classified governance on a scale from 1 to 4:

1. Exclusive responsibility of the local level. The local level has complete authority to act on this target.
2. Responsibility shared between the local and national levels. The local level has a certain authority to act on this target; however, these competencies are also shared with the national level.
3. National-level responsibility supported by the local level. The national level has the main responsibilities necessary to act on this target; however, it can delegate to the local level for implementing an action. The local level has a certain authority for ensuring action on the ground, but it does not hold decision-making power.
4. Exclusive national-level responsibility. The national level has the full authority to act on this target. The local level does not have the authority to intervene, although it can sometimes influence priorities through representations at the national level.

2.6. Localization

The final information produced in the SDGT-PG considered the role of the different levels of governance in implementing initiatives; this governance level can affect whether a target can be achieved. Combining the priority level (Figure 2) and the governance assessment, we could determine what should be considered by local and national planners and what targets can be achieved jointly, in some form of multilevel governance (Table 2).

Table 2. Initiatives to be undertaken according to the level of priority and our governance assessment. These are proposals aimed at local (Quebec City) and national (Quebec, Canada) levels.

		Level of Governance			
		Local	**Shared**	**National Supported by Local**	**National**
Urgent	**Local**	Implementation of actions	Implementation of actions	Search for opportunities	Advocacy at the appropriate governance level
	National	Financial, technical, human support	Financial, technical, human support	Actions considering local characteristics	Implementation of actions
Priority	**Local**	Implementation of actions	Implementation of actions	Search for opportunities	Search for opportunities
	National	Direct support at the local level	Direct support at the local level	Actions considering local characteristics	Implementation of actions
Medium term	**Local**	Implementation of actions	Implementation of actions	None	None
	National	Financial, technical, human support	Direct support at the local level	Actions considering local characteristics	Implementation of actions
Long term	**Local**	Search for opportunities	Search for opportunities	None	None
	National	Financial, technical, human support	Long-term planning	Long-term planning	Long-term planning
Consolidated	**Local**	Consolidation of actions already implemented	Consolidation of actions already implemented	Consolidation of actions already implemented	None
	National	None	Collaboration with the local level	Collaboration with the local level	Consolidation of actions already implemented
Non-priority	**Local**	Consolidation of actions already implemented	Consolidation of actions already implemented	Consolidation of actions already implemented	None
	National	None	Collaboration with the local level	Collaboration with the local level	Consolidation of actions already implemented
Not relevant	**Local**	None	None	None	None
	National	None	None	None	None

3. Results

3.1. Performance Assessment

Our diagnosis aimed to assess the potential achievement of the SDG targets by identifying those already considered within the city strategic documents and by assessing the degree to which they had been achieved. We found that the city documents dealt with 85 targets, representing 50% of the total SDG targets (Table 3). We noted a difference in coverage between the operational targets (71%) and the MoI (15%). No document covered the targets of SDG 14; however, five SDGs had 100% of their operational targets considered by at least one strategic document.

Table 3. SDG and targets covered by the analyzed Quebec City strategic documents.

SDG	Results Target (*n*)	Results Target Covered (*n*)	Results Target Covered (%)	MoI Targets (*n*)	MoI Targets Covered (*n*)	MoI Targets Covered (%)	Total Targets (*n*)	Total Targets Covered (*n*)	Total Targets Covered (%)
1	5	4	80	2	0	0	7	4	57
2	5	2	40	3	0	0	8	2	25
3	9	5	56	4	1	25	13	6	46
4	7	4	57	3	1	33	10	5	50
5	6	4	67	3	0	0	9	4	44
6	6	6	100	2	0	0	8	6	75
7	3	3	100	2	0	0	5	3	60
8	10	9	90	2	0	0	12	9	75
9	5	5	100	3	0	0	8	5	63
10	7	6	86	3	0	0	10	6	60
11	7	7	100	3	1	33	10	8	80
12	8	6	75	3	1	33	11	7	64
13	3	3	100	2	0	0	5	3	60
14	7	0	0	3	0	0	10	0	0
15	9	6	67	3	1	33	12	7	58
16	10	6	60	2	1	50	12	7	58
17	x	x	x	19	3	16	19	3	16
Total	107	76	71	62	9	15	169	85	50

Note: SDG 1: No poverty; SDG 2: Zero hunger; SDG 3: Good health and well-being; SDG 4: Quality education; SDG 5: Gender equality; SDG 6: Clean water and sanitation; SDG 7: Affordable and clean energy; SDG 8: Decent work and economic growth; SDG 9: Industry, innovation and infrastructure; SDG 10: Reduced inequalities; SDG 11: Sustainable cities and communities; SDG 12: Responsible consumption and production; SDG 13: Climate action; SDG 14: Life below water; SDG 15: Life on land; SDG 16: Peace, justice, and strong institutions; SDG 17: Partnerships for the goals.

Table 3 illustrates that 84 targets (50%) are not covered, including 53 MoI targets (of a possible 62). SDG targets are also unequally covered by the strategic documents (Figure 3). Dividing the analyzed documents into blocks of five (for clarity's sake in Figure 3), we noted that 46 targets were represented only within 1 to 5 documents, 15 targets were found in 6 to 10 documents, 12 targets were covered by 11 to 15 documents, only 4 targets were found in 16 to 20 documents, and 8 targets were linked to 20 or more documents. Thus, most documents covered only a limited number of SDG targets.

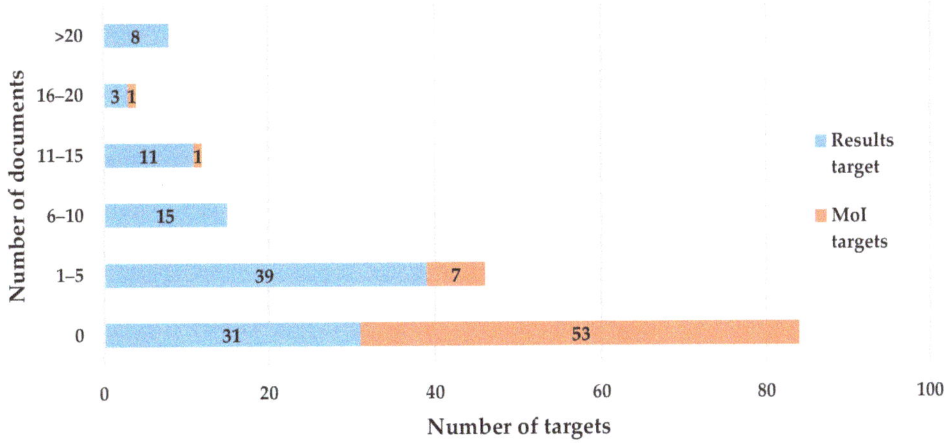

Figure 3. Number of targets covered by the strategic documents.

Of the eight most covered targets, three come from SDG 11 (Sustainable cities and communities), two from SDG 9 (Industry, innovation, and infrastructure) and SDG 16 (Peace, justice, and strong institutions), and one from SDG 3 (Good health and well-being).

We used the potential coverage of SDG targets to assess performances in the SDGT-PG. Except for SDG 2 and 14, at least 60% of the targets showed some performance in terms of potentially achieving the target (Figure 4). No target could be labeled as "achieved" because we found no indicators confirming this level of performance. Targets of SDG 6 (Clean water and sanitation) and 11 (Sustainable cities and communities) showed more than 80% of the targets had a performance ranking of 3 "in the process of being achieved" (Figure 4).

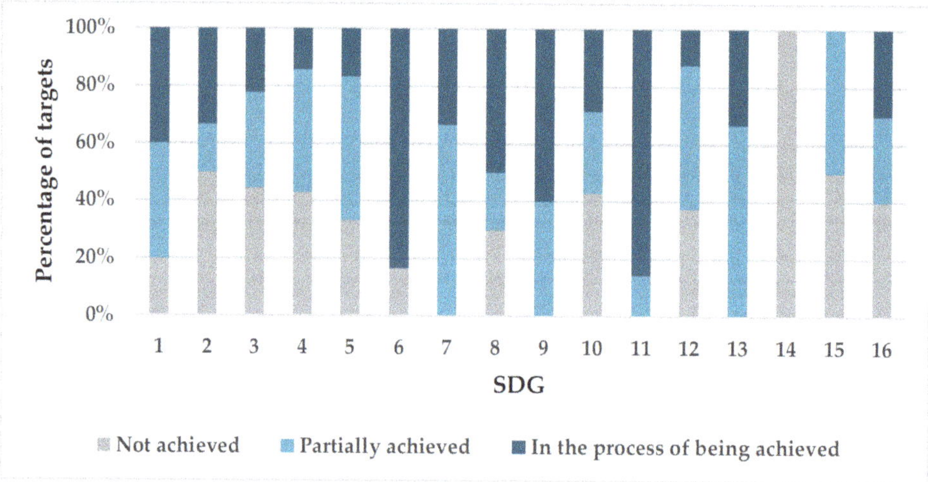

Figure 4. Distribution of the performance of the SDG targets found within the 89 Quebec City strategic documents.

3.2. *Synergies and Trade-Offs*

The city officers and our research team identified 687 potential interactions, including 638 synergies and 49 trade-offs. These interactions involve 86 targets and 16 SDGs. Table 4 shows the targets that most influenced other targets on the basis of the number of times they are the source of interaction. The table also reveals the targets most influenced by other targets according to the number of times they are impacted by other targets. All these targets exhibited positive and negative interactions. Among all the analyzed interactions, the most influencing and most influenced targets were all strongly positive (Table 4).

Table 4. Most influencing and most influenced targets. Numbers in parentheses identify the number of positive/negative interactions, respectively, in which the targets are involved. (The full cross-impact matrix is available in the Supplementary Materials).

Most Influencing Targets	Most Influenced Targets
9.5—Scientific research, technological capabilities, and innovation (50/4)	3.4—Non-communicable disease, mental health, well-being (54/2)
7.3—Energy efficiency (41/4)	10.2—Empowerment; social, economic, and political inclusion (45/5)
11.3—Sustainable urbanization, participatory planning and management (42/3)	15.1—Terrestrial and inland freshwater ecosystems (25/4)
9.4—Upgrade infrastructure; resource-use efficiency; clean technologies and industrial processes (26/4)	11.6—Environmental impact of cities; air quality; waste management (23/1)
10.2—Empowerment; social, economic, and political inclusion (28/2)	2.1—Hunger; nutritious and sufficient food (20/3)
12.8—Information and awareness for sustainable development (29/1)	13.2—Climate change measures (20/1)

Influencing targets came from various SDGs. The exception was SDG 9, for which two targets were found in the five most influencing targets (Table 4). In the context of applying the SDGs at the municipal level, we expected and noted Target 11.3 (Sustainable urbanization; participatory and integrated planning and management) to be one of the most (the third) influential targets (Table 4).

The most impacted targets came from six SDGs. The most impacted target is 3.4 (Non-communicable diseases, mental health, and well-being). We found a single target (10.2. Empowerment; social, economic, and political inclusion) in both the most influencing and the most influenced targets (Table 4).

The targets included in Table 4 have the highest positive results. In terms of negative impacts, targets 7.3, 9.4, 9.5, and 15.1 most negatively affected other targets (sum: −4). The most often negatively impacted targets were targets 8.1 (sum: −7) and 10.2 (sum: −5). The highest positive interaction was between targets 11.3 and 10.2 (sum: 5). We observed the highest negative results (sum: −2) for interactions between targets 9.4 and 3.4, between 9.5 and 10.3, and between 15.1 and 11.1.

At the SDG level, targets from SDG 9 (sum: 100) and SDG 11 (sum: 99) had the highest positive net influence on other targets (Figure 5). Targets SDG 3 (sum: 76), SDG 11 (sum: 69), SDG 8 (sum: 53), SDG 10 (sum: 52), and SDG 12 (sum: 51) were most often impacted by other targets, according to their net influence. We noted that 94.7% of net influence was positive when excluding absent relationships. The greatest net-positive influence was from SDG 11 toward SDG 10 (sum: 23). The net influence of SDG 15 toward SDG 8 was the most negative (sum: −3) (Figure 5).

	SDG 1	SDG 2	SDG 3	SDG 4	SDG 5	SDG 6	SDG 7	SDG 8	SDG 9	SDG 10	SDG 11	SDG 12	SDG 13	SDG 14	SDG 15	SDG 16	Sum
SDG 1		3	5	2	6			1	-1	7	3				1		27
SDG 2	3	2	3					-1		-1	-1	2			1		8
SDG 3					2												2
SDG 4	4	2	3	1	9	3	1	7		8	3	3	1			3	48
SDG 5	2	2	4					2		1	1					1	13
SDG 6	2		12			6		3			1	1	2	1	7		35
SDG 7	3	3	4	-1		1	1	7	3	1	7	6	7				42
SDG 8		3	3		2					4	4	4	4		4		28
SDG 9	3	1	2	4	1	6	9	12	6	2	19	15	8	1	9	2	100
SDG 10	4	5	5	2	9	1		3			5		1			4	39
SDG 11	3	3	16	3	4	1	2	7	3	23	11	7	5		4	7	99
SDG 12	1	4	8	2	1	7	4	12	3	-1	7	8	7		4		67
SDG 13			3				2	2	2	1	2	3					15
SDG 14																	0
SDG 15	1		8			6		-3	1	1	5	1	3	2	12		37
SDG 16	1			5	5			1	1	6	2	1	1			6	29
Sum	27	28	76	18	39	31	19	53	18	52	69	51	39	4	42	23	

Figure 5. Cross-impact matrix of 16 SDGs. Numbers indicate the net influence of positive and negative interactions between targets of the corresponding SDGs.

3.3. *Importance, Prioritization, and Governance*

The participants at the prioritization workshop assessed the level of importance of the 107 operational targets (Figure 6). They found all targets to be relevant. The participants considered most targets as important (56.1%), with 32 targets deemed essential (29.9%) and 15 as unimportant (14%). The SDGs having the highest percentage of essential targets were

SDG 6 (66.7%) and SDGs 5, 12, and 16 (50%) (Figure 5). SDGs 3 and 14 had the highest percentage of unimportant targets (44.4% and 57.1%, respectively).

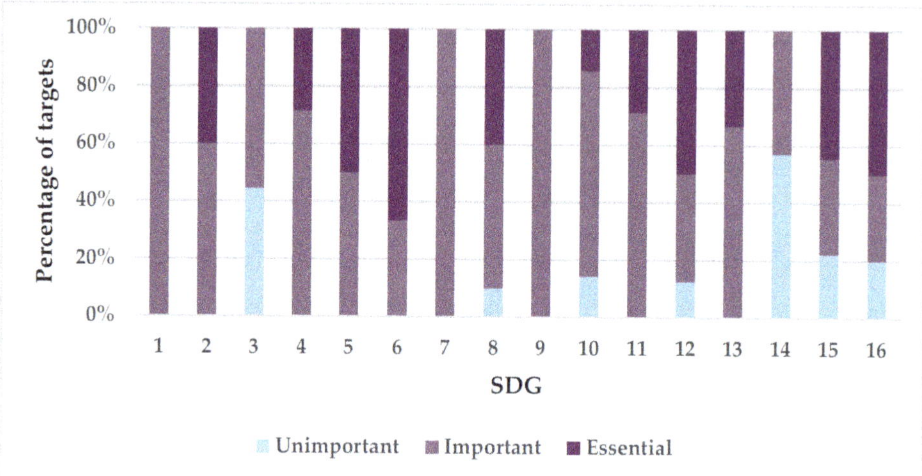

Figure 6. Distribution of the level of importance given to the SDG targets by the Quebec City round tables.

We obtained a prioritization index by crossing performance with importance. Eight targets, among eight different SDGs, were prioritized as urgent (Figure 7), whereas 27 targets were deemed a priority. We noted five priority targets in SDG 15, four in SDG 12, three each in SDGs 4, 14, and 16, two each in SDGs 2, 5, and 8, and one each in SDG 1, 10, and 13. SDGs 3, 7, 9, and 11 did not have any urgent or priority targets (Figure 6). Thirty-four targets were prioritized in the medium term and fifteen in the long term. Additionally, Quebec City needed to consolidate 23 targets. SDG 11, with four targets, and SDG 8 and 9, each with three targets, showed the most targets to be consolidated.

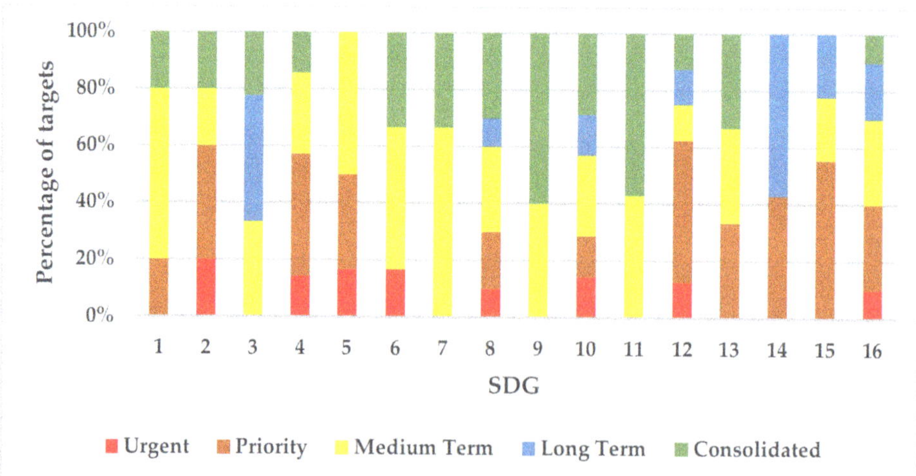

Figure 7. Prioritization levels for each SDG. (A detailed table including all targets is available under the "detailed results" tab of the SDGT-PG available in the Supplementary Materials).

The governance assessment showed that the project team members considered Quebec City to have exclusive power over six targets (5.6%) (Figure 8). These targets are found in SDGs 6, 11, and 12 (each having two targets). On the other hand, they assessed 29 targets (27.1%) as being exclusively national (provincial or federal) jurisdiction and responsibility. Among the SDG targets most associated with the national level, we noted five of the seven targets of SDG 14 (71%), two of the three targets of SDG 7 (67%), and four of the seven targets of SDG 4 and 10 (57%). Twenty-six targets (24.3%) represented a shared responsibility, and 46 targets (43%) were primarily national competence, although supported at the local level. Overall, the national level was better positioned to intervene on 75 targets (70%); for instance, the national level holds most of the authority to intervene in regard to all targets of SDGs 4 and 7 (Figure 7).

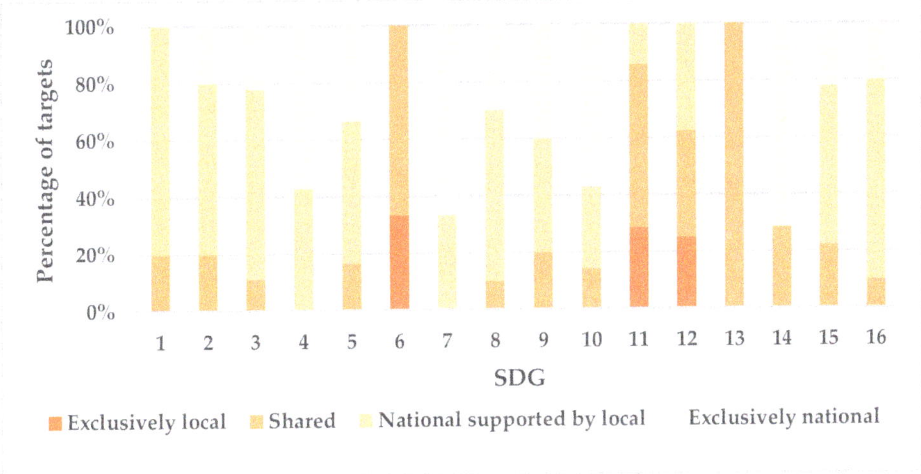

Figure 8. Distribution of the SDG targets among the levels of governance.

4. Discussion

The success of implementing the 2030 Agenda requires the mobilization of all actors at all levels. Our SDG localization approach focuses on the local level and includes an original systemic tool to identify priorities in a context of strategic planning. We used parameters found in the literature [6,20,30]; they were evaluated separately but integrated to define the priorities.

In our study case, assessing the current sustainability context for Quebec City is necessary to clarify the starting point and to develop a sustainable development strategy based on achievements [14]. The development and application of our dictionary of expressions linked to the SDG targets identified the targets considered (or not) within the city's strategic documents.

A proper analysis of performance requires contextualizing performance in terms of governance level. Local governments, depending on the effective distribution of powers in a given country, have varying levers on the SDGs. Quebec City is located in the province of Quebec and also falls within the Canadian national governance. The responsibility for municipalities resides with the provinces under the Canadian constitution. In the province of Quebec, cities have the legislative powers of development and urban planning, housing, roads, community and cultural development, recreation, urban public transport, and wastewater treatment [40]. We strongly recommend that an expert assessment of governance parameters, in accordance with the national/provincial legislative texts, be undertaken when applying an SDGT-PG.

Examining the distribution of powers among government levels allows an analysis of performance crossed with an evaluation of governance (Table 5). In this study, we observed that no exclusively local responsibility target was achieved. Among the exclusively national targets (at the provincial and/or federal level), however, 69% of the targets had not been achieved. In Canada, navigation, coasts, and inland fisheries are a federal responsibility. Five of the seven SDG 14 targets related to oceans and marine resources are exclusively a national responsibility and have not yet been achieved. The two other targets are considered as a shared responsibility. In contrast, Quebec City was on track to achieve 83% of the targets under its responsibility. These are targets of SDG 6 (Clean water and sanitation) and 11 (Sustainable cities and communities), which correspond to the fields of competence given to municipalities in provincial legislation. The other targets of exclusive local responsibility were partially achieved.

Table 5. Distribution related to the performance and level of governance.

	Exclusively Local	Shared Responsibility	National Supported by Local	Exclusively National	
The target is not achieved at all	**0%**/0%	**23%**/16.2%	**23.9%**/29.7%	**69%**/54%	100%
The target is partially achieved	**16.7%**/2.8%	**23%**/16.7%	**50%**/63.9%	**20.7%**/16.7%	100%
The target is in the process of being achieved	**83.3%**/14.7%	**53.9%**/41.2%	**26.1%**/35.3%	**10.3%**/8.8%	100%
	100%	**100%**	**100%**	**100%**	

Note: For the pairs of percentage values, the percentage in bold represents the relative distribution of target performance related to the governance level, whereas the percentage presented in normal font represents the relative distribution of the targets' level of governance related to the performance.

Agenda 2030 states that the SDGs and their targets are global and that [national] governments should define their priorities according to their particular contexts [1]. This contextualization applies to all levels of governance, from local to national. The successful implementation of SDGs requires multilevel governance implemented with communication channels that promote vertical integration [7,41]. Although cities have extremely varied contexts, they encounter common obstacles, such as issues of power [24], and can seize specific opportunities addressed by our approach. Moreover, localization allows local authorities to participate more effectively to achieve national SDGs.

4.1. Obstacles, Limitations, and Challenges of SDG Localization

The scope of the 2030 Agenda limits its localization. The formulation of targets is addressed at the national and global levels, and their text-based interpretation can have a demobilizing effect on the local-level actors. Local actors may see this agenda as being focused on global issues and, thus, they may ultimately reject the agenda outright [42]. Implementing SDGs at the local level requires localizing the targets by adjusting the labels without distorting their meaning. In our case study, the Quebec City project team modified the wording of targets, for which the scope was explicitly national, to provide a local-scale feel to the target. This adaptation increases the tangibility of targets for local actors, who must assess the importance of the targets and ensure that the targets are implemented at the appropriate—local—level. One could assume that targets explicitly mentioning a national scope would be assessed as less important or not applicable for local actors. For the MoI targets, however, adapting these targets to local contexts is difficult, as these targets often involve international partnerships for implementing the 2030 Agenda. From the governance parameter of the SDGT-PG, responsibility for the MoI targets occurs exclusively at the national level. For the sake of adaptation and contextualization, and not to give the impression to local actors that the 2030 Agenda is addressed only at the national level, we chose to exclude the MoI targets from our prioritization approach.

Localizing the SDGs involves implementing the SDGs in the logic of vertical, horizontal, and territorial integration. A siloed approach predominates, and moving toward an integrated approach is not straightforward.

Forming a group of leaders from different municipal services promoted horizontal integration. The leaders were not used to working in a multiservice group. Their collective work and dialogue broke down existing silos. The multiservice workshops greatly helped identify potential synergies and trade-offs. This horizontal integration occurred at several stages of our approach. During the diagnosis stage, our analysis of strategic documents, using the dictionary of expressions related to the SDG targets, identified the initial potential synergies. For example, we identified that the following targets touched all services: 4.4 (Skills for employment and entrepreneurship), 8.2 (Economic productivity), 9.1 (Sustainable infrastructure, economic development, well-being), 9.5 (Research, technological capabilities, innovation), 16.6 (Efficient, accountable, and transparent institutions), and 16.7 (Participation in decision-making). These shared targets do not systematically imply synergistic actions, but the diagnosis identified those actions carried out by several municipal services sharing common objectives. Our dictionary has proven to be a highly relevant and effective tool for undertaking this diagnosis.

The identification of 687 potential interactions formalized the links between city services and contributed to horizontal integration. The in-depth analysis and articulation of interactions illustrated the integrated nature of the actions of all services to members of the leader group. We observed that 92.8% of the interactions were positive by nature. This result closely matches the systemic analysis applied to the case of Sweden by Weitz et al. [31], where 96% of interactions were synergies. Referring to the classification of SDG targets to the five pillars of the 2030 Agenda (population, planet, prosperity, peace, and partnership) in Tremblay et al. [15], we observed that 83% of positive interactions (sum of +2 and greater) were linked to the same pillars. Half of the negative interactions related to different pillars. This illustrates the complexity of SDG targets and their interactions, and how the different pillars are integrated and indivisible.

The limits and challenges of vertical and territorial integration are multiple and complex. These types of integration refer to the principle of subsidiarity, "the search for the 'optimal scale of government,' [28] and the concept of multilevel governance, a system of continuous negotiation among nested governments at several territorial tiers" [43]. These limits and challenges are universal but vary depending on the context. Thus, there is not a single solution, but it is possible to provide adaptable reflections from our approach.

The actors of governance, at different scales, have variable levels of control and power over their context. This control varies from none (e.g., distribution of natural resources across the territory) to full (e.g., the adoption of policies). In addition, the actors interact according to different paradigms, at their level, in a complex system where the dominant paradigm of economic neoliberalism is omnipresent and, sometimes, underground [44–47]. It is well known that states tend to protect their powers despite the recognized importance of applying the principle of subsidiarity for implementing sustainable development [11].

The application of the principle of subsidiarity is linked directly to power issues, a very sensitive subject [24]. Local governments, to respond effectively to their sustainability challenges, must have the corresponding powers. From this perspective, Jones [48] writes, "Where national and state/provincial governments fail to act, city governments are severely limited in the implementation of [sustainability] policy." To address these issues, governments must collaborate. Using the governance assessment in the SDGT-PG, we guided local governments on the types of actions available to them on the basis of their specific governance context and target priority while also proposing actions at the national level. The terms "Search for opportunities" and "Advocacy at the appropriate governance level" apply to targets having a high priority level (urgent or priority) and whose governance is at the national level. This intersection between three parameters of the SDGT-PG helps guide the advocacy that local governments must undertake at higher levels. This observation does not guarantee success and an openness to dialogue; however, it provides guidelines for a structured argument based on an inclusive approach. The aim is to reduce what the OECD identified as the "policy gap" [49]. To achieve this, we must establish mechanisms for collaborating between the levels of governance to make the implementation of public policies relevant and effective.

Localizing the SDGs requires an integrated commitment of human and financial resources [3]. SDSN [6] observed that, despite the importance of localizing SDGs, questions regarding capacities and mobilizing resources remain unanswered. Thus, the major constraints that cities face relate primarily to their limited political and fiscal powers, their lack of access to finance, the low levels of institutional capacity, the lack of multilevel government cooperation and integration, and the difficulty in establishing multi-stakeholder partnerships [6]. Becoming aware of these constraints is, however, a necessary step. Cities can act directly on a few aspects of sustainability, but they need the collaboration and openness of higher levels of governance to tackle the ensemble of issues. Open and empowered multilevel governance is essential for localizing SDGs horizontally, vertically, and territorially within an integrated approach [50].

4.2. Opportunities

The 2030 Agenda is mobilizing an enormous quantity of resources across the globe, and actors at all levels are developing appropriate tools and approaches. The number of scientific articles having "2030 Agenda" as a keyword has increased rapidly from 44 in 2015 to 246 in 2017 to 632 in 2020 (Scopus, search results using "2030 Agenda" as a keyword, 5 November 2020). The SDGs and their targets provide a relevant framework at all scales and are internationally recognized. The principle of integration is increasingly applied, and organizations (national, local, private) increasingly choose the SDG framework for the sake of multilevel consistency.

This willingness to join the SDG movement must be supported politically. In Quebec City, the mayor undertook the process, leading to a strategy and an action plan for sustainable development. This engagement at the highest levels of local government is essential for committing all necessary resources to ensure the success of the process [48]. Thus, the mayor's office established a competent project team that mobilized stakeholders, coordinated and analyzed activities, and developed the necessary strategy. Furthermore, a team of leaders, mobilized within all of the city's administrative units—because of the support of the directors of the various units mobilized by the mayor's office—has been trained in sustainable development issues. The team members communicated the progression of the approach and raised awareness with their colleagues. They sought their views at various stages of the process [48,51,52]. This multiservice mobilization was achieved through the mayor's commitment, through a top-down approach, to provide the means for achieving the results. The presence of a city councilor of the executive committee at every stage of the process testified to this political will. Mobilization at the highest level facilitates awareness of the efforts and actions to be implemented to vertically integrate the process. In our case study, Quebec City does not hold all the necessary powers to respond to the priorities that emerged from the prioritization exercise. City officers will be obliged to develop partnerships with higher governance bodies. As the mayor is the process holder in this case study, he will feel all the more invested and convinced of the need to carry out this task and to use the right communication channels to develop a multilevel collaboration. However, it is important to reiterate that the mobilization of the mayor alone cannot guarantee a successful implementation of sustainable development. It is also essential for all stakeholders to rally and face the challenges related to sustainability.

Cities must build on existing structures and actions already underway that fit within the sustainability framework to ensure optimal localization of the SDGs [14]. Our diagnosis provides a relatively rapid portrait of the situation, an exercise that can often be tedious. In our case study, we included the diagnosis at the stage of evaluating the performance parameter of the SDGT-PG. The use of the dictionary made it possible to undertake rigorous work with a minimum mobilization of human resources. It provided a solid starting point on which to build the remainder of the process and made it possible to identify a common starting point for all actors involved.

Crises can constitute opportunities to introduce a sustainability approach. Some previous crises (climate, financial, energy, etc.) have been drivers of change. For example, the 2008 financial crisis motivated some countries to embark on a transition movement [53,54]. The COVID-19 pandemic may also turn out to be an opportunity to provide arguments that favor the implementation of a sustainable development strategy. Quebec City, as most other local and national governments, must implement a post-containment/COVID-19 recovery strategy. This recovery strategy, linked to a sustainable development strategy, could offer an opportunity to facilitate ownership of the shift and the actions proposed by the city. In terms of sustainability, however, not all crises become opportunities. As stated in the 2030 Agenda: "There can be no sustainable development without peace and no peace without sustainable development" [1]. Thus, crises such as armed conflicts remain major obstacles to sustainability.

Local governance is the closest level of government to citizens and their issues. This reality allows, in theory, to quickly implement measures to respond effectively to identified problems. The local level involves fewer actors and fewer divergent issues than at the national level. This difference could explain why differing from "business as usual" can be easier to implement at the local level [55]. For example, in the context of local actions, actors are less influenced by the dominant paradigm of neoliberalism and thus allows the emergence of approaches considered more radical when compared with "business as usual" actions [47,55,56]. Cities should support grassroots initiatives [42] and socio-ecological transition projects [57] undertaken by local community groups in their territories. These partnerships are much easier to support by local governments that are in direct contact with these groups. Leadership at the top of the city hierarchy (top-down) and support of bottom-up initiatives are not contradictory and mutually reinforce each other [24]. In this sense, Quebec City has opened a dialogue with local partners from various civil society organizations with the objective of identifying challenges, issues, and opportunities, as well as proposals for action.

The identified limitations and opportunities routinely brought us back to the need for multilevel governance to ensure implementation of the 2030 Agenda [7,11]. The national level of governance, although holding most of the powers (Figure 7), must be aware that the national level is not always the most appropriate level in regard to local actors and issues [28]. The motivation of local governments can be hampered by the lack of collaboration of higher governance bodies [42]. The evaluation of the governance parameters shows higher authorities must collaborate with local governments. As Meuleman and Niestroy state [24], the issues and contexts differ at all levels, and a lack of integration and collaboration can lead to failure. A multilevel governance approach that relies on collaboration and cooperation will help promote vertical integration and policy coherence [50]. Our analyses identified targets representing opportunities to build such collaborations, allowing local and national authorities to optimize their contributions for achieving the goals of the 2030 Agenda.

5. Conclusions

Our approach aligns with the best practices for localizing SDGs and includes the concepts of contextualization, localization, systems approach, and integration. Although we apply this approach to the local level, it is flexible and adjustable enough to be applied at all levels of governance. Our approach provides a procedure that empowers sustainability actors in line with vertical and horizontal integration through capacity building, awareness, and direct participation, a procedure that, to our knowledge, has not been provided in previous studies focused on the local level.

Each application of our approach should be contextualized, as the opportunities and limitations differ from place to place. In our case, we were limited by a lack of data; the indicators of the SDG targets had yet to be assessed. Therefore, it was impossible to accurately assess performance. We stated that they were potential performances, and we remain conservative in our assessments by not describing any targets as being fully achieved.

The systemic tools and approach presented in our study will help planners develop strategies and action plans for implementing the 2030 Agenda. Although our approach is complete, it can only be implemented with a mobilization at the highest level and with the involvement of stakeholders who represent the complexity of the system in which the agenda is being implemented. SDG localization faces other challenges, in particular the adaptation of SDG tools and approaches to the private sector, where each particular sector comprises its challenges, contexts, opportunities and specific scopes of organizations governance. Future research could help define, as in the present study, good practices in localizing the SDGs, and methodologies for adapting the 2030 Agenda to the private sector.

Supplementary Materials: The following are available online at https://www.mdpi.com/2071-1050/13/5/2520/s1, Table S1: List of analyzed documents, Table S2: Dictionary linked to the content of the 2030 Agenda, Table S3: Matrix of links between SDG targets and the analyzed Quebec City strategic documents inspired by the Rapid integrated assessment (RIA), Table S4: Cross-impact matrix, Table S5: Sustainable development goal target prioritization grid of Quebec City.

Author Contributions: Conceptualization, D.T., C.V., and S.G.; SDGT-PG methodology, O.R., D.T., and C.V.; WordStat, D.T.; Validation, D.T., J.-F.B., O.R., S.D., S.G., and C.V.; Formal analysis, D.T., S.G., and S.D.; Writing—original draft preparation, D.T.; Supervision, J.-F.B. and C.V.; Project administration, C.V. All authors commented all the sections and reviewed the manuscript. All authors have read and agreed to the published version of the manuscript.

Funding: This research received no external funding.

Institutional Review Board Statement: Not applicable.

Informed Consent Statement: Not applicable.

Data Availability Statement: Data is contained within the article or Supplementary Material.

Acknowledgments: The authors acknowledge the contribution of the project team and the group of city leaders. The authors thank the project director, Yohan Maubrun, for his leadership and openness. The authors acknowledge the support of the Organisation Internationale de la Francophonie and its subsidiary body, l'Institut de la Francophonie pour le Développement Durable, for their support in the development of systemic sustainability analysis tools since 2012. The lead author thanks the Université du Québec à Chicoutimi and the Chaire en éco-conseil de l'Université du Québec à Chicoutimi for their financial support. We also thank Murray Hay of Maxafeau Editing Services.

Conflicts of Interest: The authors declare no conflict of interest.

References

1. United Nations. Transforming Our World: The 2030 Agenda for Sustainable Development; Resolution Adopted by the General Assembly on 25 September 2015. (A/RES/70/1). 2015. Available online: http://www.un.org/ga/search/view_doc.asp?symbol=A/RES/70/1&Lang=E (accessed on 8 December 2016).
2. Allen, C.; Reid, M.; Thwaites, J.; Glover, R.; Kestin, T. Assessing national progress and priorities for the Sustainable Development Goals (SDGs): Experience from Australia. *Sustain. Sci.* **2020**, *15*, 521–538. [CrossRef]
3. Krellenberg, K.; Bergsträßer, H.; Bykova, D.; Kress, N.; Tyndall, K. Urban sustainability strategies guided by the SDGs-A tale of four cities. *Sustainability* **2019**, *11*, 1116. [CrossRef]
4. Zinkernagel, R.; Evans, J.; Neij, L. Applying the SDGs to cities: Business as usual or a new dawn? *Sustainability* **2018**, *10*, 3201. [CrossRef]
5. United Nations Development Programme. *A Multi-Dimensional Focus for the 2030 Agenda*; United Nations Development Programme: New York, NY, USA, 2017; p. 14.
6. Kanuri, C.; Revi, A.; Espey, J.; Kuhle, H. *Getting Started with the SDGs in Cities: A Guide for Stakeholders*; Sustainable Development Solutions Network: New York, NY, USA, 2016; p. 110.
7. Graute, U. Local Authorities Acting Globally for Sustainable Development. *Reg. Stud.* **2016**, *50*, 1931–1942. [CrossRef]
8. Kharrazi, A.; Qin, H.; Zhang, Y. Urban big data and sustainable development goals: Challenges and opportunities. *Sustainability* **2016**, *8*, 1293. [CrossRef]
9. Klopp, J.M.; Petretta, D.L. The urban sustainable development goal: Indicators, complexity and the politics of measuring cities. *Cities* **2017**, *63*, 92–97. [CrossRef]
10. United Nations; Department of Economic and Social Affairs; Population Division. *World Urbanization Prospects: The 2018 Revision*; United Nations: New York, NY, USA, 2019; p. 126.

11. Fenton, P.; Gustafsson, S. Moving from high-level words to local action—Governance for urban sustainability in municipalities. *Curr. Opin. Environ. Sustain.* **2017**, *26–27*, 129–133. [CrossRef]
12. Hoornweg, D.; Hosseini, M.; Kennedy, C.; Behdadi, A. An urban approach to planetary boundaries. *Ambio* **2016**, *45*, 567–580. [CrossRef] [PubMed]
13. Parnell, S. Defining a Global Urban Development Agenda. *World Dev.* **2016**, *78*, 529–540. [CrossRef]
14. Gustafsson, S.; Ivner, J. Implementing the Global Sustainable Goals (SDGs) into Municipal Strategies Applying an Integrated Approach. In *Handbook of Sustainability Science and Research*; Leal Filho, W., Ed.; Springer International Publishing: Cham, Switzerland, 2018; pp. 301–316. [CrossRef]
15. Tremblay, D.; Fortier, F.; Boucher, J.-F.; Riffon, O.; Villeneuve, C. Sustainable development goal interactions: An analysis based on the five pillars of the 2030 agenda. *Sustain. Dev.* **2020**, *28*, 1584–1596. [CrossRef]
16. United Nations Development Programme; UN-Habitat; GTF of Local and Regional Governments. *Localizing the Post-2015 Development Agenda- Dialogues on Implementation*; United Nations: New York, NY, USA, 2015.
17. Riffon, O. Une typologie pour l'analyse des représentations du développement durable des instruments de mise en oeuvre à l'échelle territoriale. In *Les Instruments de L'action Publique et les Dispositifs Territoriaux*; L'Harmattan: Paris, France, 2016; pp. 43–58.
18. Barnett, C.; Parnell, S. Ideas, implementation and indicators: Epistemologies of the post-2015 urban agenda. *Environ. Urb.* **2016**, *28*, 87–98. [CrossRef]
19. Bibri, S.E.; Krogstie, J. Smart sustainable cities of the future: An extensive interdisciplinary literature review. *Sustain. Cities Soc.* **2017**, *31*, 183–212. [CrossRef]
20. Osborn, D.; Cutter, A.; Ullah, F. *Universal Sustainable Development Goals—Understanding the Transformational Challenge for Developed Countries*; Stakeholder Forum: London, UK, 2015; p. 25.
21. Hamdouch, A.; Depret, M.-H. Sustainable development policies and the geographical landscape of the green economy. Actors, scales and strategies. *Finisterra Revista Portuguesa de Geografia* **2012**, *47*, 49–80. [CrossRef]
22. Villeneuve, C.; Tremblay, D.; Côté, H.; Bonfils, S.; Prescott, J. *La Gouvernance du Développement Durable Dans la Francophonie*; Institut de l'énergie et de l'environnement de la Francophonie: Québec, QC, Canada, 2012.
23. Nilsson, M.; Chisholm, E.; Griggs, D.; Howden-Chapman, P.; McCollum, D.; Messerli, P.; Neumann, B.; Stevance, A.-S.; Visbeck, M.; Stafford-Smith, M. Mapping interactions between the sustainable development goals: Lessons learned and ways forward. *Sustain. Sci.* **2018**, *13*, 1489–1503. [CrossRef] [PubMed]
24. Meuleman, L.; Niestroy, I. Common But Differentiated Governance: A Metagovernance Approach to Make the SDGs Work. *Sustainability* **2015**, *7*, 12295–12321. [CrossRef]
25. Finnveden, G.; Gunnarsson-Östling, U. Sustainable development goals for cities. In *Connecting the Dots by Obstacles? Friction and Traction Ahead of SRIA Urban Transitions Pathways*; Bylund, J., Ed.; JPI Urban Europe: Brussels, Belgium, 2016.
26. Hansson, S.; Arfvidsson, H.; Simon, D. Governance for sustainable urban development: The double function of SDG indicators. *Area Dev. Policy* **2019**, *4*, 217–235. [CrossRef]
27. Nilsson, M.; Griggs, D.; Visbeck, M. Map the Interactions between Sustainable Development Goals. *Nature* **2016**, *534*, 320–322. [CrossRef] [PubMed]
28. Piattoni, S. Multi-level Governance: A Historical and Conceptual Analysis. *J. Eur. Integr.* **2009**, *31*, 163–180. [CrossRef]
29. Rivera, M. Political Criteria for Sustainable Development Goal (SDG) Selection and the Role of the Urban Dimension. *Sustainability* **2013**, *5*, 5034–5051. [CrossRef]
30. Allen, C.; Metternicht, G.; Wiedmann, T. Prioritising SDG targets: Assessing baselines, gaps and interlinkages. *Sustain. Sci.* **2019**, *14*, 421–438. [CrossRef]
31. Weitz, N.; Carlsen, H.; Nilsson, M.; Skånberg, K. Towards systemic and contextual priority setting for implementing the 2030 Agenda. *Sustain. Sci.* **2018**, *13*, 531–548. [CrossRef] [PubMed]
32. Villeneuve, C.; Tremblay, D.; Riffon, O.; Lanmafankpotin, G.; Bouchard, S. A Systemic Tool and Process for Sustainability Assessment. *Sustainability* **2017**, *9*, 1909. [CrossRef]
33. United Nations Development Programme. *Rapid Integrated Assessment (RIA)—To facilitate Mainstreaming of SDGs into National and Local Plans*; United Nations Development Programme: New York, NY, USA, 2017; p. 93.
34. Provalis Research. *WordStat*; Provalis Research: Montreal, QC, Canada, 2019.
35. Provalis Research. *WordStat 7—User's Guide*; Provalis Research: Montreal, QC, Canada, 2015; p. 199.
36. Lucas, P.L.; Kok, M.T.J.; Nilsson, M.; Alkemade, R. Integrating Biodiversity and Ecosystem Services in the Post-2015 Development Agenda: Goal Structure, Target Areas and Means of Implementation. *Sustainability* **2014**, *6*, 193–216. [CrossRef]
37. Le Blanc, D. Towards Integration at Last? The Sustainable Development Goals as a Network of Targets. *Sustain. Dev.* **2015**. [CrossRef]
38. International Council for Science. *A Guide to SDG Interactions: From Science to Implementation*; International Council for Science: Paris, France, 2017; p. 239.
39. Weitz, N.; Persson, A.; Nilsson, M.; Tenggren, S. *Sustainable Development Goals for Sweden: Insights on Setting a National Agenda*; Stockholm Environment Institute: Stockholm, UK, 2015; p. 57.

40. Ministère des Affaires Municipales et de L'HABITATION. Régime Municipal Général. Available online: https://www.mamh.gouv.qc.ca/organisation-municipale/organisation-territoriale/organisation-territoriale-municipale/regime-municipal-general/#:~{}:text=Les%20responsabilit%C3%A9s%20de%20la%20municipalit%C3%A9,assainissement%20des%20eaux%20us%C3%A9es%2C%20etc (accessed on 16 October 2020).
41. United Nations Development Program; Office of the High Commissioner for Human Rights. *Global Thematic Consultation on Governance and the Post-2015 Development Framework—Consultation Report*; UNDP and OHCHR: New York, NY, USA, 2013.
42. Lo, K. Urban carbon governance and the transition toward low-carbon urbanism: Review of a global phenomenon. *Carbon Manag.* **2014**, *5*, 269–283. [CrossRef]
43. Marks, G. Structural policy and multilevel governance in the EC. In *The state of the European Community*; Lynne Riener: Boulder, CO, USA, 1993; pp. 391–410.
44. Boehnert, J. The Green Economy: Reconceptualizing the Natural Commons as Natural Capital. *Environ. Commun. J. Nat. Cult.* **2016**, *10*, 395–417. [CrossRef]
45. Brand, U. Green Economy—The Next Oxymoron?: No Lessons Learned from Failures of Implementing Sustainable Development. *Gaia* **2012**, *21*, 28–32. [CrossRef]
46. Kosoy, N.; Brown, P.G.; Bosselmann, K.; Duraiappah, A.; Mackey, B.; Martinez-Alier, J.; Rogers, D.; Thomson, R. Pillars for a flourishing Earth: Planetary boundaries, economic growth delusion and green economy. *Curr. Opin. Environ. Sustain.* **2012**, *4*, 74–79. [CrossRef]
47. Wanner, T. The New 'Passive Revolution' of the Green Economy and Growth Discourse: Maintaining the 'Sustainable Development' of Neoliberal Capitalism. *New Political Econ.* **2015**, *20*, 21–41. [CrossRef]
48. Jones, S. Climate Change Policies of City Governments in Federal Systems: An Analysis of Vancouver, Melbourne and New York City. *Reg. Stud.* **2013**, *47*, 974–992. [CrossRef]
49. Charbit, C.; Michalun, M. *Mind the gaps: Managng Mutual Dependence in Relations among Levels of Government*; OECD: Paris, France, 2009; p. 189.
50. Corfee-Morlot, J.; Kamal-Chaoui, L.; Donovan, M.G.; Cochran, I.; Robert, A.; Teasdale, P.-J. *Cities, Climate Change and Multilevel Governance*; OECD: Paris, France, 2009; p. 125.
51. Betsill, M.M. Mitigating Climate Change in US Cities: Opportunities and obstacles. *Local Environ.* **2001**, *6*, 393–406. [CrossRef]
52. Burch, S. In pursuit of resilient, low carbon communities: An examination of barriers to action in three Canadian cities. *Energy Policy* **2010**, *38*, 7575–7585. [CrossRef]
53. Li, J.L.; Lin, B.Q. Green Economy Performance and Green Productivity Growth in China's Cities: Measures and Policy Implication. *Sustainability* **2016**, *8*, 947. [CrossRef]
54. Mathews, J.A. Green growth strategies—Korean initiatives. *Futures* **2012**, *44*, 761–769. [CrossRef]
55. Loiseau, E.; Saikku, L.; Antikainen, R.; Droste, N.; Hansjürgens, B.; Pitkänen, K.; Leskinen, P.; Kuikman, P.; Thomsen, M. Green economy and related concepts: An overview. *J. Clean. Prod.* **2016**, *139*, 361–371. [CrossRef]
56. Hobson, K. 'Weak' or 'Strong' Sustainable Consumption? Efficiency, Degrowth, and the 10 Year Framework of Programmes. *Environ. Plan. C Gov. Policy* **2013**, *31*, 1082–1098. [CrossRef]
57. Audet, R.; Segers, I.; Manon, M. Expérimenter la transition écologique dans les ruelles de Montréal: Le cas du projet Nos milieux de vie! *Lien social et Politiques* **2019**, 224–245. [CrossRef]

sustainability

MDPI

Article

Ecological Footprint as an Indicator of Corporate Environmental Performance—Empirical Evidence from Hungarian SMEs

Áron Szennay [1,2], Cecília Szigeti [3,*], Judit Beke [4] and László Radácsi [5]

1 Department of Finance, Faculty of Finance and Accountancy, Budapest Business School, 1149 Budapest, Hungary; szennay.aron@uni-bge.hu
2 Doctoral School of Regional and Economic Sciences, Széchenyi István University, 9026 Győr, Hungary
3 Department of International and Theoretical Economics, Kautz Gyula Economics Faculty, Széchenyi István University, 9026 Győr, Hungary
4 Department of International Economics, Faculty of International Management and Business, Budapest Business School, 1165 Budapest, Hungary; lisanyi.endrene@uni-bge.hu
5 Department of Management, Faculty of Finance and Accountancy, Budapest Business School, 1149 Budapest, Hungary; radacsi.laszlo@uni-bge.hu
* Correspondence: szigetic@sze.hu

Abstract: Small- and medium-sized enterprises (SMEs) play a significant role in the national economies of the EU member states. This economic activity has an inevitable environmental impact; however, environmental performance indicators are mostly measured at larger companies. Since the ecological footprint (EF) is a suitable measure of unsustainability, this paper considers it as a measure of the environmental impact of SMEs. An EF calculator for SMEs was developed that is freely available online, and it is a methodological innovation per se. Our previous research projects highlighted that the calculator must be easy-to-use and reliable; therefore, the calculator considers only the common, standardizable, and comparable elements of EF. Our results are based on validated ecological footprint data of 73 Hungarian SMEs surveyed by an online ecological footprint calculator. In order to validate and test the usefulness of the calculator, interviews were conducted with respondents, and results were also checked. The paper presents benchmark data of ecological footprint indicators of SMEs obtained from five groups of enterprises (construction, white-collar jobs, production, retail and/or wholesale trade, and transportation). Statistical results are explained with qualitative data (such as environmental protection initiatives, business models, etc.) of the SMEs surveyed. Our findings could be used as a benchmark for the assessment of environmental performance of SMEs in Central- and Eastern Europe.

Keywords: ecological footprint; environmental performance of SMEs

Citation: Szennay, Á.; Szigeti, C.; Beke, J.; Radácsi, L. Ecological Footprint as an Indicator of Corporate Environmental Performance—Empirical Evidence from Hungarian SMEs. *Sustainability* **2021**, *13*, 1000. https://doi.org/10.3390/su13021000

Received: 26 November 2020
Accepted: 16 January 2021
Published: 19 January 2021

1. Introduction

There is a broad consensus around the need and usefulness of indicators and metrics to define the planetary boundaries. Humanity's demand on resources has been expanding, which has a significant impact on the Earth system; therefore, many researchers now believe that this era can be considered as a new geological epoch, the so-called Anthropocene [1]. The World Overshoot Day, calculated by Global Footprint Network (GFN), is a high-level and easy-to-understand indicator of global (un)sustainability, since it "marks the date when humanity's demand for ecological resources and services in a given year exceeds what Earth can regenerate in that year" [2]. Since 1970, this date occurs before 31st December each year, and, since the beginning of the 2010s, it lands around the 1st of August. This figure means that in the 2010s, humanity used up approximately 1.7 times more resources each year than the ecosystems of the Earth can regenerate. Although environmentally friendly (i.e., "green") consumption habits and technologies are becoming more common, recent studies show that even conscious consumers change their habits occasionally (e.g., during holiday) [3]. Considering this, it is not surprising that Mathis Wackernagel [4] called our economy the largest Ponzi scheme ever. However, as a result of the COVID-19 pandemic and the lockdown measures introduced in the developed and emerging world, the Overshoot Day landed on 22nd August in 2020.

Environmental sustainability (with special regard to the reduction of greenhouse gas emission and the increase of renewable energies) is one of the headline targets the Europe 2020 strategy of the European Union (EU) [5]. Since Europe's 25 million small- and medium-sized enterprises (SMEs) play a significant role in the economies of EU member states, their contribution to sustainable development is also crucial. SMEs make up over 99% of all enterprises in all EU countries, they generate around two-thirds of all jobs and account for more than half of EU's GDP [6]. Evidence shows that both the regulatory stakeholder pressure and organizational stakeholder pressure positively influence green production practices, corporate reputation, and the environmental performance of manufacturing SMEs [7], which means that the sustainability efforts of both the EU as a whole and the individual member states have a positive impact on the attitudes of SME managers towards sustainability. This finding is supported by evidence from the energy sector, i.e., debt increases the value of SMEs in countries with strong environmental commitment, which makes it possible to facilitate growth with additional external capital [8]. Italian evidence highlights, however, that decision-makers of SMEs "have a high school diploma mainly used bank loans or overdrafts as compared to those that received formal training" [9]. Nonetheless, firms with external capital must maintain financial capacity to repay it, which might create significant problems in case of a crisis situation [10], and capital structure considerations may also play a crucial role [9,11,12]. Another aspect is that a large share of SMEs are family businesses that make up between 57 and 66 percent of the enterprises with 3 and 99 employees in Hungary [13]. Evidence shows that Hungarian family businesses have better chances of survival and create higher value added than non-family businesses [14].

Experience has shown that, although several managers of SMEs are interested in metrics on environmental performance, their businesses/companies cannot afford paying for comprehensive environmental audit and advisory; therefore, they do not have enough experience in selecting the most appropriate measures. Our results suggest that the ecological footprint (EF) is a suitable metric for SMEs because (1) it is easy to understand, therefore making it easy for even managers who do not have enough relevant expertise to use it; (2) the calculation is standardizable, therefore capable of providing performance metrics at a low cost or even for free; and (3) quantitative performance indicators allow them to support the selection of the most appropriate projects or measures to enhance corporate environmental performance (CEP). Our aim was to develop an easy-to-use EF calculator for SMEs which could measure the common elements of corporate environmental impacts reliably. Based on experiences with carbon footprint calculators, it has been found that there is a trade-off between accuracy and simplicity. A calculator that measures the EF of SMEs is needed because, whereas large enterprises have sufficient resources to make unique calculations, it can be difficult for SMEs to find resources and expertise [15]. The results of standardized calculations can be complemented with unique items (e.g., material consumption or more sophisticated data on meals) or longitudinal assessment of CEP can be conducted. Based on the results of our previous analyses, the usefulness and accuracy of the calculator developed was validated, the results were discussed with the respondents, and we made attempts to improve the calculator [16]. Nevertheless, lacking benchmark data can be considered as the most critical problem. Therefore, this paper aimed to calculate sectoral comparative benchmark data.

This paper is structured as follows: Chapter 2 summarizes the concept of the EF and its potential role in measuring of corporate environmental performance. At the end of the chapter, some examples of sectoral EF calculations are presented. The third chapter gives an account of the methodology and the sample used, while the fourth chapter summarizes our results.

2. Theoretical Framework

2.1. The Concept of Ecological Footprint

The ecological footprint (EF) concept was developed by Mathis Wackernagel and William E. Rees [17] in 1996. Since the introduction of the concept, the EF has been used to measure environmental sustainability both at a global level and of individual consumption, as well [18–22]. Nonetheless, other indicators could also be used for measuring environmental sustainability [23–25], but it is only the EF that indicates the upper limit of growth properly [26]. The GFN started its National Footprint Accounts (NFA) program in 2003 based on Wackernagel's calculations, and, since then, the EF calculation methodology framework is regularly updated [27]. The most recent update, which contains data sets for most countries and the world from 1961 and 2017, was published in 2020 [28].

The indicator represents the size of land needed for humanity at a given level of technological development to satisfy its needs and absorb waste generated. Compared to other indicators of environmental impact, the most important advantages of the EF are the following: the EF is easy to understand, and it is relatively easy to determine the upper limit of sustainable consumption.

According to the concept of GFN, EF considers six land types: built-up land, forest products, grazing land, cropland, fishing ground, and carbon. Resource usage is expressed and measured by land usage, which are standardized with the help of equivalence factors (EQF) in global hectares (gha)—globally comparable hectares. This conversion number serves as a tool to compare different land types (e.g., cropland, forest, etc.). Since productivity of the particular land types may show regional differences, an adjustment-specific yield factor (YF) is applied [29].

Besides the spread of spatial calculations [30–34], corporate calculations were also introduced. The principles of corporate EF calculations were developed by Nicky Chambers and her colleagues in 2000 [35]. Although the concept of EF calculation was developed by examining (un)sustainability at a macro-level, it is equally useful at a micro-scale, for example, for corporations or other organizations. EF calculations could help corporations to find intervention fields [36] where environmental measures are the most effective, i.e., a particular amount of money spent has the greatest positive impact on corporate environmental performance.

A clear sign of global unsustainability of CO_2 emissions is that, although the usage of all land types has been increasing since the Industrial Revolution, the increase in carbon usage had the most significant role. Carbon usage grew from 43.8% to 59.9% of total land usage between 1961 and 2018, while it has an annual growth rate of 2.54%, the second highest among the land types [37].

2.2. Ecological Footprint as a Possible Corporate Environmental Performance Indicator

The usage of natural resources of business operations has an obvious impact. The concept of environmental performance attempts to measure and manage such impacts. Trumpp et al. [38] reviewed the related literature and identified 16 articles that give a definition of corporate environmental performance (CEP). Since 5 articles refer to the definition of International Organization for Standardization (ISO) standard 14031, and they capture the most important aspects of the 11 other definitions, the authors argue that "the ISO definition provides an encompassing and parsimonious definition". The ISO standard defines environmental performance as the "measurable results of an organization's management of its environmental aspects" [39]. However, the exact and comparable measurement of CEP is not easy because the ISO definition is "fuzzy enough to impose no clear conceptual boundaries" [40].

According to Jung et al., environmental performance measures can be grouped into five categories [41], where general environmental management (GEM) represents the strategic level, while the other four categories (input, process and operation, output, and outcome) are operational. Input measurement considers the raw material (for example, water, timber, metals, etc.) and energy (electricity, fossil fuels, etc.) consumption, while output measures reveal desirable outputs (energy or pollutant savings) and undesirable outputs, for example, emission of air, water, or even land pollutants. As Schultze and Trommer summarize, these two measures refer to "companies' physical interactions with the natural environment" [42]. Process measures deal with optimization of corporate operations to enhance CEP, i.e., the increase in material efficiency and raising awareness of employees and suppliers. Outcome measures concern financial outcomes of the actions taken (for instance, avoided costs, fines, penalties, or even cost savings) and non-financial outcomes, which comprise mainly stakeholder relations, for example, complaints, lawsuits, or reputational issues [41].

We argue that the EF can be considered as an input/output environmental performance measure, since it focuses on the resources (raw material and energy consume, built-up land, etc.) that are consumed in business operations. Furthermore, we argue that the EF is a suitable tool to measure and manage CEP because ecological footprint:

(1) is a well-known and easy-to-understand measure of environmental sustainability;
(2) is a quantitative indicator and is measured on a ratio scale, therefore providing adequate data to create key performance indicators (KPIs);
(3) is a reliable indicator because calculations are based on scientifically proven data, such as carbon emission factors of electricity grid or fossil fuels, local food consumption, etc.; and
(4) calculations can be standardized through online calculators, therefore providing a low-cost solution for small- and medium-sized companies.

Although standardized calculations and methodologies of EF calculators can be considered as an advantage, especially for SMEs and individuals, Harangozó and Szigeti found that online corporate carbon footprint calculators may have validity and reliability issues, even in the case of the simplest business operations [15]. The authors suggested that the reliability of online EF calculators can be enhanced with more detailed input data and using local data (e.g., electricity mix). Furthermore, while corporate carbon footprint calculations are more commonly used among SMEs than EF calculations [43], understanding further aspects of EF brings new insights to improving environmental performance at the SME level.

2.3. Impact of Environmental Performance on Financial Performance

Although some authors suggested that the EF can be reduced at low or no cost [44], further engagements consume scarce corporate resources (e.g., financial funds, human resources, managerial attention, etc.). Since these resources could be used for other projects with net present value, companies will engage only in environmental projects the benefits of which exceed their costs. The link between sustainability and corporate financial performance (CFP) is an empirically well-studied area (see References [45–52]). Meta analyses (e.g., References [53–55]) mostly showed a positive relationship. Although there is no consensus on which indicators measure sustainability the best, we have found no study that used EF as a proxy. We suggest, however, that EF could be a suitable indicator of CEP because, (1) as we mentioned above, the EF has some advantages over other indicators; and (2) the EF is measured in ratio scale, therefore making the link between CEP and CFP examinable with more sophisticated methods than in case of other proxies measured by dummy variables (e.g., certificates, non-financial disclosures, etc.).

According to the theoretical model of Schaltegger and Synnestvedt [56], up to a point, environmental efforts pay off (see point A in Figure 1); after that, marginal benefits will be decreasing. Nonetheless, further environmental protection efforts may be confirmed because the economic performance will be higher than at the starting point up to point B in Figure 1. Two other consequences are as follows: (1) due to managerial skills, attitudes towards and the ignorance of environmental performance may vary at a given level of economic performance; and (2) several factors (e.g., change of consumer attitude, technological development, etc.) may allow to implement further environmental protection efforts, i.e., it causes the curve to shift right (see dashed line in Figure 1).

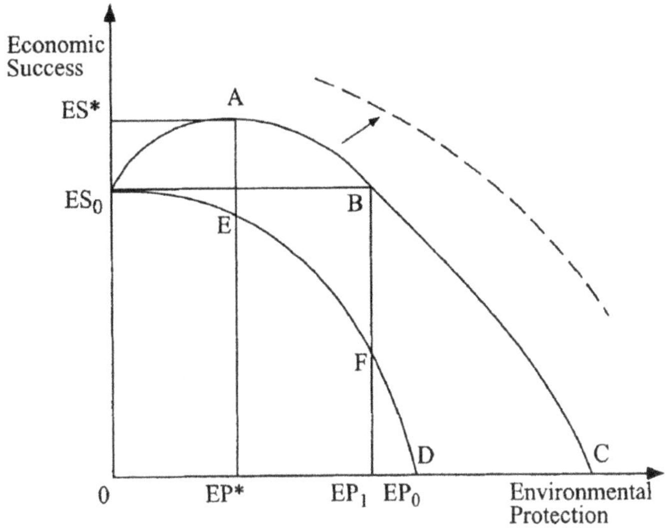

Figure 1. Possible relations between corporate environmental protection and economic success. Source: Reference [56].

By analyzing a sample of 4186 companies in OECD countries, empirical evidence on the positive relationship between environmental protection efforts and financial performance has been found [57]. Furthermore, Zhang et al. [58] provide a more sophisticated version of the model by adding the effects of environmental uncertainty. The authors suggest that environmental uncertainty may influence both costs and benefits of CEP through several factors. Their empirical findings show that the link between the corporate environmental performance (CEP) and the CFP is "steeper and of a lower plateau in higher levels of environmental uncertainty characterized by high dynamism, low munificence, and high complexity".

2.4. Sectoral Average of EF

As it was mentioned earlier, EF was developed to calculate environmental impacts of larger areas (regions, states, countries, etc.) and individuals or their households. In addition, the EF concept was complemented with other, specific calculations to determine sustainability of industrial branches or companies, among others [44]. Although ecological and carbon footprint calculations may be suitable tools for measuring both environmental and economic improvements and related reporting [36], however, one of the main limitations of corporate footprint calculations is the lack of benchmark data; namely, there are no industrial or sectoral averages available to assess the calculated footprint value. Recent research [59–62] aims to fill this gap and to provide guidance for both advisors and managers to assess CEP. To highlight both methodological approaches and impacts of different business models on EF values, in this subsection, we provide a brief insight on the results from three different specific EF calculations.

Mining is one the most CO_2 intensive sectors; thus, there is a legitimate demand on calculating total EF and optimizing it. Murakami et al. [59] have found that underground mines (1) have significantly lower EF for built-up land due to their smaller land-use change, and (2) fossil fuel consumption is also much lower due to their electrification; therefore, the EF could be decreased by using renewable energy sources.

Residential homes have a rather high EF in the EU. Energy consumption of households makes 26.1 percent of total final energy consumption in the EU, out of which heating is the largest portion (63.6%) [63]. Residential buildings have an average energy intensity of 180 kWh/m^2, but it shows significant differences among countries, even when they are located in the same climate zone [64]. Another aspect that studies have shown is the high variability of emissions associated with construction and operation of buildings during their life cycle [65]. Since the Energy Performance of Buildings Directive requires all new buildings to be nearly zero-energy by the end of 2020 in the EU [66], EF minimalization measures should focus on the construction phase. Incorporating EF figures in construction cost databases could support in optimization of both environmental impact and costs of construction. A case study from Andalusia (Spain) highlights that the substitution of traditional construction units with lower EF solutions could result in 18% reduction of the EF, while the total cost increased only by 7% [60]. Using recycled materials (e.g., wood, concrete, steel) could reduce the EF significantly [61].

Since Hungary is an export-oriented, open, and small market economy, industrial parks can be considered as important engines of economic growth and regional development (see References [67,68]). In a case study from China [62] researchers claim that through eco-industrial transformation, EF of HETDA industry park of China can be reduced by 15.9 percent [62]. Nevertheless, other studies have shown that most eco-industrial parks are at a very early stage of development [69].

3. Methodology

A mixed methodology was used in this study. On the one hand, an online ecological footprint calculator was developed according to the special needs of the SME sector. A brief outline of the calculator can be found in the appendix. On the other hand, with a special regard to EF, we conducted interviews and mini case studies to gain deeper understanding of the unique features of SMEs operating in different sectors.

Both the monetary and employment figures are standardized. First, although financial data was collected in local currency (Hungarian forint, HUF) in the survey, results are expressed in euros. Since survey data considers both 2018 and 2019, an arithmetic average of daily exchange rates of the European Central Bank was applied (322.0932 HUF/EUR). Second, all employment data are expressed in full-time equivalents.

3.1. Calculation of EF

The Table 1 cites only articles in which figures, methodology, etc., were directly used in the calculator developed.

Although material usage was part of a previous version of the calculator, later it was excluded from the formula due to the fact that the 500+ materials we employed in the explorative phase could not be standardized in a proper way [16].

Table 1. Element of ecological footprint (EF) calculated, their short description, and calculation method.

Element of EF	Description	Calculation Method	Literature
EF_{meals}	Food consumption during work time, calculated on the base of Hungarian national average values.	Equation (1)	Mózner [70]
$EF_{water\ consumption}$	Water consumed by employees during work time. Industrial water consumption is excluded.	Equation (2)	Chambers et al. [35]
$EF_{built\text{-}up\ area}$	Total area of non-water absorbent surfaces.	Equation (3)	Lin et al. [29]
$EF_{electricity\ consumption}$	Electricity consumption from electricity grid, included heating and boiling with electric devices.	Equation (4)	IEA [71] DEFRA 2018 [72]
$EF_{heating\ and\ boiling}$	Heating and boiling with fossil fuels, e.g., natural gas, coal, or wood.	Equation (5)	DEFRA 2018 [72]
$EF_{transportation}$	All transportation-related EF, including commuting (both public transport and vehicles owned by employees or by the enterprise), transportation of goods, using of corporate cars, flying, etc., petrol, gasoline, and gas consumption of equipment (e.g., generators) are included.	n/a	DEFRA 2018 [72]

The EF of meals was calculated on the basis of Hungarian average values of people's food consumption [70] (see Equation (1)). Average values do not take into consideration food consumption exceeding the minimal human needs (e.g., alcohol or candy consumption, import goods, etc.); therefore, they provide a rather lower estimate than the real figures. To achieve more accurate results, different EF factors were used for both females and males, as well as the characteristics of jobs (i.e., white collar or blue collar). Since the abovementioned values reflect the total food consumption of a given year, we assumed that employees have *n* working days a year, and they consume *i* percent of their meals at the workplace, where *n* and *i* values are given by the SMEs surveyed for each employee category. Calculation of the EF of food consumption was as follows:

$$EF_{meals} = \frac{n_{female}}{365} \times i_{female} \times \sum E_{job} \times EF\ factor_{job} + \frac{n_{male}}{365} \times i_{male} \times \sum E_{job} \times EF\ factor_{job}, \quad (1)$$

where:

n—number of working days of both female and male employees,

i—percent of at workplace consumed meals,

E—number of employees at a given job type (e.g., white collar or blue collar), and

EF factor—EF factor of each job type (e.g., white collar or blue collar).

The EF of food consumption is one of those EF elements which could differ significantly among regions [73]. An EF calculation on food consumption conducted by a Polish research team showed a much larger EF per capita figure. (It is interesting to note that Poland is another Central Eastern European country and EU member state.) The higher number is partly due to methodological considerations.

Spanish and Chilean EF values on food consumption, both of them based on Food and Agriculture Organization (FAO) of the United Nations data, show significant differences too, 0.97 and 1.43 gha per person, respectively [61].

According to our methodology, the EF of water consumption calculates with the EF of building and maintenance of water pipelines, sewage, and wastewater treating facilities. Since exact measures are not available, we assumed that the EF of water consumption is a function of employee number (see Equation (2)).

$$EF_{water} = \left(E_{female} + E_{male}\right) \times EF\ factor_{water}, \quad (2)$$

where:

E—number of both female and male employees; and

EF factor—EF factor of water consumption.

The EF of built-up area was calculated on the base of buildings' ground floor and other covered and non-water absorbent (e.g., asphalt or concrete) surface (see Equation (3)).

$$EF_{built-up} = \left(S_{building} + S_{other}\right) \times EF\ factor_{built-up}, \quad (3)$$

where:

S—covered surface, both ground floor of buildings and other non-water absorbent surfaces, in square meters; and

EF factor—EF factor of built-up area.

The EF of electricity consumption is based on carbon intensity figure (264 g CO_2e/kWh 2015) of International Energy Agency (IEA) [71]. CO_2e (carbon dioxide equivalent) is a term for describing different greenhouse gases in a common unit. For any quantity and type of greenhouse gas, CO_2e signifies the amount of CO_2 which would have the equivalent global warming impact. This value was adjusted from CO_2e to CO_2 figures by the British organization called Department for Environment, Food and Rural Affairs (DEFRA) database 2018 [72] in order to determine carbon intensity values in CO_2/kWh instead of in CO_2e. After that we added estimated impacts of energy generation and losses of electricity transmission and distribution. Although the renewable energy generation of enterprises was taken into consideration, its EF factor was determined as 0. The EF of electricity consumption was calculated as follows:

$$EF_{electricity} = El_{grid} \times EF\ factor_{electricity} + El_{renewable\ generated} \times 0, \quad (4)$$

where:

El—electricity consumed (i.e., bought from the electricity grid or generated by the enterprise); and

EF factor—EF factor of electricity consumption.

The calculation of EF of heating and boiling is based on carbon intensity factors of DEFRA database 2018 [72]. It includes the usage of different fossil energy sources, e.g., natural gas or even burning coal.

$$EF_{heating\ and\ boiling} = \sum FES_i \times EF\ factor_i, \quad (5)$$

where:

FES—fossil energy source (e.g., megajoules of natural gas or tonnes of wood logs); and

EF factor—EF factor of specific fossil energy source.

Besides heating and boiling, transportation and the related carbon footprint generally makes up the largest portion of EF [74]; therefore, our online EF calculator provides the following options to determine the EF:

(1) usage of different fuel types (i.e., petrol, gasoline, LPG), if accurate analytical records are available;
(2) mileage of vehicles of different fuel types (kilometers a year) and average fuel consumption (liters per 100 km);
(3) mileage of different category and fuel type of cars and small vans;
(4) number and average distance of trips in case of taxi and air travel; and
(5) an average of daily distance in case of public transport (underground, tram, bus).

Since SMEs in general use several different transportation modes, only the first two calculation methods are mutually exclusive. All carbon intensity factors are based on the DEFRA 2018 database [72].

3.2. The Sample

Enterprises in our sample were required to have the following attributes:

(1) It is a small- or medium-sized company, defined by the Commission of the European Communities [75], namely has less than 250 employees and its turnover is less than €50 million or its balance sheet total is less than €43 million.
(2) Energy consumption of corporate activities can be separated from other activities, e.g., private home of managers and/or owners.
(3) Managers and/or owners are willing to participate in the survey.

Data was collected from three sources: (1) SMEs known from our professional network or from our university networks; (2) commercial and industrial chambers in Hungary were asked to send calls for survey to their member companies, and we participated in some of their events; and (3) students were asked to assist with our study. Mini case studies were conducted about most of the companies surveyed to gather additional qualitative data.

Companies were filtered out from our sample as an outlier when one or more figures varied significantly from other companies of the same group and we had no plausible explanation for this (e.g., equipment used, working processes, etc.).

Anecdotical evidence suggest that the SMEs of different business activities may have similar EF. Therefore, a preliminary qualitative analysis was conducted to classify SMEs on the basis of the determining factors of their ecological footprint, i.e., based on the attributes of their CEF. This is inevitably different from statistical classifications (i.e., NACE in the EU or SIC in the USA). We suggest that a more detailed and more accurate result could be achieved by analyzing a larger database. For example, white-collar jobs have similar environmental impact, regardless of whether the enterprise is involved in bookkeeping, software development, civil engineering planning, or even fashion design. The ecological footprint of white-collar jobs is determined mainly by (1) the conditions of the property used (place, size, insulation, effectiveness, and usage of air conditioning and heating, etc.), (2) commuting habits of employees and home office opportunities, (3) number and length of business trips and vehicles used, and (4) the number of employees.

The study focuses on the following five groups of SMEs (see Table 2):

Table 2. Classification of SMEs analyzed.

Name of Group	Common Sense	Related Subsection
construction	Extensive use of machines, heavy-duty vehicles. EF is determined mostly by fossil fuel consumption.	Section 4.1
white-collar jobs	Knowledge-intensive activities, moderate land use, equipment with low consumption (e.g., laptops, plotters, etc.). Vehicle usage is limited for passenger cars and only for field visits or commuting. EF is rather balanced among determining factors.	Section 4.2
production	Technology-intensive activities, significant usage of equipment and land. EF is determined mostly by energy and fossil fuel consumption, but built-up land usage and food consumption are also significant.	Section 4.3
retail and/or wholesale trade	Significant land use (buildings and parking lots), moderate use of equipment (e.g., refrigerators). Moderate vehicle usage. EF is determined significantly by heating and boiling; fuel consumption could be significant in case of home delivery or other vehicle usage.	Section 4.4
transportation	Extensive use of trucks and other resource usage is negligible. EF is determined most of all by gasoline consumption.	Section 4.5

Variation of EF among group of enterprises can be explained by several coexisting factors:

(1) The operation of SMEs may differ. For example, the EF will be greater if a retail store transports goods with its own van and/or provides home delivery for costumers, or if an engineering office must make trips for its field works.

(2) Manager's attitudes towards sustainability may vary significantly. Some managers attempt to engage in environmentally friendly projects (e.g., energy efficient equipment, solar panels, etc.), while others do not.

(3) The organization culture may also be different.

One of the limitations of the EF calculator is that it ignores all the factors that are beyond the control of companies. Accordingly, financial performance is measured by an adjusted value added, which is calculated on the available accounting data as the sum of personnel costs, amortization, and after-tax profit. Adjustment had to be made because of a simplified tax type eligible only for small companies. If a company chooses this tax type, it substitutes corporate tax and social contributions of employment. Since personnel costs of companies of different types are directly not comparable, we chose after-tax profit instead of pre-tax profit. We suggest that these kinds of calculations provide more comparable results among the analyzed SMEs but have the limitation that all value-added figures presented show an underestimation of real values.

4. Results

Our sample consists of 73 SMEs from the five groups. Four out of the five groups have 15–20 valid items, while the smallest sub-sample (transportation) comprises only 4 items. This can be explained by the relative simpleness of the sector; the EF of these SMEs is determined almost completely by fuel consumption (liters of diesel per 100 km). Detailed results are presented in the following subsections. For detailed numerical information see Tables 3 and 4.

Table 3. Descriptive statistics.

			Construction	White-Collar Jobs	Production	Retail and Wholesale Trade	Transportation
	Valid cases		17	17	15	20	4
specific EF (global hectares/employee)	Mean		1.25	0.46	1.47	1.10	20.15
	95% Confidence Interval for Mean	Lower Bound	0.87	0.32	0.85	0.73	17.00
		Upper Bound	1.62	0.60	2.08	1.47	23.30
	5% Trimmed Mean		1.20	0.43	1.42	1.06	20.20
	Median		0.93	0.44	1.21	0.81	20.56
	Std. Deviation		0.72	0.27	1.11	0.79	1.98
eco-efficiency (global hectares/th. EUR)	Mean		0.089	0.051	0.067	0.088	1.055
	95% Confidence Interval for Mean	Lower Bound	0.065	0.029	0.033	0.050	0.410
		Upper Bound	0.113	0.074	0.100	0.126	1.701
	5% Trimmed Mean		0.086	0.047	0.064	0.079	1.040
	Median		0.076	0.041	0.047	0.071	0.918
	Std. Deviation		0.047	0.043	0.061	0.081	0.406
specific value added (th EUR/employee)	Mean		15.40	15.29	32.98	17.24	20.64
	95% Confidence Interval for Mean	Lower Bound	11.94	7.23	14.04	12.64	11.79
		Upper Bound	18.85	23.34	51.93	21.84	29.49
	5% Trimmed Mean		14.99	13.01	27.78	16.83	20.74
	Median		14.96	11.38	22.97	15.78	21.57
	Std. Deviation		6.71	15.67	34.21	9.83	5.56

Table 4. Correlations.

Activity			Specific EF (gha/empl)	Eco-Efficiency (gha/th EUR)	Specific Value Added (th EUR/empl)	Activity	Specific EF (gha/empl)	Eco-Efficiency (gha/th EUR)	Specific Value Added (th EUR/empl)	Activity	Specific EF (gha/empl)	Eco-Efficiency (gha/th EUR)	Specific Value Added (th EUR/empl)
construction	specific EF (gha/empl)	Pearson Correlation	1	0.778 **	0.177	white-collar jobs	1	0.524 *	−0.042	production	1	0.788 **	−0.209
		Sig. (2-tailed)		0.000	0.497			0.037	0.878			0.001	0.472
		N	17	17	17		16	16	16		14	14	14
	eco-efficiency (gha/th EUR)	Pearson Correlation	0.778 **	1	−0.408		0.524 *	1	−0.507 *		0.788 **	1	−0.378
		Sig. (2-tailed)	0.000		0.104		0.037		0.045		0.001		0.182
		N	17	17	17		16	16	16		14	14	14
	specific value added (th EUR/empl)	Pearson Correlation	0.177	−0.408	1		−0.042	−0.507 *	1		−0.209	−0.378	1
		Sig. (2-tailed)	0.497	0.104			0.878	0.045			0.472	0.182	
		N	17	17	17		16	16	16		14	14	14
retail and wholesale trade	specific EF (gha/empl)	Pearson Correlation	1	0.379	0.245	transportation	1	0.591	−0.406				
		Sig. (2-tailed)		0.099	0.298			0.409	0.594				
		N	20	20	20		4	4	4				
	eco-efficiency (gha/th EUR)	Pearson Correlation	0.379	1	−0.543 *		0.591	1	−0.960 *				
		Sig. (2-tailed)	0.099		0.013		0.409		0.040				
		N	20	20	20		4	4	4				
	specific value added (th EUR/empl)	Pearson Correlation	0.245	−0.543 *	1		−0.406	−0.960 *	1				
		Sig. (2-tailed)	0.298	0.013			0.594	0.040					
		N	20	20	20		4	4	4				

**. Correlation is significant at the 0.01 level (2-tailed). *. Correlation is significant at the 0.05 level (2-tailed).

4.1. Construction

Activities of construction enterprises in our sample, ranging from civil engineering, structural architecture, and some special construction firms (e.g., planning, implementing solar panels and other electric equipment on buildings, installing shading equipment, etc.), are also present. They have an average EF of 1.25 gha/employee (confidence interval (CI): 0.87–1.62), an eco-efficiency of 0.089 gha/thousand EUR adjusted value added (CI: 0.065–0.113), and specific value added of 15.4 thousand EUR/employee (CI: 11.94–18.85). Positive correlation between eco-efficiency and specific EF ($p < 0.01$) shows that more eco-efficient construction also has lower the EF per employee figures. Significant correlations between other variables could not be identified.

The EF of construction enterprises is determined mostly by the consumption and efficiency of vehicles and other equipment used. Our mini cases show that managers mostly aimed to reduce fuel consumption; therefore, vehicles are regularly replaced by more efficient ones, private vehicle use is restricted, and employees are collected by a company vehicle. It is interesting to note, however, that the prestige of driving a car is of great importance for many people, and they drive to work even if the commuting distance is less than a few kilometers. Nevertheless, a moderate vehicle use may be allowed in most construction enterprises, since the second half of 2010s is marked with a shortage of trained and experienced professionals. Another issue is that, although there is governmental aid for purchasing battery electric vans or cars, the managers interviewed are concerned about the higher price and the lack of experience; therefore, only a small car that was used for the everyday corporate errands was to be replaced.

If the company has a larger office building, it is often retrofitted or is even equipped with solar panels.

4.2. White-Collar Jobs

White-collar jobs include mostly financial and accounting services (bookkeeping, tax advisory services, auditing, etc.), but engineering, education, or even software development enterprises are present in the sample. The group has the smallest environmental impact—an average of 0.46 gha/employee (CI: 0.32–0.60) and average eco-efficiency of 0.051 gha/thousand EUR adjusted value added (CI: 0.029–0.074), while the average specific value added is less than in other sectors, 15.29 thousand EUR/employee (CI: 7.23–23.34). Results of the correlation analysis show that (1) more eco-efficient enterprises have significantly lower specific EF figures ($p < 0.05$) and (2) higher specific value added ($p < 0.05$). This latter result means that engagement in environmental protection measures and/or project may be profitable.

Since working in an office is a human capital-intensive activity, its EF is determined mostly by the energy-efficiency of the buildings used and by the commuting practices and working trips of the workforce. While the former figure can easily be reduced by insulation and/or renovation of the buildings, by using energy-efficient lightning or even by implementing solar panels, reducing the latter figure is a more complicated issue. On the one hand, the COVID-19 pandemic showed that personal contacts can be at least partly substituted by online meetings, but working trips could be necessary in some cases; for example, engineers must visit working fields or even cultural determinations may require personal meetings. On the other hand, the prestige of commuting by car and/or living in urban agglomerations may influence the habits of employees. Furthermore, employees mostly use their own cars; therefore, it is out of the managers' control. Based on our findings, we recommend promoting more sustainable ways of commuting. For example, when it is feasible, businesses should provide shower and changing facilities for cyclists in the workplace, but biking events and/or actions may influence commuting habits, as well. Of course, financial stimuli could also be used, for example, cutting contributions on commuting with a car and providing benefits for public transport usage instead.

4.3. Production

Producer companies in our sample are very diverse—they range from manufacturing spices, wooden toys for playgrounds to producing vehicles. The EF figures of these activities differ substantially. Specific EF is EF 1.47 gha/employee (CI: 0.85–2.08) on average in this group, while eco-efficiency is favorable, 0.067 gha/thousand EUR (CI: 0.033–0.100), and specific value added is 32.98 thousand EUR/employee (CI: 14.04–51.93) due to the higher adjusted value added. Just as in the case of construction enterprises, correlation analysis shows a significant relationship only between eco-efficiency and specific EF ($p < 0.01$).

Production is technology-intensive, so EF is also highly determined by working processes and equipment used. Our mini cases show that companies attempt to implement both up-to-date working processes and efficient equipment, but EF figures are influenced significantly by other factors, such as industrial specialties, level of market competition, managerial attitudes, and governmental and/or EU grants.

4.4. Retail and Wholesale Trade

Retail and/or wholesale trade companies range from pharmacies and other fast-moving consumer goods (FMCG) stores to wholesale of electronic components or even veterinary items. The most significant difference among companies is the following: (1) transportation and/or home delivery of goods with own vehicle or by a third party; and (2) special storage needs of goods sold (e.g., storage of frozen or chilled goods have much higher energy consumption than of recyclable waste). Specific EF of the sector is on average 1.10 gha/employee (CI: 0.73–1.47), and eco-efficiency is 0.088 gha/thousand EUR (CI: 0.050–0.126), while specific added value lies at 17.24 thousand EUR/employee (CI: 12.64–21.84). We found significant correlation between eco-efficiency and specific value added ($p < 0.05$). It means, as in the case of office activities, that more eco-efficient enterprises have generally higher added value per employee; namely, there is a positive relationship between corporate environmental performance and value-added creation.

Our cases reveal that companies of the clusters sector have similar challenges as of offices, namely energetical characteristics of the buildings used.

4.5. Transportation

The fifth group is transportation, which is the most EF-intensive sector in our analysis. Specific EF figure of transportation companies is 20.15 gha/employee (CI: 17.00–23.30), which is 16 times higher than of construction companies. Average eco-efficiency is 1.055 gha/thousand EUR (CI: 0.410–1.701), and specific value added is 20.64 thousand EUR/employee (CI: 11.79–29.49). Similar to other groups, our correlation analysis identifies significant relationship between eco-efficiency and specific value added ($p < 0.05$), establishing a positive connection between corporate financial and environmental performance.

Our cases show that there are four main routes to reducing the EF: (1) increasing the efficiency of vehicle technology, which means not only lower consumption in relative terms (liters per 100 km), but highway tolls and maintenance costs are significantly lower, as well; (2) monitoring fuel consumption could mitigate misuse of tanked fuel and provide data for route optimization; (3) route optimization could decrease mileage of trucks, which means lower consumption in absolute terms (liters per trip); and (4) using lower-carbon fuels (e.g., hydrogen).

5. Conclusions and Discussions

The paper aimed to develop an easy-to-use EF calculator for SMEs which could measure the common elements of corporate environmental impacts reliably. Results are based on a sample of 73 corporate EF calculated by an online EF calculator; thus, identical approach and methodology was assured.

Our results primarily have practical implications, as they show that it is feasible to develop an EF calculator for SMEs which can provide reliable figures and is easy-to-use. As anecdotical evidence suggested, SMEs can be classified on the basis of the determining factors of their ecological footprint. Considering their EF figure calculated with a standardized methodology, benchmark data could be also calculated to measure CEP. Using a larger sample, a more detailed classification and more accurate benchmark data could be provided.

According to our results, there is significant and negative correlation between eco-efficiency (EF/value added) and added value per capita in some groups (office activities, retail and/or wholesale trade, transportation). Similarly, significant correlation cannot be found in other analyzed groups (construction, production). These findings suggest that CEP does not influence the financial performance of the analyzed SMEs negatively; rather, there is a positive link in the case of some groups. A possible explanation for the difference of sectors' results may be that production and construction are both highly technology intensive sectors; therefore, environmental protection measures are either too expensive (e.g., more advanced production technology) or there is no available solution (e.g., heavy duty vehicles with electric powertrain).

It is remarkable that transportation enterprises have much higher EF figures than enterprises from other groups. On the one hand, it highlights the importance of locality [76]. On the other hand, transportation connect participants of the value chains. It means that activities with significantly different EF figures in a value chain are separated into several enterprises, so it would seem that CEP of a specific enterprise might be high, but actually only a more environmentally intensive element of the value chain is outsourced to it.

Our results have four main limitations. First, there is no widely used and accepted methodology of conducting easy-to-use and reliable EF calculator for SMEs; therefore, we could not lean on former experiences or calculators that we could have used for the development and testing of our calculator. To provide reliable results, only the common elements of corporate EF were taken into consideration. Although EF of material usage might make up a significant proportion of corporate EF, we suggest that the number and diversity of materials would make the calculator too complex and complicated to use. Second, the sample used in the analysis is small and does not represent the real environmental performance of Hungarian SMEs. Furthermore, we suggest that the sample is positively biased because companies with higher environmental performance are more willing to participate in the survey. Third, although the most financial data were validated on the basis of disclosed financial reports, and we adjusted them according to findings of qualitative methods, firm-specific parameters (e.g., part-time employment, tax optimization, accounting policies, business models, etc.) could significantly influence the results. Fourth, our calculator does not consider material usage of companies; thus, the provided EF values are consistently underestimated.

The results presented in this article show a transition phase between individual and mass calculations; therefore, our future research aims to provide more accurate benchmark data on EF values of SMEs based on a larger sample size. This step would make it possible to conduct more sophisticated analyses using moderating variables (such as corporate governance [77,78], family businesses [79], developed, emerging, and transitional countries, etc.). Another aspect could be to complement the data set with other sectors, for example, with services, agriculture, etc., and to compare EF values of SMEs based in different countries.

Author Contributions: Conceptualization and methodology, C.S.; data collection, Á.S., J.B., C.S.; validation and formal analysis, Á.S.; writing—original draft preparation, Á.S.; writing—review and editing, J.B., L.R.; supervision and project administration, C.S., L.R. All authors have read and agreed to the published version of the manuscript.

Funding: This research was supported by a grant from the Higher Education Institutional Excellence Program of the Hungarian Ministry of Innovation and Technology to Budapest Business School (NKFIH-1259-8/2019). The APC was funded by Budapest Business School—University of Applied Sciences.

Institutional Review Board Statement: Not applicable.

Informed Consent Statement: Not applicable.

Data Availability Statement: The data are not publicly available because respondents did not permit data usage of third parties.

Conflicts of Interest: The authors declare no conflict of interest.

References

1. Lewis, S.L.; Maslin, M.A. Defining the Anthropocene. *Nature* **2015**, *519*, 171–180. [CrossRef] [PubMed]
2. Global Footprint Network Earth Overshoot Day. 2020. Available online: https://www.overshootday.org/ (accessed on 17 November 2020).
3. Anciaux, A. "On Holidays, I Forget Everything ... Even My Ecological Footprint": Sustainable Tourism through Daily Practices or Compartmentalisation as a Keyword? *Sustainability* **2019**, *11*, 4731. [CrossRef]
4. Wackernagel, M.; Pearce, F. Day of Reckoning. *New Sci.* **2018**, *239*, 20–21. [CrossRef]
5. European Commission. Europe 2020—Overview. Available online: https://ec.europa.eu/eurostat/web/europe-2020-indicators (accessed on 20 November 2020).
6. Eurostat. Statistics on Small and Medium-Sized Enterprises—Statistics Explained. Available online: https://ec.europa.eu/eurostat/statistics-explained/index.php/Statistics_on_small_and_medium-sized_enterprises#General_overview (accessed on 26 November 2020).
7. Baah, C.; Opoku-Agyeman, D.; Acquah, I.S.K.; Agyabeng-Mensah, Y.; Afum, E.; Faibil, D.; Abdoulaye, F.A.M. Examining the Correlations between Stakeholder Pressures, Green Production Practices, Firm Reputation, Environmental and Financial Performance: Evidence from Manufacturing SMEs. *Sustain. Prod. Consum.* **2021**, *27*, 100–114. [CrossRef]
8. Cariola, A.; Fasano, F.; La Rocca, M.; Skatova, E. Environmental Sustainability Policies and the Value of Debt in EU SMEs: Empirical Evidence from the Energy Sector. *J. Clean. Prod.* **2020**, *275*, 123133. [CrossRef]
9. Rossi, M.; Lombardi, R.; Siggia, D.; Oliva, N. The Impact of Corporate Characteristics on the Financial Decisions of Companies: Evidence on Funding Decisions by Italian SMEs. *J. Innov. Entrep.* **2016**, *5*, 2. [CrossRef]
10. Giacosa, E.; Culasso, F.; Mazzoleni, A.; Rossi, M. A model for the evaluation trends performance in small and medium enterprises. *Corp. Ownersh. Control* **2016**, *13*, 389–402. [CrossRef]
11. Rossi, M. Capital Structure of Small and Medium Enterprises: The Italian Case. *IJGSB* **2014**, *6*, 130. [CrossRef]
12. Rossi, M.; Lombardi, R.; Nappo, F.; Trequattrini, R. The Capital Structure Choices of Agro-Food Firms: Evidence from Italian SMEs. *IJMP* **2015**, *8*, 172. [CrossRef]
13. Kása, R.; Radácsi, L.; Csákné Filep, J. Családi vállalkozások definíciós operacionalizálása és hazai arányuk becslése a kkv-szektoron belül. *Statisztikai Szemle* **2019**, *97*, 146–174. [CrossRef]
14. Sági, J.; Chandler, N.; Lentner, C. Family Businesses and Predictability of Financial Strength: A Hungarian Study. *Probl. Perspect. Manag.* **2020**, *18*, 476–489. [CrossRef]
15. Harangozó, G.; Szigeti, C. Corporate Carbon Footprint Analysis in Practice—With a Special Focus on Validity and Reliability Issues. *J. Clean. Prod.* **2017**, *167*, 1177–1183. [CrossRef]
16. Szigeti, C.; Szennay, Á.; Lisányi Endréné Beke, J.; Polák-Weldon, J.R.; Radácsi, L. Challenges of Corporate Ecological Footprint Calculations in the SME Sector in Hungary: Case Study Evidence from Six Hungarian Small Enterprises. In *Agroecological Footprints Management for Sustainable Food System*; Banerjee, A., Meena, R.S., Jhariya, M.K., Yadav, D.K., Eds.; Springer: Singapore, 2021; pp. 345–363. ISBN 9789811594960.
17. Wackernagel, M.; Rees, W. *Our Ecological Footprint—Reducing Human Impact on the Earth*; New Society Publishers: Gabriola, BC, Canada, 1996; ISBN 978-0-86571-312-3.
18. Wackernagel, M.; Rees, W.E. Perceptual and Structural Barriers to Investing in Natural Capital: Economics from an Ecological Footprint Perspective. *Ecol. Econ.* **1997**, *20*, 3–24. [CrossRef]
19. Wackernagel, M.; Onisto, L.; Bello, P.; Callejas Linares, A.; Susana López Falfán, I.; Méndez García, J.; Isabel Suárez Guerrero, A.; Guadalupe Suárez Guerrero, M. National Natural Capital Accounting with the Ecological Footprint Concept. *Ecol. Econ.* **1999**, *29*, 375–390. [CrossRef]
20. Van den Bergh, J.C.J.M.; Verbruggen, H. Spatial Sustainability, Trade and Indicators: An Evaluation of the 'Ecological Footprint'. *Ecol. Econ.* **1999**, *29*, 61–72. [CrossRef]
21. Monfreda, C.; Wackernagel, M.; Deumling, D. Establishing National Natural Capital Accounts Based on Detailed Ecological Footprint and Biological Capacity Assessments. *Land Use Policy* **2004**, *21*, 231–246. [CrossRef]
22. Čuček, L.; Klemeš, J.J.; Kravanja, Z. A Review of Footprint Analysis Tools for Monitoring Impacts on Sustainability. *J. Clean. Prod.* **2012**, *34*, 9–20. [CrossRef]
23. Galli, A.; Wiedmann, T.; Ercin, E.; Knoblauch, D.; Ewing, B.; Giljum, S. Integrating Ecological, Carbon and Water Footprint into a "Footprint Family" of Indicators: Definition and Role in Tracking Human Pressure on the Planet. *Ecol. Indic.* **2012**, *16*, 100–112. [CrossRef]
24. Zhu, Y.; Jiang, S.; Han, X.; Gao, X.; He, G.; Zhao, Y.; Li, H. A Bibliometrics Review of Water Footprint Research in China: 2003–2018. *Sustainability* **2019**, *11*, 5082. [CrossRef]

25. Helka, J.; Ostrowski, J.; Abdel-Razek, M.; Hawighorst, P.; Henke, J.; Majer, S.; Thrän, D. Combining Environmental Footprint Models, Remote Sensing Data, and Certification Data towards an Integrated Sustainability Risk Analysis for Certification in the Case of Palm Oil. *Sustainability* **2020**, *12*, 8273. [CrossRef]
26. Toth, G.; Szigeti, C. The Historical Ecological Footprint: From over-Population to over-Consumption. *Ecol. Indic.* **2016**, *60*, 283–291. [CrossRef]
27. Borucke, M.; Moore, D.; Cranston, G.; Gracey, K.; Iha, K.; Larson, J.; Lazarus, E.; Morales, J.C.; Wackernagel, M.; Galli, A. Accounting for Demand and Supply of the Biosphere's Regenerative Capacity: The National Footprint Accounts' Underlying Methodology and Framework. *Ecol. Indic.* **2013**, *24*, 518–533. [CrossRef]
28. Getting to Know the National Footprint and Biocapacity Accounts—Global Footprint Network. Global Footprint Network. Available online: https://www.footprintnetwork.org/) (accessed on 20 October 2020).
29. Lin, D.; Hanscom, L.; Murthy, A.; Galli, A.; Evans, M.; Neill, E.; Mancini, M.S.; Martindill, J.; Medouar, F.-Z.; Huang, S.; et al. Ecological Footprint Accounting for Countries: Updates and Results of the National Footprint Accounts, 2012–2018. *Resources* **2018**, *7*, 58. [CrossRef]
30. Liu, P. Investigación sobre la huella ecológica del turismo: El caso de Langzhong en China. *Obs. Medioambient.* **2019**, *22*, 245–263. [CrossRef]
31. Danish; Ulucak, R.; Khan, S.U.-D. Determinants of the Ecological Footprint: Role of Renewable Energy, Natural Resources, and Urbanization. *Sustain. Cities Soc.* **2020**, *54*, 101996. [CrossRef]
32. Kovács, Z.; Harangozó, G.; Szigeti, C.; Koppány, K.; Kondor, A.C.; Szabó, B. Measuring the Impacts of Suburbanization with Ecological Footprint Calculations. *Cities* **2020**, *101*, 102715. [CrossRef]
33. Lu, W.-C. The Interplay among Ecological Footprint, Real Income, Energy Consumption, and Trade Openness in 13 Asian Countries. *Environ. Sci. Pollut. Res.* **2020**, *27*, 45148–45160. [CrossRef]
34. Zambrano-Monserrate, M.A.; Ruano, M.A.; Ormeño-Candelario, V.; Sanchez-Loor, D.A. Global Ecological Footprint and Spatial Dependence between Countries. *J. Environ. Manag.* **2020**, *272*, 111069. [CrossRef]
35. Chambers, N.; Simmons, C.; Wackernagel, M.; Simmons, C.; Wackernagel, M. *Sharing Nature's Interest: Ecological Footprints as an Indicator of Sustainability*; Routledge: London, UK, 2000; ISBN 978-1-315-87026-7.
36. Tóth, G.; Szigeti, C.; Harangozó, G.; Szabó, D.R. Ecological Footprint at the Micro-Scale—How It Can Save Costs: The Case of ENPRO. *Resources* **2018**, *7*, 45. [CrossRef]
37. Global Footprint Network. Open Data Platform. Available online: http://data.footprintnetwork.org/#/analyzeTrends?type=EFCtot&cn=5001 (accessed on 20 October 2020).
38. Trumpp, C.; Endrikat, J.; Zopf, C.; Guenther, E. Definition, Conceptualization, and Measurement of Corporate Environmental Performance: A Critical Examination of a Multidimensional Construct. *J. Bus. Ethics* **2015**, *126*, 185–204. [CrossRef]
39. ISO 14031:2013(En), Environmental Management—Environmental Performance Evaluation—Guidelines. Available online: https://www.iso.org/obp/ui/#iso:std:iso:14031:ed-2:v1:en (accessed on 21 October 2020).
40. Dragomir, V.D. How Do We Measure Corporate Environmental Performance? A Critical Review. *J. Clean. Prod.* **2018**, *196*, 1124–1157. [CrossRef]
41. Jung, E.J.; Kim, J.S.; Rhee, S.K. The Measurement of Corporate Environmental Performance and Its Application to the Analysis of Efficiency in Oil Industry. *J. Clean. Product.* **2001**, *9*, 551–563. [CrossRef]
42. Schultze, W.; Trommer, R. The Concept of Environmental Performance and Its Measurement in Empirical Studies. *J. Manag. Control* **2012**, *22*, 375–412. [CrossRef]
43. Csutora, M.; Harangozó, G. Twenty Years of Carbon Accounting and Auditing—A Review and Outlook. *Soc. Econ.* **2017**, *39*, 459–480. [CrossRef]
44. Wackernagel, M.; Beyers, B. *Ecological Footprint—Managing Our Biocapacity Budget*; New Society Publishers: Gabriola, BC, Canada, 2019; ISBN 978-0-86571-911-8.
45. Stanwick, P.A.; Stanwick, S.D. The Relationship between Corporate Social Performance, and Organizational Size, Financial Performance, and Environmental Performance: An Empirical Examination. *J. Bus. Ethics* **1998**, *17*, 195–204. [CrossRef]
46. Tsoutsoura, M. *Corporate Social Responsibility and Financial Performance*; University of California at Berkeley: Berkeley, CA, USA, 2004.
47. Santis, P.; Albuquerque, A.; Lizarelli, F. Do Sustainable Companies Have a Better Financial Performance? A Study on Brazilian Public Companies. *J. Clean. Product.* **2016**, *133*, 735–745. [CrossRef]
48. Crifo, P.; Diaye, M.-A.; Pekovic, S. CSR Related Management Practices and Firm Performance: An Empirical Analysis of the Quantity–Quality Trade-off on French Data. *Int. J. Product. Econ.* **2016**, *171*, 405–416. [CrossRef]
49. Rokhmawati, A.; Gunardi, A.; Rossi, M. How Powerful Is Your Customers' Reaction to Carbon Performance? Linking Carbon and Firm Financial Performance. *Int. J. Energy Econ. Policy* **2017**, *7*, 85–95.
50. Chen, Y.-C.; Hung, M.; Wang, Y. The Effect of Mandatory CSR Disclosure on Firm Profitability and Social Externalities: Evidence from China. *J. Account. Econ.* **2018**, *65*, 169–190. [CrossRef]
51. La Rosa, F.; Liberatore, G.; Mazzi, F.; Terzani, S. The Impact of Corporate Social Performance on the Cost of Debt and Access to Debt Financing for Listed European Non-Financial Firms. *Eur. Manag. J.* **2018**, *36*, 519–529. [CrossRef]
52. Taliento, M.; Favino, C.; Netti, A. Impact of Environmental, Social, and Governance Information on Economic Performance: Evidence of a Corporate 'Sustainability Advantage' from Europe. *Sustainability* **2019**, *11*, 1738. [CrossRef]

53. Orlitzky, M.; Schmidt, F.L.; Rynes, S.L. Corporate Social and Financial Performance: A Meta-Analysis. *Organ. Stud.* **2003**, *24*, 403–441. [CrossRef]
54. Van Beurden, P.; Gössling, T. The Worth of Values—A Literature Review on the Relation between Corporate Social and Financial Performance. *J. Bus. Ethics* **2008**, *82*, 407. [CrossRef]
55. Margolis, J.D.; Elfenbein, H.A.; Walsh, J.P. *Does It Pay to Be Good... And Does It Matter? A Meta-Analysis of the Relationship between Corporate Social and Financial Performance*; Social Science Research Network: Rochester, NY, USA, 2009.
56. Schaltegger, S.; Synnestvedt, T. The Link between 'Green' and Economic Success: Environmental Management as the Crucial Trigger between Environmental and Economic Performance. *J. Environ. Manag.* **2002**, *65*, 339–346. [CrossRef]
57. Harangozó, G.; Kerekes, S.; Zsóka, Á. Environmental Management Practices in the Manufacturing Sector—Hungarian Features in International Comparison. *J. East Eur. Manag. Stud.* **2010**, *15*, 312–347. [CrossRef]
58. Zhang, Y.; Wei, J.; Zhu, Y.; George-Ufot, G. Untangling the Relationship between Corporate Environmental Performance and Corporate Financial Performance: The Double-Edged Moderating Effects of Environmental Uncertainty. *J. Clean. Product.* **2020**, *263*, 121584. [CrossRef]
59. Murakami, S.; Takasu, T.; Islam, K.; Yamasue, E.; Adachi, T. Ecological Footprint and Total Material Requirement as Environmental Indicators of Mining Activities: Case Studies of Copper Mines. *Environ. Sustain. Indic.* **2020**, *8*, 100082. [CrossRef]
60. Freire-Guerrero, A.; Alba-Rodríguez, M.D.; Marrero, M. A Budget for the Ecological Footprint of Buildings Is Possible: A Case Study Using the Dwelling Construction Cost Database of Andalusia. *Sustain. Cities Soc.* **2019**, *51*, 101737. [CrossRef]
61. González-Vallejo, P.; Muñoz-Sanguinetti, C.; Marrero, M. Environmental and Economic Assessment of Dwelling Construction in Spain and Chile. A Comparative Analysis of Two Representative Case Studies. *J. Clean. Product.* **2019**, *208*, 621–635. [CrossRef]
62. Fan, Y.; Qiao, Q.; Xian, C.; Xiao, Y.; Fang, L. A Modified Ecological Footprint Method to Evaluate Environmental Impacts of Industrial Parks. *Resour. Conserv. Recycl.* **2017**, *125*, 293–299. [CrossRef]
63. Eurostat. Road Accident Fatalities—Statistics by Type of Vehicle—Statistics Explained. Available online: https://ec.europa.eu/eurostat/statistics-explained/index.php/Road_accident_fatalities_-_statistics_by_type_of_vehicle (accessed on 16 July 2020).
64. European Commission. EU Buildings Factsheets. Available online: https://ec.europa.eu/energy/eu-buildings-factsheets_en (accessed on 19 November 2020).
65. Ibn-Mohammed, T.; Greenough, R.; Taylor, S.; Ozawa-Meida, L.; Acquaye, A. Operational vs. Embodied Emissions in Buildings—A Review of Current Trends. *Energy Build.* **2013**, *66*, 232–245. [CrossRef]
66. *Directive 2010/31/EU of the European Parliament and of the Council of 19 May 2010 on the Energy Performance of Buildings*; European Union: Brussels, Belgium, 2010.
67. Lux, G. *Újraiparosodás Közép-Európában*; Studia Regionum; Dialóg Campus Kiadó: Budapest, Hungary, 2017; ISBN 978-615-5376-94-8.
68. Fekete, D.; Rechnitzer, J. *Együtt Nagyok. Város és Vállalat 25 éve*; Studia Regionum; Dialóg Campus Kiadó: Budapest, Hungary, 2019; ISBN 978-615-6020-17-8.
69. Gibbs, D.; Deutz, P. Reflections on Implementing Industrial Ecology through Eco-Industrial Park Development. *J. Clean. Product.* **2007**, *15*, 1683–1695. [CrossRef]
70. Mózner, Z.V. Sustainability and Consumption Structure: Environmental Impacts of Food Consumption Clusters. A Case Study for Hungary. *Int. J. Consum. Stud.* **2014**, *38*, 529–539. [CrossRef]
71. International Energy Agency. *Energy Policies of IEA Countries—Hungary 2017 Review*; International Energy Agency: Paris, France, 2017; p. 176.
72. DEFRA. Greenhouse Gas Reporting: Conversion Factors 2018. Available online: https://www.gov.uk/government/publications/greenhouse-gas-reporting-conversion-factors-2018 (accessed on 15 July 2020).
73. Świąder, M.; Szewrański, S.; Kazak, J.K.; Van Hoof, J.; Lin, D.; Wackernagel, M.; Alves, A. Application of Ecological Footprint Accounting as a Part of an Integrated Assessment of Environmental Carrying Capacity: A Case Study of the Footprint of Food of a Large City. *Resources* **2018**, *7*, 52. [CrossRef]
74. Csutora, M. Az ökológiai lábnyom számításának módszertani alapjai. In *Az Ökológiai Lábnyom Ökonómiája*; Csutora, M., Ed.; Aula Kiadó: Budapest, Hungary, 2011; pp. 6–16.
75. Commission of the European Communities. *Commission Recommendation of 6 May 2003 Concerning the Definition of Micro, Small and Medium-Sized Enterprises (Text with EEA Relevance) (Notified under Document Number C(2003) 1422)*; Commission of the European Communities: Brussels, Belgium, 2003.
76. Harangozó, G.; Kovács, Z.; Kondor, A.C.; Szabó, B. A budapesti várostérség fogyasztási alapú ökológiai lábnyomának változása 2003 és 2013 között. *Ter. Stat.* **2019**, *59*, 97–123. [CrossRef]
77. Rossi, M.; Nerino, M.; Capasso, A. Corporate Governance and Financial Performance of Italian Listed Firms. The Results of an Empirical Research. *Corp. Ownersh. Control* **2015**, *12*, 628–643. [CrossRef]
78. Widyaningsih, I.U.; Gunardi, A.; Rossi, M.; Rahmawati, N.A. Expropriation by the Controlling Shareholders on Firm Value in the Context of Indonesia: Corporate Governance as Moderating Variable. *IJMFA* **2017**, *9*, 322. [CrossRef]
79. Gjergji, R.; Vena, L.; Sciascia, S.; Cortesi, A. The Effects of Environmental, Social and Governance Disclosure on the Cost of Capital in Small and Medium Enterprises: The Role of Family Business Status. *Bus. Strategy Environ.* **2021**, *30*, 683–693. [CrossRef]

MDPI
St. Alban-Anlage 66
4052 Basel
Switzerland
Tel. +41 61 683 77 34
Fax +41 61 302 89 18
www.mdpi.com

Sustainability Editorial Office
E-mail: sustainability@mdpi.com
www.mdpi.com/journal/sustainability

www.ingramcontent.com/pod-product-compliance
Lightning Source LLC
Chambersburg PA
CBHW040811250926
51850CB00007BA/1138
9783036529516